Intelligent Robotic Process Automation:

Development, Vulnerability and Applications

Tanupriya Choudhury
University of Petroleum and Energy Studies, Dehradun, India

Jitendra Rajpurohit
Symbiosis International University, India

Ketan Kotecha
Symbiosis Centre for Applied Artificial Intelligence, Symbiosis International University, India

Ming Yang
Kennesaw State University, USA

Sachi Nandan Mohanty
VIT-AP University, India

IGI Global
Scientific Publishing
Publishing Tomorrow's Research Today

Vice President of Editorial	Melissa Wagner
Director of Acquisitions	Mikaela Felty
Director of Book Development	Jocelynn Hessler
Production Manager	Mike Brehm
Cover Design	Phillip Shickler

Published in the United States of America by
IGI Global Scientific Publishing
701 East Chocolate Avenue
Hershey, PA, 17033, USA
Tel: 717-533-8845
Fax: 717-533-7115
Website: https://www.igi-global.com E-mail: cust@igi-global.com

Library of Congress Cataloging-in-Publication Data

Names: Choudhury, Tanupriya, editor. | Rajpurohit, Jitendra, 1981- editor. | Kotecha, Ketan, editor. | Yang, Ming, 1975- editor. | Mohanty, Sachi Nandan, editor.
Title: Intelligent robotic process automation : development, vulnerability and applications / edited by Tanupriya Choudhury, Jitendra Rajpurohit, Ketan Kotecha, Ming Yang, Sachi Mohanty.
Description: Hershey, PA : Engineering Science Reference, [2025] | Includes bibliographical references and index. | Summary: "The book tries to provide a complete insight into RPA to its readers. The initial chapters provide the basic introduction to RPA systems, their development and related considerations. Next few chapters are dedicated to integrating RPA systems with Artificial Intelligence and Cloud for capability enhancement. Further, the book covers few emerging domains such as Healthcare, Security, Education and IoT as use-cases for RPA. The book is likely to equip its users with all basic information required to plan a higher degree, pursue research or gain expertise in RPA or any associated field"-- Provided by publisher.
Identifiers: LCCN 2024038682 (print) | LCCN 2024038683 (ebook) | ISBN 9798369343654 (h/c) | ISBN 9798369349847 (s/c) | ISBN 9798369343661 (ebook)
Subjects: LCSH: Process control. | Automatic control. | Robotics. | Automation. | Artificial intelligence--Industrial applications.
Classification: LCC TS156.8 .I56555 2024 (print) | LCC TS156.8 (ebook) | DDC 670.42/7563--dc23/eng/20241206
LC record available at https://lccn.loc.gov/2024038682
LC ebook record available at https://lccn.loc.gov/2024038683
British Cataloguing in Publication Data
A Cataloguing in Publication record for this book is available from the British Library.

Table of Contents

Detailed Table of Contents

governance operations. The next chapter will also illustrate how seriously those strong governance frameworks, robust testing processes, and ongoing monitoring are taken to protect the RPA deployments. With AI integration, the ever more complex theme of remote work, and a changed regulatory scene, the future directions for the same are discussed. A proactive attitude towards security will keep the organizations safe from challenges by the RPA system and will help them gain all the benefits of automation with risks minimized.

This chapter deals with growth and deployment, changing business operations with Intelligent RPA. It brings together the traditional automation with AI, machine learning, and cognitive technologies in handling complex processes to enhance the efficiency of operations and reduce costs. The chapter outlines the critical enablers of Intelligent RPA, the benefits attached to it during implementation, the challenges that are met during its implementation, and finally the future trends defining its direction. As such, it examines the management, operational, social, and ethical effects of Intelligent RPA adoption for organizations, emphasizing transparency, workforce transformation, and ethical governance. Further, as enterprises transition to Intelligent RPA, they will be forced to balance the celebration of technological accomplishment with people-oriented means to sustainable value creation. Lastly, the paper concludes that Intelligent RPA can eventually alter the work nature in almost every industry while factoring consideration of the ethical and social implications of automation.

This chapter delves deeper into the inclusion of generative AI in RPA and the transformative possibility it brings regarding business operation. Issues of companies pertaining to the operational dimension and their evolution from being rule-based to AI-powered systems will also be discussed. This part of the chapter will focus on how generative AI can transform decision-making processes through a better quality of decision-making support and generate content, which would prove to be an instrumental aspect for application in future scenarios. Additionally, the chapter elaborates on other significant advantages of this integration - efficiency and adaptability - as well as describes some challenges and limitations for organizations, including ethical concerns and those workers displaced by this new trend. The future RPA trends driven by generative AI - namely, hyper-automation and cognitive automation - are also explored in the book.

The integration of cloud computing with Robotic Process Automation (RPA) changes the very course of the way business is conducted, bringing about improvements in business operations efficiency, scalability, and adaptability. This chapter introduces the reader to the architecture of cloud-based RPA and discusses its benefits: that is, a lowered cost of operations and rapid deployments. Discussions include the rise of cloud-native RPA platforms and tools, as well as the importance placed on security and compliance in the cloud environment. On the other side, the chapter also analyses other issues and limitations organizations may face when implementing cloud-enabled RPA, including issues with data privacy and integration complexity among others. Outlook on cloud-enabled RPA is also reviewed in terms of implications for business and the workforce. Based on this, this chapter is specifically looking at various considerations in enabling insight into the means through which organizations can transform their processes through cloud-enabled RPA while propelling innovation into the future.

What started as a solution to automate repetitive, rules-based tasks has expanded into RPA, incorporating artificial intelligence technologies such as machine learning, natural language processing, and computer vision, and promoting a move to intelligent process automation. This integration allows RPA to discern unstructured data, make contextual judgments, and learn and change. Conversational AI enables end-to-end process automation, ensuring human-robot cooperation, while artificial intelligence-driven analytics ensure process optimization. With the

healthcare, ultimately reducing the chances of human errors and improving access to quality healthcare for the rural population.

This chapter elaborates on how NLP pushes RPA to tasks beyond rule-based activities, to areas involving complex processing such as document analysis, real-time decision-making, automation of customer services, etc. Key areas of enhancement pertain to the integration of sentiment analysis, chatbots, and intelligent extraction of data. Technical challenges and general implications of such integration are also covered in the chapter, like infrastructure requirements, avoiding biases, and responsible deployment. Changes in the workforce and equitable access to technology are two examples of managerial and social impacts. Understanding the potential and limitations of NLP-enhanced RPA will position an organization strategically to employ such technologies and optimize their operations toward innovation.

A solution for an industry-specific problem by implementing the poka-yoke technique in a laser printer system. This automation is designed to detect the keyway slot from the component and then marking the QR code of that component with specific number of the component. The primary goals of this solution in industry are to minimize human intervention and reduce product rejection rates. The same component is fitted into three different machines, each component has the keyway slot at a different angle. To address this, used photoelectric sensors to detect the keyway slot and then relay this information to printer to print the final result on the

Preface

INTRODUCTION

In an age where digital transformation continues to shape the global business and technological landscape, Robotic Process Automation (RPA) stands at the intersection of innovation and operational efficiency. This book, *Intelligent Robotic Process Automation: Development, Vulnerability and Applications*, is a timely contribution to understanding the multifaceted roles RPA plays across industries, from banking and finance to healthcare and cybersecurity. With the increasing integration of artificial intelligence, machine learning, and cloud computing, traditional automation is evolving into intelligent, adaptive, and context-aware systems, creating new opportunities and raising critical challenges.

The goal of this volume is to provide a comprehensive and authoritative exploration of RPA's development, vulnerabilities, and emerging applications. The collected chapters address both foundational and advanced topics, including deployment strategies, security concerns, tool analysis, and domain-specific use cases. Whether discussing the transformative role of RPA in rural healthcare systems or the integration of natural language processing to enhance automation capability, each chapter offers unique insights backed by real-world research and experimentation.

This book is intended for a wide range of readers, including researchers, practitioners, system architects, decision-makers, and graduate students. It will serve not only as an academic resource but also as a practical guide for organizations seeking to implement or scale intelligent RPA solutions. As automation continues to reshape the nature of work, understanding its development, vulnerabilities, and future implications has never been more important.

CHAPTER OVERVIEW

Chapter 1: RPA Testing, Deployment and Scalability

This chapter delves into the testing and deployment strategies required for robust RPA implementation and explores scalability challenges. It emphasizes the social, ethical, and managerial implications of deploying RPA in modern enterprises.

Chapter 2: RPA in Banking and Finance

Focused on the financial sector, this chapter outlines how RPA enhances accuracy, compliance, and customer service while navigating issues like legacy integration and data security.

Chapter 3: RPA for Secure Systems and Networks

Exploring the intersection of automation and cybersecurity, this chapter covers how RPA can bolster security protocols, automate threat response, and integrate with AI to build intelligent defense mechanisms.

Chapter 4: Intrinsic Vulnerabilities of RPA Systems

This chapter identifies the key vulnerabilities in RPA systems—such as poor architecture and third-party dependencies—and proposes governance models and monitoring strategies to mitigate risk.

Chapter 5: Intelligent Robotic Process Automation (RPA) Development

A deep dive into the evolution of RPA into intelligent systems, this chapter discusses the merging of AI with automation and the resulting impact on operational efficiency, workforce transformation, and ethical deployment.

Chapter 6: Generative AI as a Tool for RPA

This chapter examines how generative AI supports enhanced decision-making, dynamic content generation, and adaptability in RPA, while also addressing ethical concerns and workforce implications.

Chapter 7: Exploiting Cloud Computing in RPA

The integration of cloud platforms with RPA is explored, highlighting benefits like rapid deployment and scalability, alongside challenges such as data privacy and compliance in distributed environments.

Chapter 8: Intelligence in RPA

Focusing on the role of AI technologies like machine learning and natural language processing in enhancing RPA, this chapter outlines the movement toward intelligent process automation and the future of autonomous systems.

Chapter 9: Analysis of RPA Development Tools

This chapter provides a comparative review of leading RPA tools, analyzing their features, integration capabilities, and cost structures, and offering guidance for tool selection based on organizational needs.

Chapter 10: Intelligent Robotic Process Automation for Healthcare Service Efficacy at Grassroot Level

This chapter addresses the application of intelligent RPA in rural healthcare systems, highlighting how automation can improve service delivery, optimize resource allocation, and enhance patient outcomes.

Chapter 11: Natural Language Processing for Capability Enhancement of RPA

Exploring the integration of NLP into RPA, this chapter focuses on sentiment analysis, data extraction, and intelligent decision-making, while considering the societal and managerial implications of such advancement.

Chapter 12: Design and Development of Automatic QR Code Printing Using PLC and Machine Learning

An industry-specific case study, this chapter showcases how automation, PLCs, and machine learning streamline manufacturing workflows, reduce rejection rates, and ensure precise component tracking.

Chapter 13: Analyzing and Comparing Trajectory Planning Algorithms for Robotic Manipulators

This chapter investigates various trajectory planning algorithms used in robotic manipulators, examining their efficiency, jerk limitations, and adaptability to high-speed environments.

Chapter 14: Advancements in Examination Systems with IP-Based Robo Invigilator Techniques

Focusing on educational assessment, this chapter introduces intelligent invigilation systems that leverage RPA and computer vision to ensure integrity and efficiency in examination processes.

Chapter 15: Intelligent RPA in Education Management Systems

The final chapter explores the potential of intelligent RPA in academic administration, including student data processing, personalized learning environments, and real-time analytics for educators and institutions.

CONCLUSION

The chapters in this volume collectively underscore the transformative impact of intelligent robotic process automation across industries and domains. From its ability to streamline repetitive tasks to its evolving role in enabling hyper-automation, RPA is redefining operational efficiency, data handling, and workforce dynamics. The convergence of AI, machine learning, and cloud computing with RPA has created an ecosystem capable of addressing increasingly complex problems with greater agility and precision.

However, with innovation comes vulnerability. As the technology matures, so too do the associated risks—data breaches, system misconfiguration, ethical dilemmas, and socio-economic disruptions. These challenges are not insurmountable, but they require proactive governance, continuous oversight, and a commitment to responsible AI practices. The insights provided throughout this book are meant to serve as both a warning and a guidepost for organizations at any stage of RPA adoption.

Ultimately, the future of intelligent RPA depends not only on technological advancement but also on human-centric design and ethical integration. By understanding the technical, strategic, and philosophical dimensions of automation, we can ensure that RPA serves as a tool for sustainable growth, equitable progress, and enhanced human capability. We hope this book equips readers with the knowledge, perspective, and tools to participate in shaping that future.

Chapter 1
RPA Testing, Deployment, and Scalability

Sushree Sonam Mohapatra
College of Engineering, Bhubaneswar, India

Shashank Mittal
https://orcid.org/0000-0002-8666-5323
O.P. JIndal Global University, India

Sakshi
https://orcid.org/0009-0007-8467-8079
Lovely Professional University, India

Thi Minh Ngoc Luu
https://orcid.org/0000-0002-5972-7752
International School, Vietnam National University, Hanoi, Vietnam

ABSTRACT

This chapter focuses on navigating the complex RPA landscape, with testing, deployment, and scalability in focus. RPA is transforming the face of business operations to be more efficient, precise, and cost-effective. However, this involves the surmounting of many challenges, which include such processes as complexity, legacy systems, and workforce adaptation. The approach in the chapter is to discuss effective testing strategies, critical deployment considerations, and methods for ensuring scalability within an RPA initiative. Besides, it discusses the hands-on, management, moral, and social impacts of RPA, within which safe use of automation practices is covered. As this application of automation continues to advance, organizations must be responsive and attentive to latest trends and challenges in order to maximize gains

DOI: 10.4018/979-8-3693-4365-4.ch001

through automation with sustainable work culture.

INTRODUCTION TO ROBOTIC PROCESS AUTOMATION (RPA)

RPA stands for technological shift in business practices, which is literally automation of repetitive tasks or activities based on a rule, and which were once performed by human beings. Such a technology employs application of software robots, commonly called "bots" in performing tasks such as data input, transaction processing, and report generation, as if a human would deal with these digital systems. RPA works toward making business processes efficient by removing mundane work from the system, reducing errors, and liberating the employees to higher-value creative and decision-based work (Aguirre & Rodriguez, 2017).

RPA does greatly benefit the organizations across all the industries to be implemented. Businesses will save significantly by automating routine work as once it gets deployed, RPA requires minimal human intervention. Other than reducing the cost, the speed and accuracy provided by RPA improve productivity and enhance the quality of outputs. These strengths make RPA particularly attractive in industries that work with high volumes of data and must be processed reliably and in real time. These include finance, healthcare, retail, and manufacturing.

However, big promises from RPA can be realized only if organizations take a strategic approach in its implementation. Testing and deployment are major indicators of the success of long-term performance for the RPA projects. Testing of the RPA bots will be strong and sure, so that they could be put to proper use in many processes and environments. Thus, disruptions in daily operations would be reduced. Deployment refers to configuring and integrating RPA bots into live environments so they could interact with existing systems. Scalability is about the ability to scale up an RPA solution across an organization, handling increased workloads, with the addition of more bots or processes without loss in performance or reliability.

Testing RPA is quite crucial to validate that the bots do get work done, manage exceptions properly, and integrate well with other applications. RPA bots work on multiple systems often, so there are high chances of running into unforeseen issues that may suddenly come crashing into the workflow. Thorough testing will come across these potential issues way before deployment; hence, transitions will happen more smoothly without risking many operational problems. Since the business environment has become very complex, RPA test processes also need automation to make the testing cycle more effective.

The deployment phase, though technologically oriented, essentially encompasses huge strategic planning. Organizations should thus determine whether to deploy their RPA as a cloud or on-premises, depending on their infrastructures as well as their security and compliance requirements (Fernandez & Aman, 2021). For instance, security and compliance in banking, health and other industries become a significant concern as bots may hold sensitive information, and hence, managing and controlling those bots to ensure safe handling of that information becomes a preeminent concern.

Scalability is also one of the critical considerations with regard to RPA. Even as the business grows, so does its demand on its automation systems. RPA solutions need to scale to support increased transaction volumes, more complex processes, and even additional departments or regions. The architecture of RPA should be flexible and adaptable with regard to growth scenarios that might unfold without affecting performance. Scalability is always important in the long term because it enables an organization to leverage the highest value possible from its RPA investments.

RPA becomes a potential business decision change mainly in the course of business operations by automating repetitive work processes. Success within the RPA implementation requires good planning, testing, strategic deployment, and maintaining a focus on scalability. In case of a correct RPA implementation, this technology can lead to sizeable and substantial gains toward efficiency, accuracy, and organizational performance.

RPA TESTING FUNDAMENTALS

Testing during the implementation of RPA is the most essential thing because it ensures that bots work as required without disrupting business processes. Given the fact that RPA bots are in contact with all the various applications, systems, and environments, problems characteristic of such environments require a proper method of testing. Proper testing minimizes the risk presented by bot malfunction, ensuring that automation proceeds smoothly and integrates seamlessly with other workflows (Smeets et al., 2021).

Importance of Testing in RPA

Most business-critical processes that communicate with several applications, from ERP to CRM systems, can be automated using RPA bots. These bots are, therefore, highly important to the execution of fundamental business processes, necessitating comprehensive testing to ensure that the bots are not only accurate but reliable as well. Any form of error in the performance of bots may lead to incorrect data entry or failure of a particular process as well as delay in operations. RPA testing assures that the bots behave similarly in all scenarios and do not produce exceptions; thus, the risk of operational failure will be at a bare minimum (Ibrahim, 2018).

Types of RPA Tests

RPA testing is very much a necessity because bots need to work flawlessly and with accuracy across different systems and workflows. Several types of tests are conducted that test the different functionalities of a bot, how it interacts with other systems, and its overall performance (Cernat et al., 2020). Here is a comprehensive elaboration of the major types of RPA tests:

1. Functional Testing

Functional testing is one of the aspects of testing in RPA; it is performed to ensure that the bot is doing the same thing that it was designed to do. In RPA, people keep assigning such tasks to bots as those are mundane and routine, which include data entry, form fill-ups, or retrieval of information from an email. Hence, functional testing checks that the bot performs these activities correctly and precisely across all the scenarios.

For example, in an RPA bot where the data from the invoices will be taken out and updated in an ERP, functional testing will ensure that the bot correctly

- Identifies the exact data on the invoice that should be transferred
- Input the correct data fields in the corresponding fields in the ERP system
- Handle different formats of invoices, whether PDF or XML.

It typically develops test cases that mimic the actual environment in which the bot is likely to be operated in a live setting, ensuring that the steps the bot has been programmed to execute are in line with expected outputs for each given task and will cover typical and edge cases in equal measure.

2. Integration Testing

Normally, RPA bots work upon facing varied systems, software, and platforms in an organization. Through integration testing, it will ensure that the bot will communicate well with the external systems and transfer information from one system to another without any kind of errors or disruptions.

For example, if a bot fetches customer information from a CRM system (Rababah et al., 2011) and utilizes the same information to update financial records, it would check all of these things:

- The bot is accessible to both systems without hitch
- Information passed from one system to the other is reliable and unchanged.

- No system crashes or bottlenecks when the bot interacts.

It becomes more focused in complex enterprise environments where bots are required to move around a diverse set of software solutions. It ensures the proper integration of systems and applications and how the bot can handle any variation in the way data is received or processed in various platforms.

3. Performance Testing

Performance testing is an evaluation of how an RPA bot behaves under fluctuating conditions. For instance, it can be done at high workloads or complex transactions. It encompasses the testing of efficiency and speed in processing data at scale or taking care of running tasks that take long hours to execute. Performance testing is crucial as it verifies that a bot is operating at optimum capacity without slowdowns or failure in a company's operational demand.

Key metrics for performance testing:

- Response Time: The amount of time a bot takes to complete a specific number of tasks.
- Throughput: Number of tasks a bot can do within a certain period.
- Scalability: Inability of a bot to continue to deliver levels of performance as workload builds up.

Using the above example, if one came up with a bot that was to process several hundred financial transactions within a limited period, performance testing is ensured to handle the transactions without any delay and the bot errors even when the processing peaks rise.

4. Security Testing

Any environment that bots have access to sensitive or confidential data necessitates security testing. RPA robots most of the time handle personal information, financial records, or even proprietary business data. This somehow makes bots a target of potential security breaches if not well safeguarded.

Security testing validates the following:

- The bot has only accepted systems and data with which it can interact
- The data handled by the bot is stored, transmitted, or processed in a secured manner

- Ensures adherence to the organization's cybersecurity and data protection policies

This type of testing typically involves simulating attacks or unauthorized access attempts to validate that the bot's interaction is secure. In addition, security testing ensures that the bots follow the regulatory standards, mainly in industries where the data protection law is quite tight, such as healthcare HIPAA and finance GDPR.

5. Regression Testing

Regression testing is important since bot processes evolve with bot upgrades. This ensures that such changes in bot configurations or even systems it communicates with may not introduce new faults or influence existing functionality. So, after having modified its processes, rules, or the underlying software systems, the bot can perform its original tasks.

For example, suppose that the application UI with which the bot interacts varies. This ensures that regression testing determines whether it may recognize and act with the new UI without failing.

6. Usability Testing

While RPA bots are mainly backend automation tools, some bots would need to interact with end-users, such as in attended automation. Even in such scenarios, usability testing makes sure that those bots are easy to interact with and will not provide a non-technical user with a grueling experience.

Example: an attended RPA bot can assist a customer service representative by auto-filling data fields in customer engagements. User experience testing ensures that the interface offered to the representative is intuitive, friendly to use, and enables fluid human-bot interaction ().

7. Exception and Error Handling Testing

If a bot handles its exceptions and errors properly, it may continue uninterrupted in a situation that otherwise would have broken the application flow. Testing for exception and error handling tests the functionality in which a bot can respond correctly when things go wrong, for example, for example:

- Handling missing or incorrect data
- Managing a system downtime or connectivity issues
- Recovery of application crashes or timeouts

Therefore, exception handling is the most significant testing for resilience-whether the bot can tackle these situations. A well-tested exception will ensure that it either retries correcting the task or report the problem to human operators for further review and solution.

Each type of RPA test is critical for guaranteeing that bots provide consistent and reliable performance within complex and multi-system environments. These tests, from functional and integration testing to performance, security, and usability assessments, cover every detail related to the behavior of bots. A well-crafted testing strategy for RPA will help organizations identify risks at an early stage in their implementation and ensure there is a much smoother deployment, resulting in higher operational efficiency (Santos et al., 2020).

Common Challenges in RPA Testing

Despite such benefits of RPA testing, a number of problems also present during the process of performance of the bots. The hardest thing is the integration of bots across multiple applications; each application has different interfaces and interaction models. Bots have to work well with these diverse systems without revealing any potential variation of their behavior during testing.

Another key challenge is handling dynamic changes in the interfaces of users. Most RPA bots rely on structured paths to interact with applications, and if an application's interface changes, it may make the bot break down. Bots must, therefore, be able to withstand minor changes in interfaces without malfunctioning and testing should be able to simulate such changes to verify the resilience of bots.

This is also a significant challenge to maintain test data. A test environment requires clean, consistent data sets which can mimic real-world cases, at times very challenging to create and manage. Testing will not reflect the bots' performance in production when improper test data is used.

Tools and Frameworks for RPA Testing

Various tools and frameworks are also designed to help organizations streamline the testing process of RPA. Many RPA vendors have inbuilt testing tools, through which teams can automate and manage test cases on the same platform they use for bot development. Such tools can automate the repetitive, monotonous test tasks, thereby saving more time and effort on the human end.

Apart from the standard ones mentioned earlier, there are several specialized testing frameworks that may be integrated with existing RPA tools to achieve more coverage of the test. A more intense testing environment could be used for RPA bots by simulating an application environment using tools such as Selenium and UFT (Dechamma & Shobha, 2020).

Another trend observed in RPA projects is that CI and CD pipelines are being increasingly utilized to automate testing at each stage of development. Pipelines ensure bots keep getting tested as changes keep occurring and, thus, there are no last-minute errors at the time of deployment.

Testing assumes a very important role in ensuring that the automation initiatives are successful. It has ensured that the functions of the bot are integrated, working correctly, and free from security loopholes to deliver RPA solutions with consistent and reliable results. Though integration complexity and UI-related changes are some risks possible, modern tools and frameworks can help offset the risks and pave the way for the successful roll-out of RPA deployments.

STRATEGIES FOR EFFECTIVE RPA TESTING

An RPA project can only be deemed successful by an effective, detailed testing strategy. The modern RPA solution is complex, as it may be related to many systems and applications. Bots developed need to be error-free, but this alone cannot ensure the reliability of the bots. Therefore, organizations need to adopt effective strategies that will minimize risks while guaranteeing the bot's reliability. Here are some of the key strategies in conducting effective RPA testing:

Develop Comprehensive Test Cases

The comprehensive test cases are developed with possible scenarios that a bot may face in effective RPA testing. Test cases should at least involve common process flows but also edge cases and exceptions (El-Attar & Miller, 2010). For example, if a bot is charged to process information for customers, the test cases should address missing or incorrect data, varied forms of data, and system timeouts.

Negative and positive test cases are required. In the case of positive testing, the bot is tested to ensure that it adheres to the requirements in ideal conditions. Negative testing shows how the bot would act in case of a failure such as improper inputs or system crashes. Having thorough test cases therefore ensures the organization identifies matters early, so the bot is always ready to face the real challenges.

Test in Multiple Environments

RPA bots generally interact with multiple systems, platforms, and applications. Hence, the testing must be done across different environments. Testing in multiple environments will surely guarantee that bots behave uniformly irrespective of the infrastructure on which they are located. For instance, testing has to be done on a number of browsers, different kinds of operating systems, or software versions that the bot communicates with. Also, bots should be tested both in production and non-production environments to identify probable issues that may not appear only under certain conditions.

Besides that, organizations should simulate the most nearly to real-world conditions when testing. That is simulation network latency, server down time, data format variation for the sake of smoothness in handling variation without causing process interruptions.

Automate RPA Test Processes

Since RPA testing involves many repeated scenarios, the process of automation is quite an efficient enhancement in their ability to do so. Automated testing allows large-scale tests to be performed more rapidly and accurately than manual testing would permit. Indeed, use of these tools is for creating and executing test cases, verifying results, and identifying errors at different stages of bot development.

Automated testing also facilitates CI/CD for RPA projects (Hannonen, 2020). Organizations now ensure that bots run continually, with continuous updates and changes so that there is little chance of the bots released that have not tested. The implementation is faster and more reliable.

Leverage Test Data Management

Test data management is an important part of a comprehensive RPA testing plan. To model the scenario truthfully, test data should represent a diverse mix of inputs that bots would otherwise face in real-life production environments. Test data management incorporates the creation, maintenance, and organization of datasets that make testing realistic without risking live data.

Organizations should utilize either anonymized or synthetic data for testing, especially in industries dealing with sensitive information like financial records or personal identification details. This will help avoid getting into any compliance issues while the bot is adequately tested. In addition to this, refresh test data over time so that the bot is exposed to new data sets that reflect changing business needs.

Monitor and Track Bot Performance

The necessity to track and monitor bot performance in the testing phase would help identify areas of bottlenecks and inefficiency. One can closely monitor information concerning response time, throughput, and resource utilization while making sure that the bot acts as it should. If there is a compromise in a bot's performance at high workloads or in some circumstances, optimization and more testing may be necessary.

Organizations should also implement automated monitoring tools that measure bot performance during the duration of the testing cycle to ensure performance trends and potential issues are detected before reaching production. Monitoring also ensures bots remain scalable given that, at some point, they are given larger workloads over time.

Incorporate Exception Handling Testing

Perhaps one of the most important aspects of RPA is the way that bots handle exceptions and errors. Good exception handling means the capability of a bot to deal with unexpected scenarios without altogether breaking the process. This becomes an area in which extreme rigor must be exercised to test how bots respond to different failures, say system crash, data discrepancy, or network issue (Sinha & Harrold, 2000).

Important to include in the workflow of a bot is automated error recovery. For example, this would allow a bot to retry when it cannot retrieve data from a third-party system due to connectivity problems or escalate the problem to human operators. Exception handling tests should simulate these scenarios and show that a bot can recover gracefully with no interference in the business.

Continuous Feedback and Collaboration

Testing an RPA application calls for collaboration from business analysts, developers, testers, and end-users as a team (Madakam et al., 2019). Every one of them brings their own insight of what might go wrong during the process and where improvements can be brought about. Such feedback loops in teams become an effective way for early detection of such looming challenges that might surface during the testing, ensuring optimum bots with regard to technical performance as well as business requirements.

It further ensures that the functionality of the bot matches the real-life use cases by including end-users in the testing process. They provide usability feedback on the tool and ensure that the bot works as it is designed to do in the operations of everyday life. This collaboration facilitates better shifting towards production, hence minimizing the probability of post-deployment issues.

A good testing approach should be in place for any RPA project to be successful. Thorough test cases, testing on multiple environments, automated test processes, proper management of test data, performance monitoring, and exception handling raise the possibility that RPA bot will be as expected in the organization. Continuous feedback among teams is needed to create more reliable and effective RPA systems. All these approaches would reduce the risk of failures and improve overall efficiency if implemented appropriately on RPA Deployments.

RPA DEPLOYMENT: KEY CONSIDERATIONS

Tremendous planning and attention to detail will be involved in the implementation of an effective RPA (Kyheröinen, 2018). An effective implementation of RPA ensures that, at the very least, the bots perform their tasks as expected and are seamlessly integrated into the existing systems and processes with no disruption in core operations. Achieving all this involves several considerations prior to, during, and after deployment.

Clear Definition of Objectives and Scope

Before launching a project, an RPA project needs to have well-defined objectives and scope (Chaugh et al., 2022). This will include identification of specific processes to be automated, what you aim to achieve with automation for example, become more efficient or reduce errors or cut costs and a clear outline of the outcomes. This will help heavily prevent scope creep.

The other basic building block, here, is determining KPIs by which to measure success in deployment. This should line up with wider organizational strategic goals and be tied into metrics such as the time saved, errors reduced, or productivity increased. With defined goals and measurable KPIs, business organizations can then determine how effective their RPA deployment is and change tactics as needed.

Infrastructure Readiness

A very important deployment is that the infrastructure needs to be adequate to support the automation. In some way, this can be considered an evaluation on the compatibility of existing systems towards RPA tools and assure the infrastructure can take in the added capacity that Bots will have brought in a process. In this case, if an RPA bot was set to automate the processing of high-volume transactions, then the system has to be adequate enough for the processing with an increased data volume without affecting its performance.

Organizations should also provision the needed technical resources. These include server capacity, network bandwidth, and data storage, among others. At times, upgrading or scaling up of the infrastructure might be necessary to support the bots, depending on the environment. For instance, the presence of multiple systems and applications that interact with each other requires upgrading or scaling an infrastructure. Cloud-based infrastructure often offers more scalability and flexibility to allow for adjustments to the system as the RPA needs change over time (Kunduru, 2023).

Data Security and Compliance

Data security is also considered a concern in the deployment of RPA, especially when the bot interacts with sensitive and confidential data. Organizations have to ensure that any automated activity is in compliance with relevant data protection laws and with the internal guidelines of security. For instance, it may belong to the healthcare, finance, or government sectors, which all have very strict regulations, such as HIPAA or the GDPR, that need to be applied in its deployment (Schmidt, 2020).

RPA bots should be properly controlled so sensitive information is not accessed without authority. This can be achieved through role-based access and only permitting interactions with the data needed to accomplish a job. In certain deployments, organizations would also consider encrypting data during transit and storage and maintain an audit trail for every bot interaction to ensure all activity can be traced and accounted for.

Change Management and Stakeholder Buy-In

Most organizations that apply RPA expect changes in current workflows, and the workers can feel uneasy about being replaced or having their routines changed. Effective change management, therefore, dictates that organizations should engage with the stakeholders at the initial stages, which may involve the workers who will be losing jobs due to automation. Letting them be included in the process of implementation can mitigate most of the issues by communicating the benefits of RPA. Organizations end up building a collaborative rather than a resisting culture (Wallace, 2022).

Training programs must be offered to introduce employees to working with bots and new roles in the automated environment. It is essential to educate employees on how RPA will supplement human work, not replace them, in securing employee buy-in and assuring a very smooth transition.

Pilot Testing and Phased Rollout

It also minimizes the risks of failure and is very helpful in the rollout of the bot pilot. In this, a pilot deployment will provide an organisation with a chance to test the bot in a controlled environment often through a small, non-critical process, thereby allowing the RPA team to pick up on any underlying issues with the automation, hone it, and gather feedback from users to scale up the solution.

Following the pilot, it will be crucial for an organization to consider phased roll-out. That is, where bots are rolled out in phases and not all at one go. It will allow gradual roll-out this way, which will cause less friction and an opportunity to quickly adjust due to unforeseen problems that may arise (Bower & Walker, 2007). The rolling out of automation in a phased manner ensures that organizations can continue measuring performance, solving problems, and improving the latter rollouts.

Bot Monitoring and Maintenance

Even after deployment, without continuous monitoring and maintenance, the process is not sustainable in the long term. RPA bots must be updated or modified periodically as business processes change, application interfaces change, or new regulatory requirements arise. Continuous monitoring ensures that the bots are performing according to expectation and that identification of problems can be quickly made and corrected.

Bot health monitoring systems should be implemented by organizations and should monitor performance metrics, error rates, and resource usage (Ashok & Gopikrishan, 2023). Major problems that have the potential to cause operations may require alerts from the RPA team such as system failure and suspicious behavior. Monitoring aside, bots should also undergo maintenance regularly to update scripts, address bugs, and optimize performance.

Scaling for Future Growth

Now that there is a small pilot RPA in place, the organization needs to think and plan how it is going to scale its automation initiative to other areas of its business. Scaling requires careful planning for more infrastructure, additional resources, and further governance frameworks in place to be able to support a larger number of bots across several departments or processes.

The organisations should implement governance frameworks when scaling efforts, therefore, ensuring that all new automation projects serve the business goals and comply with corporate policies, include defining standard development standards, testing, deploying, monitoring and all the resources available to such a project as IT, infrastructure, etc.

Above all, it is a complex process in deploying RPA, which also includes handling a few key considerations: clear objectives to be defined; infrastructure readiness; use of the required security and compliance needs; management of change to people; pilot testing and phased rollouts; and going continuously for monitoring and maintenance (Syed et al., 2020). Long-term success is therefore provided through pilot testing and phased rollouts. If the above considerations are handled correctly, organizations can successfully deploy their RPA and create a platform for future scalability and growth in their automation initiative.

ENSURING SCALABILITY IN RPA

Scalability is one of the most crucial factors of success in the long term in Robotic Process Automation (RPA). While organizations may automate more and more processes and introduce more and more bots within their operational frameworks, their RPA solution must scale nicely. They should absorb workloads and complexity increases without losses in efficiency or stability. This requires strategic planning across several dimensions, including infrastructure, process selection, bot management, and governance (Noppen et al., 2020).

Flexible and Scalable Infrastructure

Scalable infrastructure is the very foundation of IT. Such infrastructure needs to be flexible and resilient enough to accommodate more bots as well as larger volumes in terms of workloads done through automation. Scalable infrastructure would mean that it can add more bots anytime without going into a bottleneck in performance or time on downtime (Garrison & Nova, 2017).

Organizations should utilize cloud infrastructure for scalability. With cloud environments, organizations can readily add or remove resources based on business needs. This will scale business RPA initiatives without significant upfront expenses involving hardware or servers. Hybrid cloud environments helping organizations that have specific data security or compliance requirements rely on a combination of private and public cloud infrastructures.

Organisations must also ensure that their IT systems, that is, their databases and software applications, are designed to handle the increased traffic created by bots. This requires cooperation between the IT and RPA teams in estimating what can be accommodated in a given system and what upgrades or optimization will be necessary.

Prioritizing High-Value, Scalable Processes

Not all processes are scalable to have volume without being able to go operational. For successful scaling, the scope needs to be narrowed not only to repetitive and rules-based processes but also to processible processes across departments or business units. The greatest high-volume potential is obtained from processes that can easily be replicated across a multitude of functions without much need for customization (Alimoradi et al., 2023).

For example, processing invoices from the same region would typically come under one team or group for customer service queries or new employee onboarding. For this reason, they are excellent choices for scalable automation. It is likely to yield greater returns on investment in RPA while reducing complexity in scalability by focusing on high-value processes used across the organization.

Thirdly, the process should be stable and less variable. Even though processes are not long and overly complex, many exceptions or frequent changes can make scaling harder because they must be continuously updated and regulated with bots. Thus, organizations should try to stabilize or redesign such processes before scaling their automation.

Bot Management and Monitoring

The more organizations are deploying bots, the more there is an increased need for effective bot management and monitoring. Such a central bot management system is therefore important in terms of scalability because it enables an organization to manage tens of thousands of bots from one interface, featuring scheduling bot activity, tracking performance, and automated error handling (Haban & Shin, 1990).

Good monitoring tools ensure that bots continue to work well at scale. Measured metrics, such as bot uptime, response time, and resource usage, should be monitored for bottlenecks or inefficiency. Automated alerts may then notify teams about performance degradation so teams can rapidly correct the issues and minimize disruption to business operations.

Maximum transparency and accountability should be exercised by logging and tracking audit trails of bot activities and interactions with various systems as the number of bots and their interactions increases over time.

Governance and Compliance Frameworks

Scalability demands governance frameworks to ensure uniformity, quality, and compliance across the organization in RPA. How bots operate will be an essential concern-factor in enforcing company policies, industry regulations, and security protocols as automation continues to grow. Governance frameworks must lay down guidelines clearly for bot development, deployment, and maintenance that define consistency across teams and processes (de Almeida et al., 2021).

A scalable governance model would typically have a Center of Excellence (CoE) implemented to manage RPA initiatives across the organization (Costa et al., 2022). Here, best practices are standardized, reusable automation components are developed, and all RPA projects are ensured to be aligned with business goals as well as regulatory requirements. This way, organizations can be more in control of their bots by centralizing RPA governance rather than risk potential compliance issues as scale increases with automation.

Organizations need to make their bots flexible. Bots should be adaptable as the business processes change, the systems change, or regulatory requirements change. This reduces the need for constant bot reconfigurations as the automation landscape changes, which is essential for scalability.

Standardizing Automation Components

The basis of scaling also must be on standard reusable components and modules that can easily be “plugged-in” into new bots. This promotes quicker developments and deployments of new bots in addition to maintaining consistency across the organization. Reusable components can include pre-built workflows, templates for common tasks, as well as rule libraries that can be applied across various processes (Glohr et al., 2013).

For example, an organization automating data entry for several departments could build a generic data entry module which it can use for thousands of other forms or systems. This saves time and resources building new bots from scratch, which allows the organization to scale the automation efforts quickly and efficiently.

Standards also help increase bot maintenance by the updates of a module. Updates made within a single module can propagate throughout all bots using the same module. Therefore, it means there is less downtime and increases the possibility of bots staying in line with changing business requirements.

Load Balancing and Resource Optimization

With the increase of bots, the most important fact is effective resource utilization. It requires proper load balancing because every bot should be spread evenly across all the available servers or virtual machines to optimize the procedure. If not balanced properly, some bots are likely to face bottlenecks or constraints of resources, which increases processing time and even leads to errors.

Companies can have intelligent load-balancing systems that automatically direct resources to bots based on activity and the workloads that they process. These systems monitor in real time and dynamically re-balance the distribution of bots for peak performance. Additionally, RPA platforms with built-in scalability offer better workload management with queuing task activities or shifting processing priorities at peak times (Sathya et al., 2023).

Continuous Improvement and Optimization

Scaling is not a one-time effort but rather an ongoing, continuous improvement and optimization process. During the lifecycle of RPA deployments, organizations should continuously review their automation efforts to determine areas where and how they can improve-a process involving the review of bot performance, as well as process efficiency and error rates, toward perfecting and optimizing bots for scalability (Venigandla & Vemuri, 2022).

Regular process audit of automation might also identify opportunities for further automation or improvement in bots. As organisations monitor emerging technologies and best practices, they should reflect and extend them to their RPA strategy, remaining scalable and competitive.

Ideally, the culture in the organization should appreciate continuous learning by ensuring the teams are up to date with the latest developments on RPA tools, techniques, and trends. This way, the automation initiative will remain scalable and adaptable to future needs.

Scalability in RPA can only be ensured considering infrastructure, process selection, bot management, governance, and continuous improvement. Scalability of the infrastructure and efficiency of process selection would be accompanied by strong governance frameworks and optimized resource usage in order to have wider expansion of RPA initiatives. With such embedding of automation in operations, performance, reliability, and business value over time depend highly on scalability.

CHALLENGES AND FUTURE TRENDS IN RPA

Organizations embracing RPA to streamline processes and make things more efficient are presented with several challenges that impact the potential for success and scalability of the automation effort. Coinciding with this transformation is the rapidly shifting nature of the RPA environment, with emerging trends that provide new opportunities and resolutions to these challenges. Such an organization in quest for maximizing automation's benefits should understand the obstacles and direction RPA will take.

Challenges in RPA Implementation

Some of the biggest challenges in RPA implementation arise from process complexity and variability. RPA thrives on repetitive, rule-based, process activities. Business processes are rarely simple because there might be many exceptions and decision points in a large number of process activities. Therefore, detailed mapping to understand the actual process flows is involved in automating such processes, thereby making development cycles long with costs increasing more. Sometimes, it results in diminishing returns where such processes entail too much effort to be automated, not suited for RPA.

The integration with the legacy systems is another significant challenge. Most of the companies have non-integrated or outdated IT systems, which creates difficulties in setting up RPA, as bots interact with various applications, databases, and interfaces, whereas all legacy systems may not be supporting modern RPA tools. Those integration issues cause performance bottlenecks, frequent errors, and maintenance issues, thus limiting the scalability of the automation efforts (Lamberton et al., 2017).

Change management and workforce adaptation are other challenges, especially when the workforce is considered in an organization, where employees feel that automation could destroy jobs. Employee resistance to change can be a major problem against RPA adoption and benefits of automation. It would require effective communication, training, and employee engagement strategies to facilitate smooth transition into the automated environment.

The stability and resilience of the bot are always of concern as business processes or underlying systems change. Bots must be constantly monitored and maintained so that they do not start behaving irrationally. Any change in software interfaces, updating enterprise applications, or altering regulatory requirements will cause malfunctions in bots and would lead to operational disruption if not immediately addressed.

Future Trends in RPA

This maturation in RPA in recent times has bred several trends that have the potential to fashion automation into the future. Most of them come as solutions to the current challenges and expand the erstwhile potential use cases of RPA.

The first notable kind is Intelligent Process Automation, or IPA. It brings RPA closer to AI technologies like machine learning, natural language processing, and computer vision. Using IPA, organizations can automate processes that contain elements of decision-making and unstructured data or are involving customers (Samokhvalov, 2014). This evolution introduces a qualitative shift in RPA which enables bots to execute tasks that otherwise lie far out of their capacities, like breaking down bigger datasets or comprehending customer sentiment.

Another emerging trend is hyper-automation. It involves the automated execution of end-to-end business processes through various automation tools --RPA, AI, process mining, and BPM software. Hyper-automation empowers an organization to develop tasks beyond single tasks, working toward more comprehensive and transforming automation (Haleem et al., 2021). The objective is toward a totally integrated digital workforce that completes human employees and yields sufficient operational improvement.

Cloud-based RPA is also another growing trend, ensuring increased scalability, agility, and cost efficiency. These cloud-based RPA platforms allow organizations to scale their automation initiatives fast and without requiring significant investments in infrastructure upfront. Access to advanced AI capabilities is also easier from such platforms, thus allowing businesses to more significantly leverage their cognitive automation.

This brings out yet another area to focus on: the importance of RPA and cybersecurity. Since bots deal with business data that is most sensitive and critical, ensuring stronger security measures is a must-do. In the future, RPA vendors would increase efforts in security features including advanced encryption, access control, and real-time threat detection to ensure automation remains safe and compliant with industry regulations.

RPA democratization refers to the trend of accessibility in making automation available for use even for business users who are not technical employees. The presently developed low-code or no-code RPA platforms allow business users to design and deliver automation solutions sans deep technical know-how. This democratization of RPA can accelerate its adoption within organizations and speed up everyday tasks being automated in faster and more agile ways.

RPA and future workforce are yet another trend to discuss in this context. With RPA coming to the forefront, comes the rising reality that it will not replace jobs, but redefine roles. The repetitive tasks would better be done by bots while human employees would work on other more strategic work, and more creative initiatives. New skill sets would include digital literacy as well as problem-solving to ensure that the future workforce could work effectively alongside bots. More investment will be made in reskilling and upskilling of employees to prepare for the new age of work.

While there are opportunities that RPA brings along with it to a business, challenges like process complexity, integration of systems, change management, and the maintenance of the system go along with RPA. However, emerging trends such as Intelligent Process Automation, hyper-automation, RPA based on cloud platforms, democratization of automation, etc. help overcome these challenges and widen possibilities for applying RPA. With appropriate monitoring of these trends and with a forward-thinking approach, organizations can fully exploit the potential of RPA and continue to push efficiency and innovation in their operations.

IMPLICATIONS OF RPA

These implications result from implementing RPA, bringing several benefits but also several implications that must be taken seriously by organizations. They range over practical, managerial, ethical, and social dimensions and affect deployment and integration within businesses and society in general.

Practical Implications

From the practical view, RPA will highly enhance operational effectiveness in automating repetitive, rule-based, time-consuming, and error-prone operations. That goes with processing at a faster rate, better accuracy, and also costs saved for an organization. Employees are freed up from mundane activities that way to do more on more critical and value-added things.

However, effective RPA implementations critically depend on careful planning, particularly when it comes to the selection of the correct processes to be automated (Plattfaut et al., 2022). Not everything should be automated; otherwise, complex or high-variability processes would lead to many errors and increased maintenance. Scalability and the basic needs-the necessity for bots to continuously be monitored so that they remain effective with changing business processes-must be considered.

The integration of RPA into the existing system will be difficult, especially when there are legacy systems or applications in multiple, disparate software. There is the need for the organizations to invest in the appropriate IT infrastructure so that it can also support RPA and make sure that these systems can withstand the workload from automated processes.

Managerial Implications

From a managerial viewpoint, RPA calls for a paradigm shift in how an organization works in process management, workforce dynamics, and strategic decision-making. Perhaps most importantly, it needs to have a more holistic view of operations-through an understanding of what lies in RPA as an add-on to human efforts and where it can make a huge difference.

The most important managerial implication relates to change management:. Change Management RPA brings along many changes in the workflows, and if not planned appropriately, it may lead to resistances to change in employees. Management needs to communicate the benefits of RPA towards teams; this can be through addressing concerns that may have individuals worried about losing jobs and letting employees understand how automation will free them to do more strategic work. This needs a good thought out communication plan, well taken training and development programs that would help employees learn alongside bots.

A second managerial challenge of this type is governance and oversight: the more bots being deployed in organizations, the more complex their performance, security, and compliance become. Managing such complexity could prove a necessity and there should be a Centre of Excellence (CoE) established for best practices as well as monitoring the performance of bots. Managers ensure that RPA initiatives further broader business goals and deliver measurable outcomes (Costa et al., 2022).

RPA also touches workforce planning and talent management. Where routine work is automated, the managers must reconsider how many staff members are necessary, where best to place them, which should be reskilled and how these employees in a different assignment should be redeployed.

Ethical Implications

RPA raises further ethical issues for organizations, largely on themes of employment and data privacy. One of the biggest ethical concerns is how RPA might be able to affect jobs. RPA automates many routine tasks; therefore, it can potentially eliminate some jobs, mostly those administrative or clerical in nature. This is a difficult ethical question because businesses must weigh their desire to improve efficiency against their responsibility to the worker (Beerbaum, 2022).

It requires organizations to opt for ethical automation that may include the reskilling and upskilling of affected workers, thereby allowing them to take on jobs requiring critical thinking, creativity, or customer contact. That way, it minimizes the effects of job loss due to automation and keeps the workforce employed and empowered.

Data privacy and security will also be key concerns for ethics, especially as bots interface with information that might be sensitive in nature. Organizations must, therefore, ensure that when the bots are designed or operated, they abide by the regulations of data protection, and substantial security measures exist to avoid unauthorized information access.

Social Implications

Adoption of RPA can transform labor market realities and redefine the nature of work. Once such widespread automation comes in, society would have to settle questions on the future of employment, especially for those sectors where repetitive tasks exist. Society will get a chance to rethink the human-to-machine nature of work.

RPA and economic inequality is an important social factor in this regard. Because of RPA, productivity and profitability go up, but without equal benefit distribution due to automation, a negative consequence may be inferred. The people who most likely feel the pinch of job displacement and poor conditions are lower-skill and lower-wage workers while productivity and profitability go up. This requires both businesses and governments to focus more on social policies encouraging education, digital literacy, and skills development in preparation for the changed nature of jobs in a more automated world.

On the positive side, RPA will alleviate the drudgery of mundane work and free up workers' time for meaningful work. As such, their job satisfaction and wellbeing become stronger, and a more motivated and productive workforce emanates from that source (Sharma et al., 2023).

With automation, organizations can streamline better to reduce waste and minimize resource consumption in a manner that is aptly relevant to the larger goals of sustainability and CSR. In this manner, business houses could exploit RPA for improving the efficiency of their operations and thereby contribute to reducing carbon footprint and making a more general impact on sustainability efforts.

CONCLUSION

RPA technology is the change agent for the future of business operations, promising to deliver important benefits in terms of efficiency, accuracy, and cost savings. The importance of this technology associated with RPA becomes increasingly noticeable when organizations embrace it as a critical part of their workflows. This calls for a holistic approach to putting it into practice. Informed from this chapter, careful planning and strategic management are called for in navigating the complexities and challenges surrounding RPA.

Based on the practical implications of RPA, the task is to select processes appropriate for automation because not all tasks can be amenable to robotic intervention. Organizations require investment in robust IT infrastructure and ensure a smooth integration interface with existing systems to make the fullest use of RPA. Maintenance of automated processes through continuous monitoring is critical to sustenance and adaptation over time to business needs.

From the management's side, it will force the leaders to involve in change management practices for the employees to accept and embrace. If proper, transparent communication involving the benefits of RPA and reskilling opportunities is made to the employees, then the possible apprehensions of job loss will be mitigated, and a collaborative environment where humans and bots can work in collaboration will be fostered. Governance structures through creation of Centers of Excellence, for example, will facilitate oversight and accountability in RPA initiatives.

The ethical effects of RPA transcend mere words since it is centered on the consequences it may unleash in employment and data privacy. Organizations are concerned with the ethical application of automation in staff development and compliance with the rules that govern data protection regulation. This would ensure even distribution of all the advantages of RPA.

Socially, RPA might radically alter the nature of labor, making it more of an opportunity and a challenge. With automation of routine tasks, society must adjust to this change by inducing skill building and electronic literacy. This shall prepare the workforce for a world in which work is moving fast, ensuring each employee is better prepared to grow in such a world.

The future looks very promising for RPA and even brighter in the light of trends like Intelligent Process Automation and hyper-automation, which indeed push into the forefront even more complex and integrated solutions for automation. Organisations embracing these trends and taking proactive steps in overcoming the challenges related to them will be very well positioned to capitalise on this opportunity to harness the true potential of RPA and drive innovation and growth.

RPA is, at a very basic level, something more than a technological innovation-it is a catalyst for fundamental change in business practices. By understanding and thereby dealing with the practical, managerial, ethical, and social implications of RPA, businesses may better navigate that landscape and create a more sustainable and inclusive future for work.

REFERENCES

Aguirre, S., & Rodriguez, A. (2017). Automation of a business process using robotic process automation (RPA): A case study. In *Applied Computer Sciences in Engineering: 4th Workshop on Engineering Applications, WEA 2017, Cartagena, Colombia, September 27-29, 2017* [Springer International Publishing.]. *Proceedings*, *4*, 65–71.

Alimoradi, R., Aliee, F. S., & Farahani, B. (2023, November). An Efficient Method For Selecting and Prioritizing Business Processes In Robotic Process Automation. In *2023 7th Iranian Conference on Advances in Enterprise Architecture (ICAEA)* (pp. 85-92). IEEE. DOI: 10.1109/ICAEA60387.2023.10414471

Ashok, K., & Gopikrishnan, S. (2023). Statistical analysis of remote health monitoring based IoT security models & deployments from a pragmatic perspective. *IEEE Access : Practical Innovations, Open Solutions*, *11*, 2621–2651. DOI: 10.1109/ACCESS.2023.3234632

. Beerbaum, D. (2022). Artificial intelligence ethics taxonomy-robotic process automation (RPA) as business case. *Artificial Intelligence Ethics Taxonomy-Robotic Process Automation (RPA) as Business Case (July 17, 2022).*

Bower, D. C., & Walker, D. H. (2007). Planning knowledge for phased rollout projects. *Project Management Journal*, *38*(3), 45–60. DOI: 10.1002/pmj.20005

Cernat, M., Staicu, A. N., & Stefanescu, A. (2020, November). Towards automated testing of RPA implementations. In *Proceedings of the 11th ACM SIGSOFT International Workshop on Automating TEST Case Design, Selection, and Evaluation* (pp. 21-24).

Chugh, R., Macht, S., & Hossain, R. (2022). Robotic Process Automation: A review of organizational grey literature. *International Journal of Information Systems and Project Management*, *10*(1), 5–26. DOI: 10.12821/ijispm100101

Costa, D. A. D. S., Mamede, H. S., & Silva, M. M. D. (2022). Robotic Process Automation (RPA) adoption: a systematic literature review. *Engineering Management in Production and Services, 14*(2).

de Almeida, P. G. R., dos Santos, C. D., & Farias, J. S. (2021). Artificial intelligence regulation: A framework for governance. *Ethics and Information Technology*, *23*(3), 505–525. DOI: 10.1007/s10676-021-09593-z

Dechamma, P. R., & Shobha, N. S. (2020). A review on robotic process automation. *International Journal of Research in Engineering. Science and Management*, *3*(5), 237–244.

El-Attar, M., & Miller, J. (2010). Developing comprehensive acceptance tests from use cases and robustness diagrams. *Requirements Engineering*, *15*(3), 285–306. DOI: 10.1007/s00766-009-0088-6

Fernandez, D., & Aman, A. (2021). Planning for a successful Robotic Process Automation (RPA) project: A case study. [JIKM]. *Journal of Information and Knowledge Management*, *11*(1), 103–117.

Glohr, C., Kellermann, J., & Dörnemann, H. (2013). The IT factory: a vision of standardization and automation. In *The Road to a Modern IT Factory: Industrialization–Automation–Optimization* (pp. 101–109). Springer Berlin Heidelberg.

Haban, D., & Shin, K. G. (1990). Application of real-time monitoring to scheduling tasks with random execution times. *IEEE Transactions on Software Engineering*, *16*(12), 1374–1389. DOI: 10.1109/32.62446

Haleem, A., Javaid, M., Singh, R. P., Rab, S., & Suman, R. (2021). Hyperautomation for the enhancement of automation in industries. *Sensors International*, *2*, 100124. DOI: 10.1016/j.sintl.2021.100124

Hannonen, A. (2020). Automated testing for software process automation.

Ibrahim, S. S. N. (2018). The Role of RPA in Shaping the Future of Enterprise Systems.

Kunduru, A. R. (2023). Cloud BPM Application (Appian) Robotic Process Automation Capabilities. *Asian Journal of Research in Computer Science*, *16*(3), 267–280. DOI: 10.9734/ajrcos/2023/v16i3361

Kyheröinen, T. (2018). *Implementation of robotic process automation to a target process–a case study* (Master's thesis).

Lamberton, C., Brigo, D., & Hoy, D. (2017). Impact of Robotics, RPA and AI on the insurance industry: Challenges and opportunities. *Journal of Financial Perspectives*, *4*(1).

Madakam, S., Holmukhe, R. M., & Jaiswal, D. K. (2019). The future digital work force: Robotic process automation (RPA). *Journal of Information Systems and Technology Management*, *16*, e201916001. DOI: 10.4301/S1807-1775201916001

Noppen, P., Beerepoot, I., Van De Weerd, I., Jonker, M., & Reijers, H. A. (2020). How to keep RPA maintainable? In *Business Process Management: 18th International Conference, BPM 2020, Seville, Spain, September 13–18, 2020* [Springer International Publishing.]. *Proceedings*, *18*, 453–470.

Plattfaut, R., Borghoff, V., Godefroid, M., Koch, J., Trampler, M., & Coners, A. (2022). The critical success factors for robotic process automation. *Computers in Industry, 138*, 103646. DOI: 10.1016/j.compind.2022.103646

Rababah, K., Mohd, H., & Ibrahim, H. (2011). Customer relationship management (CRM) processes from theory to practice: The pre-implementation plan of CRM system. *International Journal of e-Education, e-Business, e-. Management Learning*, *1*(1), 22–27.

Samokhvalov, I. (2024). *Transforming Management Reporting with Intelligent Process Automation (IPA)*. Enhancing Business Analytics, Forecasting, and Decision-Making in Organizations.

Santos, F., Pereira, R., & Vasconcelos, J. B. (2020). Toward robotic process automation implementation: An end-to-end perspective. *Business Process Management Journal*, *26*(2), 405–420. DOI: 10.1108/BPMJ-12-2018-0380

Sathya, V., Jayashree, K., & Malathi, S. (2023). Robotic Process Automation (RPA) Applications and Tools for the Workforce Management System. In *Designing Workforce Management Systems for Industry 4.0* (pp. 251-264). CRC Press.

Schmidt, A. (2020). Regulatory Challenges in Healthcare IT: Ensuring Compliance with HIPAA and GDPR. *Academic Journal of Science and Technology*, *3*(1), 1–7.

Sharma, C., Bharadwaj, S. S., Gupta, N., & Jain, H. (2023). Robotic process automation adoption: Contextual factors from service sectors in an emerging economy. *Journal of Enterprise Information Management*, *36*(1), 252–274. DOI: 10.1108/JEIM-06-2021-0276

Sinha, S., & Harrold, M. J. (2000). Analysis and testing of programs with exception handling constructs. *IEEE Transactions on Software Engineering*, *26*(9), 849–871. DOI: 10.1109/32.877846

Smeets, M., Erhard, R., & Kaußler, T. (2021). *Robotic process automation (RPA) in the Financial Sector*. Springer Fachmedien Wiesbaden. DOI: 10.1007/978-3-658-32974-7

Syed, R., Suriadi, S., Adams, M., Bandara, W., Leemans, S. J., Ouyang, C., ter Hofstede, A. H. M., van de Weerd, I., Wynn, M. T., & Reijers, H. A. (2020). Robotic process automation: Contemporary themes and challenges. *Computers in Industry*, *115*, 103162. DOI: 10.1016/j.compind.2019.103162

Venigandla, K., & Vemuri, N. (2022). Autonomous DevOps: Integrating RPA, AI, and ML for Self-Optimizing Development Pipelines. *Asian Journal of Multidisciplinary Research & Review*, *3*(2), 214–231.

Wallace, E. (2022). *The Role of Organisational Change Management in Robotic Process Automation Implementation Projects* (Doctoral dissertation, Auckland University of Technology).

Chapter 2
RPA in Banking and Finance

Rohit Yadav
https://orcid.org/0000-0002-7573-8005
IILM University, Greater Noida, India

Sandeep Kumar Singh
https://orcid.org/0000-0002-1741-7254
O.P. Jindal Global University, India

Thi Minh Ngoc Luu
https://orcid.org/0000-0002-5972-7752
International School, Vietnam National University, Hanoi, Vietnam

Ali Najeeb
https://orcid.org/0000-0003-4663-9347
Qasim Ibrahim School of Business, Villa College, Maldives

ABSTRACT

Robotic Process Automation has revolutionized the banking and finance industry by automating routine, time-consuming tasks, thus increasing efficiency, accuracy, and adherence to regulatory requirements. The chapter then discusses the RPA functionalities in numerous financial processes, such as transaction monitoring, customer onboarding, and reporting towards regulatory compliance, along with the essential benefits-such as savings, reduced errors, and scalability. It further elucidates the interfacing of RPA with other advanced technologies, AI, and ML, for predicting a future of hyper-automation and more customer-centric solutions. The chapter further delineates the problems in the implementations of RPA, such as interfacing RPA with legacy systems, issues with data security, and employee adaptation. In banking and finance, RPA will be much more pronounced in deeper

DOI: 10.4018/979-8-3693-4365-4.ch002

automation capabilities, more fluid customer experiences, and a hybrid human-robot workforce. The ever-growing reliance on RPA will make its users stand out against the increasingly digital financial environment.

INTRODUCTION TO RPA IN BANKING AND FINANCE

In the present scenario, RPA is revolutionizing industries by automating standard repetitive work based on some pre-defined rules that have otherwise been done manually. For the banking and finance sectors, accuracy, adherence, and effectiveness play a very significant role. For these industries, there has emerged a revolutionary technology called RPA (Villar & Khan,2021). An organization can simplify its processes by duplicating human activity related to data typing or entry, processing transactions, and even customer communication over various digital channels using RPA. Most likely, by automating these processes, which are quite repetitive, banks and other financial institutions will cut down on errors, make their operations more efficient, and will thereby provide better quality services to their customers. This chapter explores the growing impact of RPA within the banking and finance industries and how it's shaping the future of the industry (Tekkethil et al.,2021) Probably, the high volume of transactions is the most significant driver behind the adoption of RPA in the banking and finance sector, along with many complicated regulatory requirements which must be adhered to. There is vast customer data managed by banks and financial firms processing millions of daily transactions and following strict regulations on anti-money laundering, Know Your Customer, and several other financial reporting standards. Traditionally, all this type of activity is performed manually by teams of employees. In a bank or financial institution, RPA will enable most of this work to be automated, thereby improving the velocity of processing while at the same time improving speed, accuracy, and compliance. With RPA, for instance, the KYC process can be automated to speed up the verification of customers' identities, applying minimum resources while reducing risks associated with human factors (Kumar & Balaramachandran,2018).

Beyond improving operational efficiency, RPA in banking and finance helps improve customer experience. Customers nowadays want faster and more efficient services, be it opening a new account or applying for a loan, or managing existing ones (Romao et al., 2019). Through RPA, these expectations are indeed addressed as back-end processes speed up different customer-facing services. For instance, processing of loans that used to take days or weeks can now be accomplished within hours using RPA. Automating labor-intensive work such as document verification, data entry, or credit assessment also brings banks quick service to customers. It approves loans with higher percentages of the customers satisfied and retained.

Besides that, RPA has proven itself to be the most cost-effective approach for financial institutions. Automation of routine tasks greatly reduces labor costs within the banks and forces the employees towards activities of higher value, such as strategic decision-making, relationship management, and product innovation (Lamberton et al.,2017). The business environment can change rapidly in hours, and RPA is scalable up or down on such changeovers, helping banks handle the strange spurt of transactions at any given time without necessarily expanding their headcount. Its scalability and flexibility make RPA an incredibly hot investment opportunity for big global banks and small financial institutions alike (Yarlagadda,2018).

Notably, adopting RPA in banking and finance has its own share of challenges. An institution needs to carefully integrate RPA with legacy systems, ensure that data is kept safe and private, and iron out issues connected with displacement of people. The benefits for an institution in terms of increased efficiency in operations, cost saving, improved compliance, and enhanced customer experience outweigh all these challenges. As bank as well as financial institutions are increasingly adopting RPA, in the course of events, the technology is going to be much more integral to the future of the evolution of the banking industry, opening doors and paving the way for even advanced automation as well as intelligent systems in the near future (Madakam et al., 2019).

CASES OF RPA IN BANKING

1. Customer Onboarding and KYC Automation (Global)

The most common application of RPA for the banking sector is the onboarding of customers, with Know Your Customer (KYC) compliance included within the process. These KYC regulations demand that banks identify and verify their customers against fraud, money laundering, among other illegal transactions by cross-checking the documents. This has traditionally been more of a manual verification process where documents get cross-checked, data is inputted, and verified against regulatory databases. With RPA, HSBC and other banks have streamlined a large part of the onboarding process saving considerable time in opening a new account(Romao et al., 2019). RPA bots can also access multiple systems to collect and verify data to hasten customer verification while complying with regulatory conditions in the shortest time possible. Such automation eliminates errors, expedites customers' onboarding times, and improves customer satisfaction.

2. Loan Processing and Approval (North America)

Their processing times are generally long because of the multiple sources of information required, credit assessment of the customer, and documentation. RPA systems have streamlined this for North America's banks, such as JPMorgan Chase and Bank of America. Bots using RPA can quickly pull in the details of their customers, do the credit assessment, and cross-check relevant documentation to ensure loan approvals in a significantly shorter time compared to traditional methods (Pramanik et al.,2019). This automation does aid the banks in making timely decisions on personal loans, mortgages, and business financing. It is error-free when it makes decisions. It frees the bank employees from spending too much time on dealing with minor cases that eventually require less human judgment.

3. Fraud Detection and Compliance Monitoring in Europe

The function of fraud detection and compliance is the most vital in the banking industry, especially in Europe. Its regulatory requirements are stronger compared to other parts of the world. Some major banks, including Barclays in the UK and BNP Paribas in France, have opted to use RPA solutions side-by-side with fraud detection systems. Since RPA can monitor transactions in real-time, suspicious activities are flagged by bot on specific predefined rules and patterns(Jędrzejka,2019). Hence, the compliance teams are relieved since fraudulent transactions are caught early, and action can be taken concerning them in time. RPA also assists in regulatory reporting by ensuring compliance with regulations such as the EU General Data Protection Regulation (GDPR) and also PSD2.

4. Transaction Processing and Reconciliation (Asia)

In Singapore and India, where digital transactions are on an unprecedented increase, banks face the daunting task of processing and reconciliation of millions of transactions daily. Banks in Singapore, like DBS Bank, and in India, like ICICI Bank, have utilized RPA to manage all such business operations in the most efficient manner. The RPA bot does automate all transactions, compares records across multiple systems, and resolves discrepancies in real-time. Such repetitive tasks result in minimal human interference, lesser processing times, and lesser chances of errors in banks(Verno,2019). This automation is particularly useful for high transaction volume management, such as foreign exchange, stock trading, and payment reconciliations.

5. Account Maintenance and Updates (Middle East)

The Middle Eastern region, particularly Emirates NBD and Qatar National Bank, widely utilize RPA to maintain accounts and perform recurrent updates. Common tasks involved in the cycle included renewal of customer information, password resetting, and changes of account settings, all left to customer service teams, which means error risks and delay times(Gosen, 2019).RPA bots can process such requests real-time, leaving chances of neither errors nor delay times. This ensures that the change of account occurs fast enough, hence improving the customer experience and reducing human capabilities of making complex requests for customers (Muzikante et al.,2018).

All these cases show how RPA is driving efficiencies in the world's banking operations. Work that a while back required a lot of time through manual processing is streamlined toward increased productivity, compliance, and improved service delivery to customers.

CASES OF RPA IN FINANCE

1. Invoice Processing and Accounts Payable Automation (United States)

Invoice processing and accounts payable are among the most common use cases of RPA in finance. The leading U.S. firms, General Electric (GE) and Coca-Cola, to name a few, have applied RPA to deal with the large number of invoices coming from vendors. RPA BOTs can automatically extract invoice data, validate it against purchase orders, and process payments without any human intervention(Madakam et al.,2019). This saves processing time, fully eliminates manual errors, and ensures timely payments. Accounts payable automation enables the finance departments to manage the cash flow more efficiently and prevents costly late-penalties payments.

2. Financial reporting with data aggregation (UK)

Financial reporting is a time-consuming activity that requires gathering data from various sources. Organizations in UK, such as BT and Barclays Bank, have already deployed RPA to automate their cycle of financial reporting. RPA bots can extract data from various systems, produce reports, or validate figures. It results in speeding up the time when financial close processes are executed(Jędrzejka,2019). This implies that you are always in compliance with regulatory requirement such as IFRS. Automated data aggregation and reporting enable finance teams to shift from focusing on financial trends, providing strategic recommendations, instead of being overwhelmed by the task of manually aggregating data(Gosen, 2019)..

3. Tax Compliance and Audits (Canada)

Tax compliance is a crucial role for the finance department, as it involves dealing with complex regulatory frameworks. For instance, RPA is used in Canada by Bombardier and Royal Bank of Canada (RBC) to streamline tax compliance processes. This means that financial information can be automatically mined by RPA bots, tax liabilities computed, and tax filings prepared according to Canadian tax laws(Stephan Hayes & Baker,2014). Moreover, in audits, RPA can quickly obtain historical records thus enabling auditors to scrutinize transactions and ascertain their compliance; this might reduce the risk of non-compliance and time expended on tax preparation and audit.

4. Treasury Management and Cash Flow Monitoring (Europe)

While at Europe, large multinational corporations such as Siemens in Germany and Unilever in the Netherlands, have integrated RPA in the management of treasury and for cash monitoring. The complexity of key treasury functions, such as cash forecasting, bank reconciliations, and foreign exchange transactions, makes it difficult to manage liquidity across a multiplicity of currencies and accounts (Verno,2019). RPA bots are able to monitor cash positions in real-time, initiate transfers, and optimize cash allocation across different business units. Automation improves cash flow visibility that enables treasury teams to make better decisions regarding investments, debt, and hedging.

5. Risk Management and Regulatory Reporting (Asia)

Countries within Asia, for instance, Hong Kong's HSBC as well as India's ICICI Bank, have adopted RPA to enhance the proficiency of risk management and incorporate automation in the regulatory reporting processes. RPA robots can collect information about risks coming from different sources, thus providing real-time risk appraisals for the effective decision-making of the decision-makers(Verno,2019). In addition, RPA delivers compliance to the regulatory requirements, and this includes adherence to both Basel III and local banking regulations. Bots enable financial institutions to accurately and timely report to the regulatory bodies, reducing the workload on the compliance teams due to manual work and ensuring the fulfillment of the dynamically shifting regulatory standards in the region.

These cases prove the adaptability and effect of RPA in finance functions across the world. Further, the organizations across the world will be able to reduce the operational costs, achieve high accuracy, and enhance their capability toward cash flow management, regulations compliance, and critical financial reporting.

BENEFITS OF RPA IN BANKING AND FINANCE

1. Increased Efficiency and Productivity

The first benefit of RPA in banking and finance is that this greatly increases efficiency and productivity. By automating repetitive, rule-based tasks such as data entry, transaction processing, and report generation, RPA enables processes to be completed faster and more accurately for banks and other financial institutions(Gosen, 2019). It could take days and even weeks to accomplish what is done in hours and even minutes now. The loan approval process, for one, took days to process; this was because it required data gathering from different systems. These are now processed more quickly. This efficiency in turn-around time enables financial institutions to process higher volumes of work without having to expand and hire more people, giving most of the employees more than enough time to do more strategic, value-adding work(Mazumdar & Jyoti, 2019) .

2. Operational Cost Savings

RPA is relatively cost-effective for financial institutions, enabling them to cut down operational costs. Banks can save on labor costs and efficiency loss resulting from human error due to the automation of activities that involved intervening in the past. More importantly, RPA generally requires less infrastructure investment as it rides on existing systems and does not require a large overhaul of their IT structure (Kaya et al.,2019). This reduction in overhead labor time leads to fewer errors and, subsequently, less rework, which further becomes cost-cutting. For example, a large bank can reduce the invoicing processing cost when utilizing the RPA because it automates the workflows from receipt to payment while inflicting the minimum amount of human intervention(Jędrzejka,2019).

3. High Accuracy and Compliance

Accuracy is very crucial in the banking and finance sectors, and consequently, errors in data processing may lead to high regulatory penalties or financial loss. With RPA, the processes are accurate. It eliminates human mistakes when there is data entry, a report on regulations, and a reconciliation of transactions. This mainly helps in compliance with complicated regulations such as AML or KYC norms. Bots provide assurance regarding the consistent execution of processes. Processes will be carried out according to the rules and standards predefined, and real-time reports can be submitted to the regulatory authorities. This will help financial institutions reduce the risk of non-compliance and penalties while ensuring they meet stringent regulatory deadlines(Devarajan, 2018)

4. Enhanced Customer Experience

In today's scenario of competitive banking, the experience a customer has defines their bank. RPA will help facilitate banks and financial institutions' ability to improve their customer experience by streamlining those processes that have direct impacts on customers. For example, automating account openings, loan processing, and customer queries means that banks are better placed to respond to customer needs quickly and efficiently. This results in faster resolution of complaints, less wait time for problems to be fixed, and overall better provision of services(Mazumdar & Jyoti, 2019) . Through the integration of RPA into customer service platforms, banks would be able to offer 24/7 support, level out high volumes of inquiries, and use automation on multiple choices through chatbots to give more personalized services that benefit the customers.

5. Scalability and Adaptability to Regulatory Changes

The financial sector is constantly evolving due to new regulations facing institutions which must adapt fast to such changes to remain compliant. RPA offers the necessary scalability and adaptability to such changes. In contrast to months or even years long implementation of traditional IT solutions, several RPA bots can be configured quickly to adapt to new regulatory requirements or increased transaction volumes(Kaya et al.,2019). This flexibility allows banks to scale the operations without major investments in new infrastructure or staff. For instance, during peak seasons or market fluctuations, peaks in workload become easy for handling by

RPA, thus running their operations smoothly and not necessarily hiring additional personnel for such times.

Summarily, the implementation of RPA in banking and finance implies a package of benefits, ranging from cost saving and accuracy enhancement to superb provision of services to customers and fulfilment of the statutory requirements. Financial institutions can hence optimize their operations, respond promptly to changing market conditions, and ascertain a competitive advantage while implementing RPA.

CHALLENGES AND RISKS OF IMPLEMENTING RPA IN FINANCIAL SERVICES

1. Integration with Legacy Systems

The core challenge in deploying RPA in financial services is the integration of the new system with legacy structures. Most banks and financial institutions today use infrastructure originally created years ago and not at all designed to cope with modern automation technologies. Legacy systems rarely provide for interoperability, which can make it difficult for RPA bots to extract, process, or input data without significant customization. This could, therefore, lead to long implementation time and higher costs(Kamat, 2019). IT teams must spend a significant amount of time and resources for RPA solutions to be compatible with the system already in place. Further, operational discontinuities or failures may also result from inadequate integration and defeat the efficiency promised by RPA solutions.

2. Data Privacy and Security Vulnerabilities

One of the major issues surrounding the implementation of RPA in financial services is handling sensitive financial information. As these RPA bots offer access to such essential customer information and transactional data, they have a highly probable chance of being attacked by hackers or suffering from a data breach(Kochhar et al.,2019). Any vulnerabilities that could be found in the automated system would mean penetration into private financial information from unauthorized and unauthorized sources, thereby creating regulatory fines, impairment of reputation, and loss in customer trust. It will be the task of financial institutions to ensure strong security-it must ensure encryption, role-based access control and continuous monitoring -so that the system cannot compromise privacy of data (Strömberg,2018).In addition, financial institutions would need to pay attention to achieving compliance with Data protection, such as GDPR and CCPA.

3. Initial High Costs and Complex Implementation

While RPA is often sold as a cost-reducing tool, installing RPA to an organization is itself a costly and complicated process, especially for large financial organizations that have a high degree of operational heterogeneity(Strömberg,2018). The installation costs of RPA include licensing fees and the infrastructural upgrades in place to support the system. The one-time RPA bots further increase this list of costs since they are configured to accept various forms of processes. Not on this list, however, is hiring or training in-house RPA experts, which increases the implementation costs of RPA(Gosen, 2019). In other cases, the financial operations are so complex that a total reengineering of the processes needs to be done in order for automation to be deployed effectively. For organizations whose work processes are quite complex, the time and resources that go into mapping and automation lead to a slower ROI than expected.

4. Job Loss and Employee Resistance

The reasons for automation cause anxiety about job loss and inspire resistance among employees because automating tasks people have typically performed gives grounds for such apprehension. This has occurred right here in the financial services industry, where RPA has placed most of work processes, such as data entry, customer service, or transaction processing, under its control(Kamat, 2019). It might instill fear in the workers on whether their job is at risk, with the resultant adverse effects of no buy-in from staff, who may not be eager to embrace RPA technologies and who is not enrolled in necessary training programs. Leading this kind of resistance through proper communication from the top will involve a clear messaging on how RPA can free employees away from repetitive tasks, which will allow them to focus on higher-value responsibilities(Strömberg,2018). A very holistic approach in change management will be needed to alleviate employee concerns and develop a culture that supports automation.

5. Maintenance and Bot Management

RPA bots, once implemented, need to be maintained and managed regularly to ensure that they continue working properly. The rules and processes that the RPA bot is programmed to handle will change with financial services updates (Kochhar et al.,2019). If the bots are not monitored, they may not adapt to variations of regulations, system updates, or process variations, all of which can result in operational failures or inaccuracies in data. Financial institutions need to provide continuous monitoring, reconfiguration of the bot, and system update resources that prevent all these issues. In addition, scaling RPA operations is complex, which requires a more advanced infrastructure to support thousands of bots, ensuring performance and complying.

6. Compliance and Regulatory Challenges

Financial services operate in an environment of high regulations. The use of RPA has now brought in some new compliance challenges. The regulatory bodies require transparency and accountability in financial processes, and RPA bots may complicate the audit trail. The activities of the RPA bots ought to be completely traceable, and in line with relevant regulations in banks(Strömberg,2018). In case of operations through RPA not documented clearly, regulatory penalties may be incurred. Additionally, new regulations may also be instituted by regulators as far as guiding the use of automation technologies; hence, refreshing their RPA systems from time to time will allow financial institutions to be in line with the new rules and regulations(Kochhar et al.,2019).

In a nutshell, RPA may have considerable benefits for the financial services industry, but it's no problem-free and has challenges with risks. Organizations need to tread very cautiously and effectively address the many issues of integration complexity, data security concerns, and regulatory hurdles while providing for employee resistance and ongoing maintenance requirements. Proactive exposure to these challenges helps the financial institutions realize maximum potential from RPA while keeping risks associated with RPA to a minimum.

RPA AND COMPLIANCE IN FINANCIAL SERVICES

1. Improved Compliance with Regulations in Financials with Automations

Global and regional regulations are stringent and dynamic in nature, and so top on the list for all financial institutions is regulatory compliance. Here again, the role of RPA has been instrumental for banks and other financial houses by streamlining tasks associated with compliance to ensure that the institution meets its legal and regulatory obligations efficiently(Kamat, 2019). RPA bots can automate compliance activities that typically involve repetitive actions such as AML checks, KYC verification, and transaction monitoring. RPA usage ensures that all human errors in business processes are removed and leads to good consistency in the proper execution of requirements based on the regulations(Gosen, 2019). For example, RPA can validate customer information from external databases, flagging suspicious activities so that institutions comply with the AML directives and therefore enhance their positions in dodging fines or penalties.

2. Audit Trails and Reporting

RPA provides an auditable and transparent record of all bots' activities. Auditing is quite essential in the finance industry, which is highly regulated. Financial institutions have to maintain comprehensive records of all the transactions and activities involved for audit purposes(Devarajan, 2018) RPA helps generate real-time reports and maintains a digital audit trail that captures every step taken by the bots. Such an audit trail is very valuable during regulatory audits because it provides clarity with a time stamp on the actions taken, thus supporting the compliance team and the auditors. The capacity to present regulators with clear documentation of the automated processes further strengthens the compliance posture of an institution and keeps the regulatory authority off its back.

3. Respond to Dynamic Regulation

Financial services operate in a dynamic legal environment with the regular emergence of new regulations and guidelines. A financial services firm should react quickly to these changes and make sure never to violate a regulation again. RPA can easily reconfigure the bots to suit new regulatory requirements. For example, if there is a new regulation on KYC, the rules can update an RPA bot to collect more data as requested by the rule so institutions meet the updated requirements (Van Chuong et al.,2019).For instance, this flexibility saves time and effort that goes into continuing to comply with the updated rules, allowing institutions to be compliant without having to overhaul their entire compliance infrastructure.

4. Risk Management and Monitoring Automation

Risk management is an integral part of compliance in the financial sector. It demands continued monitoring and pro-active action against threats on risk by being watchful of probable risks and mitigating them. RPA can automate processes that may be linked to risk management tasks such as real-time monitoring of transactions, flagging suspicious activity, and identifying potential compliance breaches(Kamat, 2019). For example, RPA bots can be configured such that once anomalous transaction patterns have been detected, it would alert compliance teams to intervene due to possible fraudulent or money-laundering transactions(Van Chuong et al.,2019). In addition, RPA allows for continued and real-time monitoring of various business units' compliance practices regarding regulatory compliance such that they comply with Basel III or even SOX requirements without necessarily having to do so manually.

5. Improving the Precision of Regulatory Reporting

Regulatory reporting usually entails a step-by-step procedure that is time-consuming and calls for minute attention. Financial institutions must make elaborate reports to the regulatory bodies on the form and nature of their operations: capital adequacy, liquidity ratios, as well as other exposures to risk(Renner, 2019). RPA will ensure that the output and the real-time delivery of regulatory reports increase, as multi-system extraction handles the collection, consolidation, and validation of data. Bots would see to it that the information was placed correctly and that submissions of reports would be done in time to avoid missing submission deadlines, penalties for submissions considered late or wrong (Kaya et al.,2019). Automating regulatory reporting enables financial institutions not only to realize an improved overall compliance but also free up their resources to devote to more strategic functions.

6. Compliance Risk Reduction

Failure to observe financial regulations would hold major consequences. Such penalties include heavy fines and reputational damage. The above are six major areas whereby bots' impact financial operations positively(Kochhar et al.,2019). Thus, RPA reduces such risks with compliance activities by executing it with precision and fewer delays. It reduces the dependence on human intervention, which is error-prone as well as prone to skipovers. For example, using RPA ensures that deadlines for any activity, be it tax filing or reporting, are adhered strictly without missing the deadline or failing to submit reporting fully(Mazumdar & Jyoti, 2019) . Automation of such critical compliance functions grants financial institutions a protective shield against regulatory risks while maintaining operational integrity.

RPA has extensive benefits for financial service sectors in terms of regulatory compliance. RPA enables the financial institution to be able to have a strong, proactive compliance framework through the automation of tasks relating to compliance, improved auditability, and the ability to adapt to any changes in regulations(Madakam et al.,2019). RPA not only saves operation cost but also minimizes the risk of non-compliance. It is very much possible to operate the institutions in an efficient and effective manner according to legal frameworks.

THE FUTURE OF RPA IN BANKING AND FINANCE

1. Integration of RPA with Artificial Intelligence (AI) and Machine Learning (ML)

Future scope of RPA in banking and finance is the incorporation of AI and ML technology in it. Currently, an organization takes pride in the fact that it automates repeated and repetitive tasks through RPA; hence, it's moving up to more complex decisions, along with predictive capability using AI and ML. For example, AI-based RPA would analyze customer behavior and financial data to offer customized services, detect fraud in real time, and streamline the process of credit risk assessment(Madakam et al.,2019) . Intelligent bots, in the future, will no longer be just confined to implementing simple instructions but will evolve with changing situations, thus making the financial process more efficient and accurate. This alliance of RPA with AI and ML will empower the financial institute to upgrade customer services and streamline internal processes and innovate at a faster pace.

2. Hyper-automation and End-to-End Process Automation

Hyper-automation-the leverage of multiple technologies such as RPA, AI, ML, and advanced analytics to automate end-to-end workflows-is going to mark one of the trends that are going to define the future of banking and finance. More use of hyperautomation would be seen in financial institutions in the area of end-to-end process automation-in customer onboarding, loan processing, risk management, and regulatory reporting. This would eliminate much of the human intervention involved and make the operations of banks smooth, fast, accurate, and of high quality. In this way, the financial institution will be able to deliver services at unprecedented speed, improving compliance, and minimizing costs. Hyperautomation will be a critical driver of agile financial services in response to dynamic market conditions and changing regulations(Madakam et al.,2019) .

3. Increased Focus on Customer-Centric Automation

The increased requirements for customized and seamless experience for customers will increase the customer-centric focus of RPA in banking and finance. Advanced RPA will enable financial institutes to deliver highly responsive, customized services through automation. For example, RPA bots can provide just-in-time financial advice, tailored product recommendations, and on-the-fly customer support. Banks will utilize RPA-based chatbots and virtual assistants that can comprehend multiple-layer customer queries in real time and resolve them on the go(Madakam et al.,2019) . This customer-centric approach will transform the end user experience and enhance customer loyalty and satisfaction since financial institutions will be better agile and responsive to individual needs.

4. Scalability and Cloud-Based RPA Solutions

Cloud-based RPA solutions are going to dominate the future of RPA in banking and finance. Financial institutions, in pursuit of digital transformation, will slowly drift towards cloud-based RPA platforms as they offer scalable, flexible, and cost-effective solutions. Cloud deployment makes it easy for financial institutions to scale their automation projects in a more fluid manner by expanding their RPA capabilities as needed without much investment in infrastructure. In addition, cloud-based RPA will enable faster deployment, simpler upgrades, and stronger disaster recovery. The scalability of cloud-based RPA solutions will assist financial institutions to respond promptly to fluctuating transaction volume, customer demand, and regulatory requirements more effectively, making RPA even more appealing to banks and financial organizations(Madakam et al.,2019) .

5. Enhanced Regulatory Compliance and Risk Management

Financial institutions will find it core to ensure they remain able to comply with regulations while working within defined limits to reduce risk. Future RPA: It will involve further advanced automation tools built specifically for handling change and compliance in the regulatory front. These RPA solutions will be able to monitor regulatory updates and ensure processes are up to date with the latest standards. They will generate compliance reports automatically. RPA integrated with AI: This will help institutions identify risks proactively and address errors proactively before issues arise. By managing compliance dynamically, RPA will reduce pressure on compliance teams and mitigate risks of non-compliance.

6. Collaborative Human-Robot Workforces

Going forward, much of the future for RPA in banking and finance would be dedicated to co-work between human employees and the RPA bots. Rather than supplant human workers, RPA would increasingly augment human positions as leaving high-order thinking to its human counterparts-professionals with the skills to determine strategy, solve problems, keep personal relationships, and make effective contributions(Madakam et al.,2019) . Financial institutions would bring into being “digital workers,” or software bots, alongside the human employees to ensure the completion of tasks in higher productivity and efficiency. As an example, their RPA bots would assist the banks to do routine, mundane administrative work, freeing up their employees for such interactions with clients, complex analyses, and innovative approaches towards problem-solving. This model of a collaborative workforce will enable the banks and financial firms to achieve higher degrees of operational agility and innovation while enabling more employees to take on more fulfilling and meaningful roles.

7. Expansion into New Financial Services Areas

As such, evolution in RPA technology will not only continue within traditional banking processes but also open further areas of financial services on a large scale. In the coming years, RPA will be applied to higher sophisticated tasks like investment management, wealth advisory services, and for insurance claims processing(Madakam et al.,2019) . RPA bots may be able to perform portfolio management, be able to analyze market trends, and be able to provide advice based on customer profiles. For instance, there is the insurance sector, which will witness inflows of RPA directly to simplify claims processing, risk assessment, and policies related to service delivery while reducing costs. Spreading into these new domains with RPA will enable financial institutions to innovate their services and meet the ever-changing needs of customers.

In conclusion, the future of RPA banking and finance promises revolutionary changes in efficiency, scalability, and service. Certainly, as it is interfaced with AI, through hyperautomation, and keeping a customer-centric focus for solutions, RPA is going to be at the center of driving the financial institutions' digital transformation. Banks and financial firms embracing more advanced RPA technologies will make them better companies to handle regulatory changes, scale up operations, and offer a superior customer experience.

CONCLUSION

With the coming of RPA, banking and finance has already seen a shift. Its integration is changing the sector, bringing in transformative efficiency, accuracy, and customer service. As institutes in the sector move forward further to embrace automation, RPA has progressed from just being a tool that aids in the automation of routine and repetitive processes toward an integrated holistic solution that enables smarter and more adaptive processes. RPA benefits add to the growing winds of change in the protracted banking and financial services industry toward cost savings, scalability in operations, and improved compliance-from which banks and financial firms are set to benefit on a competitive advantage.

RPA can add a chance for regulation compliance, risk management, as well as automated back-office processes, making it a key component of the future of financial services. RPA would generally lower the risks involved with manual processes because automation is entrusted with key functions like transaction monitoring, customer onboarding and regulatory reporting. When used along with advanced technologies like Artificial Intelligence (AI) and Machine Learning (ML), financial institutions shall better be in a position to offer a much more customized and responsive service so as to firm up its grip even further over a quite competitive market space.

While RPA does bring about transformative benefits, it also has its own set of challenges and risks-including integration into legacy systems, security and privacy of data, and employee acceptance. Solutions to these therefore need a strategic approach in terms of a change management plan, cybersecurity, and ongoing process monitoring. Beyond this, however, is the fact that financial institutions' RPA solutions are flexible, scalable, and compliant with changing regulatory environments and customer expectations.

Future of RPA in banking and finance: Innovation and expansion The future of RPA in banking and finance will be laden with innovation and expansion regarding hyper-automation, cloud-based solutions, and customer-centric automation of services and products. Financial institutions will provide more agile, efficient, and personalized services when they interact with the new ecosystem of hyper-automation, cloud-based solutions, and customer-centric automation. Human employees and RPA bots will work in tandem to redefine the workforce, making human employees focus on strategic, value-added tasks instead of settling for routine tasks generally handled by automation.

In general, RPA will strongly influence the processes of banks and financial institutions by reducing costs and improving customer experiences. As this technology develops and becomes part of intelligent systems, the financial services sector will enjoy further substantial improvements in performance and service delivery. The institutions that invest in this technology will be ready for the complexities of a future financial landscape and take advantage of the opportunities it offers.

REFERENCES

Devarajan, Y. (2018). A study of robotic process automation uses cases today for tomorrow's business. *International Journal of Computers and Technology*, *5*(6), 12–18.

Gosen, N. (2019). *How Robotic Process Automation (RPA) influences firm financial performance in The Netherlands* (Master's thesis, University of Twente).

Jędrzejka, D. (2019). Robotic process automation and its impact on accounting. *Zeszyty Teoretyczne Rachunkowości*, *2019*(105), 137–166. DOI: 10.5604/01.3001.0013.6061

Kamat, A. (2019). Challenges of robotic process automation adoption in banking and financial services. *International Journal of Research and Analytical Reviews*, *6*(2), 597–606.

Kaya, C. T., Türkyılmaz, M., & Birol, B. (2019). Impact of RPA technologies on accounting systems. *Muhasebe ve Finansman Dergisi*, (82).

Kochhar, K., Purohit, H., & Chutani, R. (2019). The rise of artificial intelligence in banking sector. In *The 5th International Conference on Educational Research and Practice (icerp)* (Vol. 127).

Kumar, K. N., & Balaramachandran, P. R. (2018). Robotic process automation-a study of the impact on customer experience in retail banking industry. *Journal of Internet Banking and Commerce*, *23*(3), 1–27.

Lamberton, C., Brigo, D., & Hoy, D. (2017). Impact of Robotics, RPA and AI on the insurance industry: Challenges and opportunities. *Journal of Financial Perspectives*, *4*(1).

Madakam, S., Holmukhe, R. M., & Jaiswal, D. K. (2019). The future digital work force: Robotic process automation (RPA). *Journal of Information Systems and Technology Management*, *16*, e201916001. DOI: 10.4301/S1807-1775201916001

Mazumdar, A. C., & Jyoti, A. (2019). Automation of Financial Services Using Artificial Intelligence with Human Touch. *International Journal of Modern Engineering & Management Research.*

Muzikante, I., Kumar Agarwal, M., & Turner, J. (2018). Robotic process automation risk analysis in circle K business centre. *Bachelor thesis, Riga Technical University, Riga.*

Pramanik, H. S., Kirtania, M., & Pani, A. K. (2019). Essence of digital transformation—Manifestations at large financial institutions from North America. *Future Generation Computer Systems, 95*, 323–343. DOI: 10.1016/j.future.2018.12.003

Renner, M. (2019). Software robot-based automation of financial administration's processes.

Romao, M., Costa, J., & Costa, C. J. (2019, June). Robotic process automation: A case study in the banking industry. In *2019 14th Iberian Conference on information systems and technologies (CISTI)* (pp. 1-6). IEEE. DOI: 10.23919/CISTI.2019.8760733

Stephan Hayes, R., & Baker, R. (2014). A participant observation study of the resolution of audit engagement challenges in government tax compliance audits. *Qualitative Research in Accounting & Management, 11*(4), 416–439. DOI: 10.1108/QRAM-02-2013-0003

Strömberg, K. (2018). *Robotic Process Automation of office work: benefits, challenges and capability development* (Master's thesis).

Thekkethil, M. S., Shukla, V. K., Beena, F., & Chopra, A. (2021, September). Robotic process automation in banking and finance sector for loan processing and fraud detection. In *2021 9th international conference on reliability, infocom technologies and optimization (trends and future directions)(ICRITO)* (pp. 1-6). IEEE. DOI: 10.1109/ICRITO51393.2021.9596076

Van Chuong, L., Hung, P. D., & Diep, V. T. (2019, July). Robotic process automation and opportunities for Vietnamese market. In *Proceedings of the 7th International Conference on Computer and Communications Management* (pp. 86-90). DOI: 10.1145/3348445.3348458

Verno, J. (2019). Risk management in corporates; can digitalization help improve risk management in cash management?.

Villar, A. S., & Khan, N. (2021). Robotic process automation in banking industry: A case study on Deutsche Bank. *Journal of Banking and Financial Technology, 5*(1), 71–86. DOI: 10.1007/s42786-021-00030-9

Yarlagadda, R. T. (2018). The RPA and AI automation. *International Journal of Creative Research Thoughts (IJCRT), ISSN*, 2320-2882.

Chapter 3
RPA for Secure Systems and Networks

Rohit Yadav
https://orcid.org/0000-0002-7573-8005
IILM University, Greater Noida, India

Sandeep K. Singh
https://orcid.org/0000-0002-1741-7254
O.P. Jindal Global University, India

Huy Hung Ta
https://orcid.org/0009-0008-6835-3036
International School, Vietnam National University, Hanoi, Vietnam

Damith Sanjaya Kumara Gangodawilage
https://orcid.org/0000-0001-7968-6284
Qasim Ibrahim School of Business, Villa College, Maldives

ABSTRACT

Robotic process automation (RPA) becomes a critical aspect to be included in most modern systems and networks to enhance cybersecurity by automating simple tasks, integrating workflows, and improving incident response. This chapter provides a study of the concept of RPA to implement its various dimensions of network security, such as access control and identity management data protection, compliance, and threat intelligence. The chapter also elaborates on the challenges and constraints posed for RPA adoption in network security in terms of integration and adaptability to dynamic, evolving cyber threats. Integrating AI into RPA promises to create adaptive intelligent and proactive security automation systems better suited for AI-driven RPA-the next frontier in cybersecurity. Features like predictive threat detection, real-time response, and enhanced compliance add an exciting dimension

DOI: 10.4018/979-8-3693-4365-4.ch003

to cybersecurity. And so, the chapter ends with the call for further innovation in the face of very changing threats to cybersecurity.

INTRODUCTION TO RPA IN SECURE SYSTEMS AND NETWORKS

Robotic Process Automation, shortly known as RPA, has gained significant heights across industries as it is proficient in automating repetitive, rule-based tasks with higher accuracy and efficiency. In the field of cyber security and secure systems, the tool of RPA is emerging to be critical in enhancing network security by automating some tough processes such as detection, response to threats, and compliance management. All these have made the digital landscape increasingly complex, and frequency coupled with the sophistication of cyberattacks have made traditional security approaches both outdated and ineffective. RPA comes in, wherein routine security tasks are automated to free human security professionals to work more strategically by focusing on the highest priority threats (Murugappan & Sree Kala, 2022).

RPA uses software robots, often referred to as "bots," to mimic the human interaction with the digital system, to execute its activity. These bots can pretty much perform a variety of security functions, such as observing the network traffic, suspicious activities' identification, managing system configurations, and access control protocols. Organizations can achieve a more secure network environment using RPA because it reduces the aspect of human error in its operations and hence, makes operational efficiency higher and faster in terms of threat responses. The most significant advantage of RPA is its ability to work continuously and therefore will furnish constant surveillance over high priority systems and data, which is what will save organizations from the constant danger looming on cyber threats.

As cyber threats evolve, safe systems and networks face numerous challenges ranging from ransomware attacks to data breaches and phishing schemes (Kurylets & Goranin, 2023). In most cases, human errors or delay in responding to security incidents magnify the damage that perpetrators of such threats enjoy. RPA mitigates this through the eradication of the need for human intervention in most tasks. For example, using RPA bots, software vulnerabilities can be automatically patched immediately when patches become available. This reduces the window of exposure of these systems to the risk. This rapid automation ensures that the systems are in compliance with security controls without reliance on the impermanent and cautious human interfaces.

RPA also finds a notable place in the organizational system while it comes to securing operations processes; it acts an asset to keep them secure as well. Access rights and security policies may also be implemented by configuring bots to determine which personnel are accessing sensitive data or systems in the network. Much of this capability is valuable only in large organizations with a huge complexity of infrastructures, where otherwise access control occurs far too inefficiently and with much greater opportunities to go wrong through manual approaches. RPA standardizes and streamlines these processes toward improvement of the overall security posture in the organization.

Another important function of RPA in secure systems is regulation compliance. Organizations, especially those in finance and healthcare, have large obligations over the protection and privacy of their data. When using RPA, an organization can monitor a continuous scan for risks within systems, which may automatically generate audit trails and compliance reports. This would save time and resources, in addition to low risks for non-compliance, which sometimes results in legal implications and loss of prestige.

In conclusion, RPA is changing the way secure systems and networks are dealt with in complex digital environments. The RPA automates extensive processes such as monitoring threats, access controls, and compliance checks. These bring together security efficiency with the efforts of cybersecurity professionals to focus more on the other challenging tasks. The more the digital world grows, the more vital the role RPA will play in the hardening of secure systems against cyber threats.

KEY SECURITY CONCERNS IN MODERN SYSTEMS AND NETWORKS

The dynamic nature of technology also generates parallel threats against systems and networks. From highly sophisticated cyber-attacks, which are turning out to be quite common, to the vulnerabilities inside, modern infrastructures are plagued by a wide array of security concerns. Below are some of the most pressing security concerns to address on behalf of organizations for the protection of their network and data integrity.

1. Malware and Ransomware Attacks

The most prevalent threat to security nowadays is malware, with viruses, worms, and ransomware. Of these, ransomware has been highly publicized because it locks users out of their systems or encrypts valuable data and demands a ransom for its restoration. With malware, the infection can then spread rapidly once penetrated into the network. It can also bring operations to a standstill, take sensitive information, or freeze a whole system. These threats have been observed attacking hospitals, government institutions, and multinationals, resulting in huge financial and reputational loss (Khan & Tailor, 2023).

Advanced threats, such as APTs, and fileless malware, which run without traditional footprints, would require more effort to detect and remove. Advanced measures must, therefore, be implemented: organizations would need advanced security measures along with automated solutions such as RPA to discover and respond to these threats well in time

2. Insider Threat

Probably the most significant threat in today's networks is the insider attack, whether from mischief or mistakes by employees or others with authorized access. Employees or contractors with a key position and privilege of access to confidential data can compromise the integrity of the network either consciously or unconsciously (Rawat et al., 2023). Human mistake can indeed open up the door to more severe breaches through mishandling of data or failure to adhere to security protocols.

Insider threats are also very hard to control since they originate from insider users who have access to your critical systems. RPA helps you keep insider risks at bay by enforcing granular access controls, tracking all user activities, and enforcing strict security policies across the organization.

3. Phishing and Social Engineering Attacks

Phishing attacks, mainly through emails, still constitute one of the most effective ways to break into network security. The attackers through phishing schemes target the victim to extract vital information such as log-in passwords or financial information. Of late, their attacks through social engineering strategies have become sophisticated, where the attackers use social media information to craft highly personalized phishing attacks.

Traditionally, phishing and social engineering attacks avoid much of this security through reliance on human vulnerability. There is scope for reducing this risk by using automated systems like RPA, which scan e-mails for suspicious content, flag them as potential threats, and can even train employees to identify phishing sites via exercises simulating the ones.

4. Distributed Denial of Service (DDoS) Attacks

DDoS attacks are called for with a view to flooding the system or network with excessive traffic that renders it inaccessible to the legitimate users. DDoS attacks will have grown strong and incessant as more devices continue to get connected with the Internet of Things (IoT). The attacks will disrupt normal business operations, cause down time, and increase customer confidence loss (Kurylets & Goranin, 2024).

DDoS attacks are particularly dangerous since they can attack any small and large organization, regardless of the industry that they operate in. Automated solutions like RPA can help identify patterns of unusual traffic, trigger mitigation procedures, and prevent service disruption in real-time.

5. Vulnerabilities in Cloud-Based Systems

Such systems have now come to be a prerequisite for any business with the widespread adoption of cloud computing, but cloud infrastructure brings ease and along with it vulnerabilities. Shared resources, inadequate access control, misconfigurations-all these bring cybercriminals to target cloud systems. With a shared responsibility model for cloud security, gaps in protection can easily become an epicenter such that the two sets of security measures of the user organization and of the cloud service provider may be inadequate or misaligned.

RPA can automate cloud configuration checks, enforce consistent security policies, and alert administrators of potential vulnerabilities so that the cloud infrastructure remains secure.

6. Zero-Day Exploits

It is an attack against previously unknown vulnerabilities in software or hardware. Precisely, this kind of exploit is dangerous because it can spread even before developers get time to release the patches. Cybercrime is always on these glitches and finds ways to exploit them without wasting much time. Zero-day attacks are hacking computers with unpredictable blows. Therefore, zero-day attacks pose a big security problem.

One can easily integrate RPA into systems so that software updates and patches can be automatically applied once fixes are available, thereby reducing the window of vulnerability and the damage wrought by zero-day exploits (Bhattacharyya et al., 2023).

Today's systems and networks have to worry about a great many security threats-malware and ransomware, insider threats, and vulnerabilities in the cloud-while getting more sophisticated all the time, coupled with ever-more complex IT infrastructures.

While traditional security is important, more and more, automated tools like RPA are the main detection, prevention, and response against these advanced threats. Using automation will enable organizations to further build on their defenses with more resilience against persistent cyber threats.

APPLICATIONS OF RPA IN CYBERSECURITY

Robotic Process Automation (RPA) has now become an activity that cannot be overlooked in modern cybersecurity since it enables the organization to automate certain repetitive tasks that may take quite a long period while at the same time enhancing the accuracy of the results and response speed. In cases where cybersecurity threats are increasing, RPA transforms the management of security processes by streamlining operations, minimizing human errors, and speeds up incident response. Some of the applications of RPA include:

1. Automated Intrusion Detection Response

This includes the automation of intrusion detection and response in cybersecurity. They can integrate with Intrusion Detection Systems as well as Intrusion Prevention Systems that are monitoring traffic and system logs on a network 24/7. Bots detect in real-time suspicions, anomalies, or breaches and alert security teams, or can automatically send configured responses.

For example, if a bot sends signals that an unauthorized login attempt is coming from a suspicious location, it can immediately lock the IP address or quarantine the system affected so it could be further investigated. This means that such response occurs in a fraction of time it would otherwise take a person to act against such attacks, thus reducing the window through which attackers would use such vulnerabilities. Furthermore, automated responses by RPA are any case more efficient than incident management because security professionals are freed from lesser threats, which they may have to handle personally.

2. Automated patch management

One of the basic cybersecurity practices is updating security patches on systems; however, manual patch management is very time consuming and error prone. This is something that can be automated through RPA by periodically scanning for vulnerabilities in systems and applying patches once they become available. In this way, known flaws in security can be addressed quickly, thus limiting the window of opportunity offered to cybercriminals (Badmus, 2023).

RPA bots can also direct the patching process according to the severity of the vulnerabilities and the systems affected. This thereby allows for automated patch management wherein the organizations can maintain consistency levels on all devices while avoiding potential security gaps that may eventually lead to breaches.

3. Incident Management and Threat Hunting

Cybersecurity incidents should be addressed promptly in the fastest and most efficient manner to reduce damage. RPA can automate incident management workflows by logging the incidents or categorizing them as per severity levels or assigning them to appropriate teams. It, therefore, helps to streamline the tracking, documentation, and resolution of security incidents, reduces response times, and generally enhances efficiencies to tackle incidents.

Another domain in which RPA can be helpful is in proactive threat hunting by collecting and analyzing system logs and user behavior data as well as network traffic. Bots can look for patterns or IOCs to indicate a latent threat. This facilitates organizations in identifying and mitigating breaches before they gain scale.

4. Network Monitoring and Traffic Analysis

Continuously monitoring network traffic is important to identify anomalies that could represent a cyberattack or data breach. Such bots can be programmed to monitor network traffic to point out unusual patterns and deviations in activities. Bots process enormous data much quicker than a human and, in real time, provide insights into networks (Tailor et al., 2022).

Once the evil activity is detected, RPA can automatically send alerts to the cybersecurity team, can force changes in firewalls, or can launch an investigation. When RPA monitors the entire network, monitoring improves both the sensitivity and responsiveness of finding potential security incidents so that threats can be identified promptly.

5. Access Control and Identity Management Automation

Further, RPA can be put into the implementation of safety in an organization by automatically implementing IAM functionalities such as account provisioning, access permissions settings, and revocation of accounts in case an employee leaves the organization. Managing access to various systems and applications is a time-consuming and error-prone process involving manual intervention that could expose some sensitive data to unauthorized users.

RPA bots can facilitate uniform enforcement of access policies, thereby ensuring that only approved personnel are allowed access to different systems or access various pieces of information. The bots may also check access permissions automatically and review them regularly, raising red flags over irregularities or outdated access rights that may potentially represent threats to the system.

6. Data Security and Compliance Management

Data protection and compliance - RPA brings in high-value improvements in cybersecurity. RPA takes care of automated encryption activities, secure data transfer, and backup processes ensuring that sensitive information is always safe and secure. In addition to these, RPA aids the organizations in attaining compliance with guidelines like GDPR and HIPAA by automating checks and providing reports for audits (Al Zarooni & El Khatib, 2023).

For instance, they can always monitor for compliance violations. They can assist in the generation of alerts whenever compliance is not maintained and automatically document corrective actions. The risk of penalties is therefore minimized, and organizations maintain a firm grip over compliance toward legal and industry standards.

Today, in the realm of cybersecurity, RPA stands to be the game changer by automating everything from intrusion detection and patch management to incident response and access control. It simplifies the processes surrounding security and frees time for valuable members of the cybersecurity team to be placed for strategic challenges. Cybersecurity professionals are also able to outsmart more sophisticated cyber threats with scalable and agile solutions from this technological power.

RPA FOR SECURE ACCESS CONTROL AND IDENTITY MANAGEMENT

Access control and identity management RPA for secure access control and identity management Anyone's security standpoint in most organizations entails controlling access to any critical system or sensitive data. However, with an increase in cyber threats and complicated IT infrastructures, maintaining user access and identity verification processes in a manual manner is time-consuming and error-prone. Security is brought about by this RPA solution through the automation of these tasks, but at the same time, it enhances the efficiency of operations. Through the automation of such processes, RPA ensures consistency, prohibits human errors, and reduces response time as these enhance access control and identity management, hence making them much safer and faster to process.

1. Automating User Provisioning and De-Provisioning

One of the most prominent applications of RPA in access control is automating user provisioning and de-provisioning. In large organizations, managing user accounts across multiple systems and platforms can be a huge undertaking by size. Inevitably, exercise creating accounts, giving roles, and ensuring the right access permissions can become inconsistent and threaten security in many instances (Beer Mohamed & Hassan, 2022). Even the process involved in deactivating or changing access permissions when an employee leaves the organization or changes job creates a backhaul such that systems remain vulnerable to unfettered access for extended periods.

RPA addresses these challenges by automating the process of account creation for a user, roles, and de-activation. Bots can be programmed to automatically create the user accounts and assign permission based on predefined rules at the time an employee joins the organization. If an employee leaves the organization, the RPA bots can deactivate accounts and revoke access immediately to ensure no unauthorized users access sensitive systems. It would not only have a security advantage but makes the onboarding and offboarding process be consistent and efficient.

2. Make access requests and approvals simpler

Access requests and approval workflows are pre-requisites to secure systems and applications access control. In a traditional set-up, the approval process for granting access is slow and sloppy and results in delayed productivity. Manual approvals also carry the risk of human error and open up channels for unauthorized persons as processes sometimes slip through the cracks and approvals are issued to those not authorized.

RPA can automate the entire process of request access by routing requests through predefined workflows, thereby ensuring that only authorized personnel will review and approve access permission. For instance, when an employee requests to access a specific system, an RPA bot can automatically verify whether the request is in compliance with that organization's security policy and ensure that the access permission is granted according to that employee's role. This automation minimizes delay, reduces the possibility of error, and ensures that access requests are treated in a timely and secure manner.

3. RBAC Enforcement

Role-Based Access Control (RBAC) is one of the most popular models for controlling access. It determines grantable access rights based on roles assigned to individuals within an organization. However, managing and enforcing RBAC policies is a complex issue because roles and responsibilities change very quickly in dynamic environments.

RPA can play an important role in ensuring continuous enforcement of RBAC because it can continuously monitor the roles and access levels through the system. Bots can thus ensure timely updating of permissions as roles change; nobody has access to unnecessary resources. Additionally, RPA periodically reviews existing controls for access, detects the differences or outdated permissions, and flags potential risks while automatically adjusting permissions to comply with security policies.

4. Securing Privileged Access Management (PAM)

Privileged access management refers to the control and monitoring of access to critical systems and sensitive data by those with enhanced privileges, such as system administrators. Poor management of privileged access is one of the causes for serious breaches in security since attackers usually target such accounts to achieve unauthorized control over most of the critical systems.

RPA enables the extension of PAM by removing human errors or omission in privileged account management, thus limiting the possibility of security breaches. Bots can monitor the usage of privileged accounts and generate alerts if any suspicious activity is detected without allowing an administrator to circumvent the security controls of the organization. RPA can also ensure automation of credentials rotation, thus reducing the chances of credential theft and unauthorized access (Ribeiro et al., 2021).

5. Real-Time Access Monitoring and Auditing

There exists access monitoring and auditing, several identity management strategies have to mean monitoring and auditing of access to both systems and data. Manual processes for auditing access can be very resource-intensive and prone to errors. RPA may therefore automate the process of real-time access monitoring and auditing since it provides constant oversight of who is accessing what and when.

Bots are also so powerful that they can automatically log much detail in the audit logs, including such activities as access attempts, systems utilized, and changes to access permissions. This kind of logging may provide the security team with many important insights into potential security risks and can ensure compliance with regulatory requirements. Automated auditing allows detection and addressing for potential security breaches such as unauthorized access attempt on such organizations in more timely manners, thereby enabling the organizations to respond in real time in case of a probable security breach.

RPA automates critical and time-consuming processes traditionally handled manually, such as user provisioning and access requests, enforcement of RBAC policies, and privileged access management. In doing so, it ensures that constant access control policy enforcement exists at all times while ensuring that no sensitive systems are compromised by unauthorized activities. Real-time monitoring and auditing thus enhance the organization's ability to detect and respond to potential security risks. In this increasingly complex digital landscape, RPA is a very valuable tool in running identity management processes towards strengthening cybersecurity efforts.

RPA IN DATA PROTECTION AND COMPLIANCE

Data protection and compliance are integral components of any organization's data security framework. Therefore, organizations are caught between ensuring such compliance while satisfying the rapidly increasingly higher demands of how data should be treated regarding regulations like GDPR and HIPAA. RPA plays a very effective role in automating those core processes that fall into the whole data protection and compliance deal, hence enhancing accuracy while introducing less human error and improving chances for organizations to meet most criteria from said regulations.

1. Automating Data Encryption and Masking

Data encryption and masking techniques ensure that the access of data to unauthorized parties is not possible. Although these are very important techniques, data encryption and masking appear to be a serious exercise when carried out manually and may have possibilities of errors during the process of ensuring that all sensitive information is masked correctly throughout the lifecycle of data. RPA has much to play here.

The RPA bots can be programmed to automate the encryption of sensitive data. Therefore, all data at rest and in transit will be encrypted with the correct organizational policies in place (Villar & Khan, 2021). Bots can mask out data with masking techniques and thus ensure that sensitive data is not accidentally exposed to unauthorized users while being analyzed or tested in software. Hence, automation helps prevent the mishandling of sensitive data.

2. Streamlining Compliance Audits

Compliance audits form the most important steps in checking whether an organization is adhering to the legal and regulatory requirements it is exposed to. Doing all the work manually, gathering all data, making reports, and ensuring availability of proper documents takes much time and effort. On the other hand, oversight even during the presentation process under audit will be subjected to non-compliance penalties.

RPA can also make compliance audits easier because it can auto-gather all needed information, automatically produce reports fit for audit, and adequately prepare all needed documents. Bots can conduct regular compliance checks to ensure that the organization is correctly calibrated with regulation requirements at intervals between scheduled audit slots. It not only puts less burden on the employees involved but also makes the audit process more accurate and comprehensive so that nothing gets missed.

3. Automating Data Access and Privacy Controls

It has been established that the ultimate concern in data protection is ensuring that only authorized personnel can access sensitive data. Manual control of data access could be prone to error, thereby allowing unauthorized personnel to gain access to protected information, which may result in potential breaches of data. Equally, the adherence to some privacy requirements, particularly the "right to be forgotten" under GDPR, would be quite complicated to handle manually.

RPA can apply controls regarding data access and privacy always and adapt access permissions based on the role of the employee and other similar duties. For example, if there has been a role change in an employee, RPA bots can automatically readjust access levels or cut off access to sensitive data, thereby preventing breaches of privacy regulations. Whereas in case of DSRs, if that be with regard to an erasure request, RPA can be programmed to automatically identify and safely delete all of a person's personal data from many systems for such a case while ensuring one remains compliant with the laws of privacy.

4. Automatic incident reporting and documentation

As a regulation, GDPR and its subsidiaries require the reporting of a breach or security incident to be transmitted within a certain timeframe. All steps taken by the organization for the management of the incident must be documented. Manual reporting and documentation have a propensity to delay and introduce inaccuracies in the reporting of incidents, thus exposing the firms to further compliance violations and legal implications (Hofmann et al., 2020).

RPA can automate the whole process of incident reporting and documentation. Bots can easily be configured to continuously monitor for security incident alert in real-time and send automated reports that detail all the necessary information on the breach, such as scope, affected systems, and also what's done in terms of mitigation. These reports automatically could then be sent to the applicable regulatory offices for the firm to ensure that it stays within its requirement of notifying others of a breach. Furthermore, RPA enables it to maintain a sophisticated record of everything happening before and after the hacking, thus providing voluminous audit trails for future use.

5. Maintenance of Continuous Compliance on Regulatory Updates

Compliances often are subject to a change due to the evolution of technology and due to security concerns. There will always be a new set of rules and regulations. Updating organizational policies according to these changes can easily become a very manpower-intensive task for large-scale organizations that have a complex structure to their IT infrastructure.

RPA can ensure continuous compliance by automatically tracking updates in regulatory matters and changing internal processes according to the changing requirements. For instance, if there is a newly enacted law on data privacy, RPA bots can be configured to enhance the controls on access, retention policies on data, and report preparation throughout every corner of the organization so that no human interaction is required, in order to meet the deadline and stay within time. In this manner, proactive management of compliance decreases the possibilities of falling behind changes in regulations and provides the flexibility to organizations to adapt fast to new legal requirements in place.

RPA is changing the face of data protection and compliance within an organization by automating such critical activities as encryption of data, compliance audits, access control, and reporting incidents. All this comes from RPA as benefits towards reduced burden and human error, which will ensure that the sensitive data remains protected while organizations increase compliance through rigid regulatory demands. As compliance challenges continue to rise, the potential of RPA in automating and streamlining such processes will become one more vital element to sustain robust cybersecurity frameworks.

ENHANCING THREAT INTELLIGENCE WITH RPA

It is hard for the various security teams to stay atop of most of the emerging attacks in today's cybersecurity environments where the complexity of threats and the voluminous data generated by security systems are so immense. Threat Intelligence is the gathering, analysis, and response to data related to security threats, but manual threat intelligence processes take too much time, are prone to mistakes, and not easy to scale. RPA can easily replace the current stages of threat intelligence with automated systems, speeding, accuracy, and efficiency that benefit cybersecurity efforts.

1. Automation of Collection and Aggregation of Threat Data

One of the main barriers in achieving effective threat intelligence is data collection, which often originates from various sources, such as threat feeds, security logs, and even external databases. Security analysts must go through massive amounts of data to determine what may constitute a threat, a task that may take hours or even days to accomplish, considering it is performed manually. Such a delay may leave an organization open to attacks that may have been avoided if they had reached detection much sooner.

RPA bots can gather and collate enormous threat data that it's collected from various sources. Through constant scanning of threat intelligence feeds, check on monitoring system logs, and data pullings of security tools, it may be available in real-time for security teams looking to make intelligent predictions about possible threats (Pramod, 2022). The bots may automatically look for correlations between different sources of data and identify any unknown pattern or patterns of indicators of compromise that may indicate an attack in the making. Automating the collection of such data increases the speed of threat detection, thus enabling security teams to respond to emerging threats faster and more efficiently.

2. Enhanced Threat Analysis and Correlation

Following the gathering of threat data, threat intelligence enters the process of analysis and correlation of such information to gain more meaningful insights about their work. This may be in terms of threats between actors and actors, techniques, and compromised systems. Analyzing all this information manually is very time consuming and would require much expertise, thus leading to delays in responses to the threats.

RPA can enhance the process of analysis by automating the correlations of threat data across systems. Then, bots can apply preconfigured rules and algorithms to look for anomalies or unusual patterns on that data that could indicate a security incident. In addition, further than this, RPA can be integrated with machine learning models for analysis of the detected threats on historical data bases and predictive insights. This automation of the stage helps security teams prioritize those most critical threats and focus more on high-risk vulnerability resolutions instead of efforts on low-priority alerts.

3. Automation of Threat Intelligence Sharing

Threat intelligence shared across organizations or within sectors is often much more effective as a collective means of defense against commonly opposing adversaries. Nevertheless, the benefits of shared threat intelligence are partially offset by the challenges of manual sharing, primarily when dealing with multiple organizations or a security team. In most cases, this process is extremely time-consuming and tends to miss some opportunities for early detection of attacks on a mass scale.

Through RPA, threat intelligence would thus be shared efficiently because it would make it automatically share all the relevant data to trusted partners, industry groups, or any threat intelligence platforms. The bots can be programmed so as to ensure critical information regarding threats, such as IoCs or attack signatures, is always shared in real time, and which will allow other organizations to use this information to go ahead and defend them against similar attacks. This results in automation that ensures rapid and uniform distribution of threat intelligence, thus enhancing the overall cybersecurity posture of the entire ecosystem.

4. Incident Response and Threat Mitigation

While detection and analysis are important, the ultimate purpose of threat intelligence is to facilitate swift response and reduction of the possibility of an attack. Traditional incident response operations typically consist of several manual steps-for instance, contacting teams that are informed, updating firewall rules, or quarantining systems suspected to be compromised-all of which can drag out the response.

RPA can significantly speed up incident response by automating the most critical actions undertaken in the mitigation process. For example, on occurrence of a security alarm, an RPA bot can automatically do so as-is isolation of affected systems, update the security policies, or block the suspicious IP addresses. The RPA can also generate reports and document all the work that was conducted in terms of steps during the course of the incident, allowing security teams to do more strategic thinking rather than repetitive tasks. This would not only shorten the time elapse between threat detection and mitigation but also the damage caused by a potential attack.

5. Ongoing monitoring and updating of threat intelligence

What changes continually is the cyber threats, and therefore, the static approaches to threat intelligence can become instantly old and irrelevant. An organization must constantly update its threat intelligence data in order to outsmart its attackers while also responding in real-time fashion with its defensive measures. However, doing so requires effort because it entails manually updating threat intelligence databases while constantly refining the detection rules.

RPA can continuously monitor and update systems with threat intelligence. Bots can automatically update security tools with the latest threat data, such as new IoCs, malware signatures, and attack vectors (Hong et al., 2023). Therefore, the company defenses would be updated with the latest intelligence, giving it the sense and ability to detect emerging threats. Additionally, RPA can continue scanning for new vulnerabilities within the organization's infrastructure by flagging up any new risks while suggesting the appropriate remediation steps.

New keys of automation are being fundamentally opened up in the realm of threat intelligence by ensuring that data collection, analysis, sharing, and even response can be done using RPA. Organizations across the world will be able to better their capabilities of detection and mitigation in terms of threats, speed up their response times, and increase the precision level of their threat intelligence efforts. The scale of cyber threats and the level of their complexity only continue rising, which goes to make RPA a perfect tool in en route towards applying this universally essential tool for Cyber Security strategies in organizations of today's world.

CHALLENGES AND LIMITATIONS OF APPLYING RPA IN NETWORK SECURITY

Even though RPA provides numerous benefits that are meant to improve the overall network security, it is not without the challenges and limitations which need to be considered. Applying RPA in network security faces various challenges including how intricate the automation is how easily it can fit into the rest of the used security systems. The next section highlights some of the key challenges and limitations which have been noted with the use of RPA in network security.

1. Complexity of Security Processes

Network security is also composed of highly complex, dynamic processes requiring deep contextual awareness and flexible, rapid decision-making. Processes that are too "judgment-intensive" for RPA demand advanced judgment, human intuition, or even real-time decision-making. For example, responding to a sophisticated, multi-stage cyber-attack often needs on-the-fly analysis or a flexible response strategy—something RPA is currently unable to handle.

Generally, RPA bots are rigid in their rules and workflows, which makes them somewhat less adaptable when it encounters new threats that are unknown or in the process of change from their established parameters. Consequently, even as RPA supports efficiency in many network security points, it may not altogether substitute for human analysts possessing skills to manage some of the more complex incidents and threats.

2. Integration with Legacy Systems

One of the primary challenges in implementing RPA in network security is integration with existing and often legacy systems and tools. Most organizations still use older security systems and infrastructure that might not be able to collaborate well with new RPA technologies. Quite often, such systems are very complex to integrate with RPA, which may require custom development, and will also demand additional resources to get this integrated without operational inconvenience.

Moreover, integration with many security tools and platforms -firewalls, intrusion detection systems, and vulnerability scanners -requires considerable configuration and testing. If security tools are not properly integrated, RPA bots may miss critical security tasks to fail gaps in network protection.

3. Maintenance and Scaling Issues

Though automation by RPA could streamline many network security tasks, the associated automation workflows can become expensive to maintain once a network environment changes over time. Changes in environments require their corresponding RPA bots to be updated over time. The frequency with which security threats may emerge or where there is a chance of regulatory requirement change keeps on increasing as the network infrastructure scales. As such, so also must the RPA deployment, increasing the system's complexity and cost of system maintenance (Ray et al., 2023).

Along with that, the accuracy of the RPA bot has to be checked regularly to ensure it is working as expected, especially in dynamic environments where security threats are constantly changing. If it's not checked, then unmonitored RPA bots can become obsoletes or unable to handle new threats for an organization.

4. Security Risks Connected With RPA Bots

Ironically, RPA bots themselves pose some new security risks if not adequately managed because they are automated units accessing sensitive systems and data. They are susceptible to attacks by cyber attackers. For example, in case an attacker penetrates an RPA bot, they would easily have unqualified access into the critical systems of the network that would leave the door for data breaches or interruptions in security operations.

To mitigate these risks, organizations must ensure tight security around their RPA bots by imposing strict access control, encrypting data at rest and in transit, and auditing bot activity on a regular basis. This, however, adds complexity to network security management and gives rise to a continuous requirement for monitoring.

5. Vulnerability in the Case of Sophisticated Threats

Relying on pre-defined rules comes with a vulnerability of RPA when dealing with sophisticated threats. Threat actors will always come up with new attack vectors and techniques, whereby, in some cases, it would be difficult for rule-based RPA to detect. The bots that have performed so well in handling repetitive and known threats would fail to detect sophisticated threats such as zero-day vulnerabilities or APTs.

In order to bridge this gap, companies need to integrate RPA with advanced AI/ML technologies that can detect threats subtly and respond accordingly. Integration of RPA with AI/ML, however, adds more complexity and may require specialized skills, thereby increasing the cost of implementation.

6. Resource and Skill Requirements

RPA can also weaken network security the amount of labor. However, those individuals entrusted with designing, implementing, or maintaining such automation systems are often extremely experienced professionals. It takes proficiency in automation technologies and knowledge in cybersecurity to develop an effective RPA bot for network security. This means that an investment in the training of a pool of professionals or hiring such professionals who can manage the two concepts is necessary for an organization.

Additionally, creating and managing RPA systems does consume significant amounts of resources in the setup process and also during maintenance. There is a need to devote sufficient time and resources to see that the RPA bots are working correctly, adhere to security policies, and are updated to address emerging threats. Small organizations might find it too resource intensive to deploy RPA as a tool of network security.

While RPA has quite promising futures in making network security much more efficient, on the other hand, more prone to making human work obsolete, it definitely brings its own set of challenges and limitations. Security processes are rather complex; integration with legacy systems is challenging; maintenance of bots; and the very risks inherent in security automation are all things that should be considered meticulously. On the contrary, the reason RPA cannot replace human expertize solely in network security is due to its inability to utilize most advanced threat capability and the necessities for specialized resources. Over time, the organizations will be able to strike a balance between automation and human oversight in a resilient and effective security strategy as they continue to experiment with RPA implementation.

FUTURE TRENDS: AI-DRIVEN RPA FOR ADVANCED SECURITY AUTOMATION

The increased sophistication in the realm of cybersecurity threats, in turn, asks for highly sophisticated tools and technologies to tackle them. RPA has proved its mettle in network security by automating a lot of mundane work, but its pure rule-based nature puts significant limitations on it. The future of RPA in security would, in effect, consist of its integration with AI-future development-and it is promising a big enhancement in the areas of adaptability, intelligence, and scope of automation. This would mean that AI-based RPA is probably going to change the face of advanced security automation. It will be forward-looking and very responsive with greater efficiency.

1. AI-Based Threat Detection and Response

Generally, conventional RPA systems operate based on a set rule and are therefore used to process known tasks or threats. However, they are certainly not versatile in dealing with unknown or emerging cyber threats. With the infusion of AI into RPA systems, organizations can take a step from rule-based automation towards intelligent automation. It is the ability of machine learning (ML) in AI that allows RPA bots to learn from historical data and recognize newer patterns that may indicate security threats.

For instance, AI-driven RPA can potentially increase the detection of threats based on real-time inspections by the network traffic, behavior related to the user, and system logs for anomalies not covered by traditional security methods. Intelligent bots can identify suspicious activities, trigger auto-responses, and even escalate these cases to a human analyst for monitoring when it is required. In this way, there would be a speedier and more efficient identification of advanced threats as well as mitigation with regard to zero-day vulnerabilities and sophisticated attacks from malware (Sibanyoni, 2021).

2. Predictive Security and Risk Management

Another aspect AI-powered RPA can bring into network security is the predictive capability to determine the onset of an incident just before it becomes fully formed. Predictive analytics allows the AI-driven bots to analyze information collected over time and, by running algorithms against that data, identify emerging trends to forecast potential attack vectors or vulnerabilities likely to emerge. This pushes organizations toward a proactive approach by patching systems, changing network settings, and increasing monitoring in areas identified to be higher risk.

Predictive security powered by AI can also improve the practice of risk management. AI-driven RPA can continuously evaluate and score security risks throughout the entire digital infrastructure of an organization. This allows the security teams to focus their efforts on the likelihood and severity level of an attack, hence utilizing resources optimally, maximizing less exposure to critical threats.

3. Adaptive Incident Response

Clearly, one of the biggest disadvantages of traditional RPA in security automation is that it does not keep pace with the constantly changing attack scenarios in real time. AI-driven RPA can go a long way in changing this scenario; by making bots learn from their previous experiences and changing their reactions to new challenges accordingly. For instance, instead of being strictly relegated to a standard-designed incident response plan, AI-enhanced RPA bots can dynamically alter their activities as per real-time data and context, thus improving response effectiveness.

Moreover, AI-driven RPA will also automate incident response workflows by instantly classifying threats, placing the affected systems in quarantine, and carrying out countermeasures with minimal human intervention. All these robots will be constantly learning from past incident outcomes, making their response protocols better and smoother as time passes. The dynamic nature of adjustments in responses as something unfolds will lead to reduced impact attacks and an improvement in the overall posture.

4. AI-Powered Security Analytics

The insights gained would be able to evaluate the effectiveness of security strategies and identify areas of improvement through security analytics. Thus, RPA driven by AI revolutionises security analytics via the automation of the collection and analysis of vast numbers of security data. Machine learning algorithms can thus dig much deeper into logs, alerts, and security reports in discovering hidden patterns, generating insights, and predicting future security incidents.

These analytics can present to the security teams real-time dashboards, pinpointing the most serious issues, possible vulnerabilities, and recommended actions. Moreover, AI-based RPA can automate the generation of compliance reports- a task that helps an organization meet regulatory compliance standards without forcing security teams to work overtime with such tasks. As AI continues to learn and improve in its processing of data and analytics, it stands to make them very more precise and actionable, putting them in a position to empower security teams to make proper decisions.

5. Securing the Endpoint and IoT Environment

A fundamental trend fueling the growth in the attack surface is the proliferation of connected devices–notably the usage of Internet of Things (IoT) devices. The smallest of them can be a rather hard nut to crack, considering they represent small form factors with limited computing resources. AI-driven RPA can take the lead in securing endpoints by taking care of automated monitoring and endpoint protection.

RPA bots can monitor IoT devices for unusual behavior, instantly apply security patches, and update device configurations in cases of vulnerability detection via AI. AI-driven bots will analyze the pattems of IoT traffic to detect threats early and prevent attacks in real-time before hitting the core systems. This approach will be critical as the number of IoT devices expands while enhancing the complexity of security management.

6. AI-Driven RPA for Continuous Compliance

As the legal frameworks become stringent around data privacy and cybersecurity, the issue of compliance has become one of the most critical concerns for organizations. AI-driven RPA can be set to ensure continuous compliance monitoring and ensure the organization complies with the rules set by the regulatory authorities such as GDPR, HIPAA, and CCPA. AI can augment the bots in reviewing and auditing the process to identify compliance gaps and automatically create a report showing adherence to regulatory standards.

This real-time monitoring reduces the risk of non-compliance that would fetch a huge loss in fines and reputation value. Also, AI-based RPA can guarantee that with the development of regulations, compliance procedures shall remain updated in real-time even without any human intervention. This saves time and resources and also reduces the risk of human error.

Adding AI to RPA is a great leap forward in automation security. The organizations will be able to build a much more powerful security system better designed for applications, according to AI combined with RPA efficiency at a cutting edge that is not located within old automation limits. AI-driven RPA has huge promises in most areas of cybersecurity, with much promise for leveraging threat detection, incident response, compliance, and endpoint security. As AI technologies become more mature, security automation will increasingly rely on intelligent data-driven processes that enable organizations to maintain a pace with these forever changing cyber threats.

CONCLUSION

The introduction of RPA into cybersecurity is a fundamental evolution of how organizations can manage and secure systems and networks within organizations. RPA has well-proven its capacity to streamline repetitive security tasks, enhance operational efficiency, and assist with incident responses. By automating these important but very time-consuming activities, organizations can free up their security teams to deal with more complex and strategic problems. However, while RPA has passed the proof of effectiveness, it imposes some restrictions on handling complex, dynamic threats which require constant innovation in security automation.

As shown in this book chapter, the greatest advantages for organizations in implementing RPA lie in areas like secure access controls, identity management, data protection, compliance, and threat intelligence. So long as proper implementation is included, RPA should enhance the security processes with greater accuracy, speed, and consistency. Nevertheless, issues such as complexity in integrating RPA with legacy systems, bot security, and adaptation to advanced cyber threats pose challenges that must be addressed by the organization. Despite these challenges, RPA has immense potential as a powerful tool in modern cybersecurity strategy.

The future of RPA, therefore, lies in the fusion with AI. AI-powered RPA shines bright with its avenue for enhancing both adaptability and intelligence of automation while predicting scenarios much earlier. This will cause RPA systems to transform from simple, rule-based bots to intelligent entities able to sense, respond, and even predict security threats in real time. The integration of AI with RPA will bring with it pro-active measures that help organizations stay at par with the latest sophisticated cyber-attacks.

RPA has already transformed vastly in automating network security, and the tool has now immense untapped potential because it will be the integration with AI that unlocks it. Technologies to counter cyber threats will have to advance with threats. For instance, AI-driven RPA will undoubtedly be at the very front of such advancement, and organizations are likely to explore it for advanced security automation, the enhancement of capabilities in threat detection, and compliance management. In this manner, embracing innovations will build a more secure and resilient digital infrastructure that better endures and withstands difficulties within the context of the cyberspace.

REFERENCES

Al Zarooni, L., & El Khatib, M. (2023). Robotics Process Automation (RPA) and Project Risk Management. [IJBAS]. *International Journal of Business Analytics and Security, 3*(1), 75–91. DOI: 10.54489/ijbas.v3i1.198

Badmus, A. D. (2023). Implementing Technology Manufacturing with Robotics Process Automation (RPA). *Journal of Knowledge Learning and Science Technology ISSN: 2959-6386 (online), 2*(2), 96-103.

Beer Mohamed, M. I., & Hassan, M. F. (2022, November). Adaptive security for cognitive robotic process automation in enterprise computing using AI-powered analytical engine. In *International Conference on Artificial Intelligence for Smart Community: AISC 2020, 17–18 December, Universiti Teknologi Petronas, Malaysia* (pp. 825-835). Singapore: Springer Nature Singapore. DOI: 10.1007/978-981-16-2183-3_78

Bhattacharyya, S., Banerjee, J. S., & De, D. (Eds.). (2023). *Confluence of Artificial Intelligence and Robotic Process Automation* (Vol. 335). Springer Nature. DOI: 10.1007/978-981-19-8296-5

Hofmann, P., Samp, C., & Urbach, N. (2020). Robotic process automation. *Electronic Markets, 30*(1), 99–106. DOI: 10.1007/s12525-019-00365-8

Hong, B., Ly, M., & Lin, H. (2023). Robotic process automation risk management: Points to consider. *Journal of Emerging Technologies in Accounting, 20*(1), 125–145. DOI: 10.2308/JETA-2022-004

Khan, S., & Tailor, R. K. (2023). Application of RPA in Human Security Systems in Smart Cities. In *Application and Adoption of Robotic Process Automation for Smart Cities* (pp. 28–46). IGI Global. DOI: 10.4018/978-1-6684-7193-7.ch002

Kurylets, A., & Goranin, N. (2023). Security ontology OntoSecRPA for robotic process automation domain. *Applied Sciences (Basel, Switzerland), 13*(9), 5568. DOI: 10.3390/app13095568

Kurylets, A., & Goranin, N. (2024). Design and implementation of a UMLRPAsec-extension for robotic process automation. *Electronics (Basel), 13*(4), 769. DOI: 10.3390/electronics13040769

Murugappan, K., & Sree Kala, T. (2022). An enhanced security framework for robotic process automation. In *Cyber Security and Digital Forensics* [Springer Singapore.]. *Proceedings of ICCSDF, 2021*, 231–238.

Murugappan, K., & Sree Kala, T. (2022). An enhanced security framework for robotic process automation. In *Cyber Security and Digital Forensics* [Springer Singapore.]. *Proceedings of ICCSDF, 2021*, 231–238.

Pramod, D. (2022). Robotic process automation for industry: Adoption status, benefits, challenges and research agenda. *Benchmarking, 29*(5), 1562–1586. DOI: 10.1108/BIJ-01-2021-0033

Rawat, R., Chakrawarti, R. K., Sarangi, S. K., Choudhary, R., Gadwal, A. S., & Bhardwaj, V. (Eds.). (2023). *Robotic Process Automation*. John Wiley & Sons. DOI: 10.1002/9781394166954

Ray, J. K., Sultana, R., Bera, R., Sil, S., & Alfred, Q. M. (2023). A Comprehensive Review on Artificial Intelligence (AI) and Robotic Process Automation (RPA) for the Development of Smart Cities. *Confluence of Artificial Intelligence and Robotic Process Automation*, 289-311.

Ribeiro, J., Lima, R., Eckhardt, T., & Paiva, S. (2021). Robotic process automation and artificial intelligence in industry 4.0–a literature review. *Procedia Computer Science, 181*, 51–58. DOI: 10.1016/j.procs.2021.01.104

Sibanyoni, N. A. (2021). A Blockchain-Based Robotic Process Automation Mechanism in Educational Setting. In *Industry Use Cases on Blockchain Technology Applications in IoT and the Financial Sector* (pp. 17–41). IGI Global. DOI: 10.4018/978-1-7998-6650-3.ch002

Tailor, R. K., Pareek, R., & Khang, A. (2022). Robot process automation in blockchain. In *The Data-Driven Blockchain Ecosystem* (pp. 113–125). CRC Press. DOI: 10.1201/9781003269281-8

Villar, A. S., & Khan, N. (2021). Robotic process automation in banking industry: A case study on Deutsche Bank. *Journal of Banking and Financial Technology, 5*(1), 71–86. DOI: 10.1007/s42786-021-00030-9

Chapter 4
Intrinsic Vulnerabilities of RPA Systems

Shashank Mittal
https://orcid.org/0000-0002-8666-5323
O.P. Jindal Global University, India

Shivani Dhand
https://orcid.org/0000-0002-4809-1365
Lovely Professional University, India

Thi Minh Ngoc Luu
https://orcid.org/0000-0002-5972-7752
International School, Vietnam National University, Hanoi, Vietnam

ABSTRACT

This chapter discusses the inherent weaknesses of RPA systems and provides solutions to mitigate associated risks. Increasingly, organizations are deploying RPA to improve operational efficiency. But this also comes with challenges such as security, data privacy, and regulatory compliance. Some of the key vulnerabilities involved are flaws in software architecture, human error, third-party integrations, and gaps in governance operations. The next chapter will also illustrate how seriously those strong governance frameworks, robust testing processes, and ongoing monitoring are taken to protect the RPA deployments. With AI integration, the ever more complex theme of remote work, and a changed regulatory scene, the future directions for the same are discussed. A proactive attitude towards security will keep the organizations safe from challenges by the RPA system and will help them gain all the benefits of automation with risks minimized.

DOI: 10.4018/979-8-3693-4365-4.ch004

INTRODUCTION TO RPA AND ITS EVOLUTION

RPA Robotic Process Automation is emerging as a revolutionizing force in the digital landscape and is proving to be an extremely effective tool that all companies use to automate mundane and rule-based tasks. RPA, whose initial and primary purposes are focused on even more simple processes that are less likely to demand human decision-making capabilities, has rapidly matured and become an essential factor in enterprise automation strategies(Beuerle, 2019). The ability to replicate human behavior, such as data entry, form processing, and handling transactions, has helped organizations work at maximum efficiency, reduce operational costs, and result in accuracy (Jaunat et al.,2019). At its most basic level, RPA depends on the functionality of high volume, laboring tasks with high precision, which sets more complex and strategic human workers free from such mundane activities (Agostinelli et al.,2019).

Evolution of RPA follows in the footsteps of traditional automation tools such as screen scraping and workflow automation. Early stages of RPA systems were limited in scope due to automation scope of specific individual tasks in isolated systems (Menolfi et al., 2018). The first releases highlighted efficiency improvements in back-office operations, such as payroll processing, order fulfillment, and invoice management. The technology soon flowered as companies realized their full potential in terms of their opportunity to integrate themselves into myriad legacy systems and also third-party software, not to mention third-party software, pushing the boundary of what could be automated (Kreuzer, 2014).

As a business becomes more and more dependent on technology, hence comes a growing request for tools that will allow the application of automation transparently on existing systems, especially legacy infrastructure (Willcocks et al., 2015). This was soon followed by more advanced RPA tools that could interact with many systems without the need for costly or intrusive overhauls. The emergence of cognitive RPA, which integrates AI with ML, also further developed the ability of RPA(Jaunat et al.,2019). It is now possible for cognitive RPA to process unstructured data, make highly basic decisions, and learn from history to perform far beyond mere rule-based automation(Beuerle, 2019).

Several factors have driven the progress of RPA. Perhaps more importantly, though, is the explosion of big data and the increasing digitization of business processes have created an enormous volume of tasks amenable to automation(Kreuzer, 2014). RPA has proved its malleability on everything from customer service and compliance reporting to finance and HR processes. Besides, it becomes non-intrusive in nature; then, it can work effectively atop the existing systems without any need for significant infrastructural changes because the firms do not dread its application owing to the disruption of businesses.

It, however, evolved to beyond improving technical capabilities and became an essential component of digital transformation. Companies can thus speed up their processes, minimizing errors while ensuring that all industry regulations are met. This has played a great role with the growing intelligent automation that combines RPA with artificial intelligence, NLP and other emerging technologies thus advancing possibilities further into automation of some other more complex workflows and decision making.

Today, RPA is not only an automation tool but also increasingly being embedded as part of the enterprise-wide intelligent automation system, continuously evolving and transforming the automation landscape. This helped businesses to not only automate discrete tasks but also optimize end-to-end processes, allowing them to be agile, scalable, and responsive in today's highly competitive global marketplace(Kreuzer, 2014). With the world increasingly dependent on digital solution streams for growth, RPA will continue to expand further in its role, supported by all the innate capabilities powered by AI-driven innovations and evolving use cases within industries.

UNDERSTANDING THE INTRINSIC VULNERABILITIES OF RPA

One is true that with RPA, one enjoys many benefits in terms of automation of repetitive task handling and streamlining of operations. However, it is not without its vulnerabilities. Intrinsic vulnerabilities are the inherent weak points in the design, architecture, and deployment of RPA systems, which thereby pose a risk to RPA. Such vulnerabilities may arise through several factors, such as the way in which RPA interacts with other systems, the nature of tasks being automated, and the limited intelligence of the bots itself. Understanding these intrinsic vulnerabilities is important to make sure that deployments of RPA happen in a secure, resilient, and more effective long-term manner (Osmundsen et al., 2019).

One of the main vulnerabilities that exist in RPA systems is **lack of built-in security features**. For instance, most of the RPA tools are ready-to-deploy with minimal coding, with an emphasis on quick deployment. This generally leads to security loopholes in which bots become not secured enough against unauthorized access or data breaches. For instance, RPA bots that handle personal data or banking transactions become easy prey for attacks if encryption and other access control measures are not followed. Moreover, because RPA systems typically interact with multiple platforms and databases, any vulnerabilities in those systems can also compromise the security of the bots.

The other intrinsic vulnerability of RPA is its **limited error-handling capability**. Most RPA systems rely on pre-defined rules and workflows, making them very efficient for structured, repetitive tasks. This however means that they are prone to unexpected changes or alterations in the processes they automate. Bots will most definitely fail to carry out their jobs the way they should if the data is in some way different from the norm, or if the format of an input changes. Such failures may not easily go unnoticed, and thus there can be larger system breakdowns or ill processing of data. Without strong error detection and recovery mechanisms, these vulnerabilities can weigh heavily on RPA systems' reliability.

RPA systems also rely on the **inherent vulnerabilities of the existing infrastructure**, especially legacy systems. Most organizations deploy RPA to automate tasks within legacy environments that were never designed with modern security or operational practices in mind. This exposes RPA bots to inheriting all the vulnerabilities old systems have, such as becoming obsolescent in software, unsupported operating system or security flaws that haven't been patched yet. Thirdly, the third-party applications used by RPA and APIs form a conduit for integration risks, among them having poor communication between systems, which then exposes this to data leakages or inefficiency in processes (Aguirre & Rodriguez, 2017).

Another crucial area of concern relates to **governance and oversight** of RPA systems. Since RPA bots are programmed to perform activities exactly like human beings do, they could sometimes work in directions not easily trackable or auditable. Misconfiguration of a bot or as when it strays from its intended process can prove quite challenging to trace the source if appropriate logging and monitoring systems are not in place. This ambiguity might lead to error multiplication in the longer run-in large-scale deployment where several bots are deployed parallelly. Lack of governance frameworks leads to operational risks and compliance in RPA systems

There is also an inherent weakness due to **human mistakes in the configuration of bots and the process** of its deployment. Many times, RPA systems are designed by very few teams that are technically not very sound, hence eventually leading to poor design or maybe poor implementation of the bots. Key mistakes are hardcoding passwords and other sensitive information, not updating bots following changes in processes, or not securing environments where bots work correctly. When businesses scale up their RPA usage, proper version control and change management becomes critical, or they will face inconsistencies in bot behavior and increased susceptibility to errors.

Intrinsic vulnerabilities in RPA systems arise from a bundle of several factors, including a weak security configuration, limited error-handling capabilities, dependency on legacy systems, poor governance, and human errors. And as RPA's maturity grows, understanding and dealing with these vulnerabilities will be invaluable to a business to realize all the benefits of automation without trading off security, reliability, and operational efficiency.

COMMON RPA SYSTEM ARCHITECTURE VULNERABILITIES

RPA systems have been touted for their efficiency and speed in automating various tasks. However, the architecture of the system, which makes such seamless automation possible, brings in several vulnerabilities. Such vulnerabilities, if not managed, can successfully destroy the security, scalability, and reliability of an organization's RPA deployment (Moffitt et al., 2018). Awareness of some common vulnerabilities in the architecture of RPA systems will be important to businesses wishing to protect their automated processes against all risk.

Weak security configurations in the architecture of an RPA system are undoubtedly the most common vulnerability. Primarily, RPA tools are so user-friendly that non-technical staff can compose and deploy bots with minimal coding. Although the ease of use does accelerate the speed of deploying bots, poor security practices such as lame encryption or even worse-hardcoded credentials directly into the bot's scripts-sacrificed to achieve that ease. In fact, sensitive information and financial and business proprietary information is often processed by RPA systems, making them a hotspot for cyberattacks. Without proper encryption, data processed by bots might be accessibly available to unauthorized persons, which can thus become vulnerable to breaches. Moreover, the ability of many access controls in bots to handle inappropriate levels of access can even be exploited by the insiders and outside hackers for exposure to critical business systems and data.

A common weakness to the current RPA systems is a **lack of scalability**. While RPA is very efficient for addressing certain recurring routine tasks, many of the architectures for the RPA systems are not scalable enough to deal with intricate and larger workflows. Organizations often experience bottlenecks when using RPA because their systems are overloaded by too much data or cannot handle more intricate workflows that require decision-making capabilities. The fundamental inflexibility of rule-based bots exposes them to the possibility of failure whenever there occurs deviations or exceptions within processes. This leads to a system frequently redesign and calibration to fit in new tasks, rather insidious context that could hamper the scalability of business.

The other serious architectural fragilities in RPA are **dependency on legacy systems.** Many organizations implement RPA on older IT infrastructures that contain processes that were not actually designed to be automated. The very nature of legacy systems is fragile in terms of having old software and limited support and so forth that can actually expose security flaws that can make the deployment of RPA risky. Because RPA bots work based on mimicking human action across different systems, they inherit vulnerability from those legacy environments. For example, when a bot uses a legacy system that is unchecked for the known security holes, it may become an entry point of such cyberattacks. Furthermore, the integration of RPA with legacy systems results in significant operations inefficiencies because the bots are negatively impacted by the outdated nature of the systems' performance they interact with.

The vulnerabilities also include **limited error handling and recovery mechanisms** in RPA systems. Bots depend on rigid workflows and break in case such changes or errors occur in the data they process. Failure to design or plan an RPA bot for change in data formats, alteration in application interfaces, or in case of system outages leads to disruption of workflows of an automated process. Such errors mostly remain unverified because monitoring systems and alerting mechanisms in RPA systems become unsatisfactory at times. Lame error handling allows processes to break down, creates bad data, and can, in extreme cases, cause a loss of confidence in the automation system.

Integration into other systems and applications also exposes to greater vulnerability through **API exploits**. RPA machines interact with many software and systems through APIs. If the APIs are not secured, it becomes a short cut to attack. In case the APIs used are not secured enough, an attacker may exploit them to gain access to a system without authorization, tamper with data, or even hold up operations. In addition, the vulnerabilities in third-party software or even the platforms that bots engage could be exploited and cause a potential breach of the overall RPA ecosystem.

Conclusion In sum, the RPA systems architecture poses common vulnerabilities that need to be addressed by organizations for their secure and reliable automation processes. Inadequate security configurations, nonscalable, dependent on legacy systems, poor error handling, and some risks integrated in the API are all critical issues that can expose RPA deployments to kinds of operational and security threats. Understanding and tackling these risks, therefore, ensures that organizations have strong, scalable, and secure RPA implementations that will give them all the benefits they expect from automation.

SECURITY THREATS IN RPA IMPLEMENTATIONS

Following the integration of most organizations with Robotic Process Automation (RPA) to increase efficiency and productivity, the security threats associated with RPA implementations have lately become the most critical threat. By their very nature, RPA systems interact with a variety of applications, data sources, and users. This makes them an easy potential target for malicious activities. While deploying

RPA undoubtedly promises huge efficiency gains, without proper security considerations, organizations risk breaches of data, system vulnerabilities, and compliance issues. To enhance enterprise operations safety from security risks associated with RPA deployment, one needs to understand the risks.

Unauthorized Access and Privilege Escalation

The most serious risk in the deployment of RPA would be unauthorized access and privilege escalation (Kokina & Blanchette, 2019). Since the RPA bot may require access to multiple systems, databases, and applications containing sensitive information from financial data, personal customer information to intellectual property, a compromised bot will compromise the systems accessed if they are not suitably secured. For one, bot credentials have in many cases been mishandled or even stored in some unencrypted formats and hardcoded into scripts.

Privilege escalation is another threat, whereby attackers exploit the gaps of the system to elevate their access level above what is otherwise intended. With too much permission, an RPA bot could be compromised by adversaries for unauthorized work and may cause a lot of loss or damage in terms of sensitive data. Organizations should adopt robust IAM protocols that give the robots minimum necessary permissions required for specific tasks.

Data Leakage and Confidentiality Breaches

One of the major disadvantages of RPA systems is their potential to handle vast amounts of sensitive data as part of their automated process. Regardless of whether customer records, monetary transactions, or confidential documents are involved, RPA bots can unintentionally become one-way leaks for data exposure if left unsecured. The more interaction a bot might have with unguarded data or the more transmission across insecure communication channels it makes, the more potential exposure can be expected. The worry is even more as there are specific industries where the regulations are strictly set, such as finance and healthcare sectors, whereby several legal and financial complications will come as an aftermath of data breach (Hofmann et al., 2020).

Another aspect of confidentiality breaches is if encryption and data masking are not included in the deployment of bots. For example, if a company deems to release sensitive information like social security numbers, credit card information, or files with medical information is not masked or encrypted, then possibly it will reach unauthorized personnel from inside or outside the company. Encryption and strong data protection protocols must be included in RPA environments to minimize data breaches as well as privacy violation violations.

Bot Tampering and Malicious Code Injection

Another important security risk in RPA deployment is the possibility of bot tampering and malicious code injection. Bots are largely developed to perform a set of scripted instructions. However, when attackers gain access to the source code or configurations of the bot, they can be able to modify it to perform malicious acts like changing payment instructions, diverting payment to another account, or even leaking confidential information. Depending on specific cases, the attackers will inject some malware into the bot; this will allow the bot to carry out unauthorized operations like ransomware attacks or distributing malware on the network.

Bot tampering is often extremely challenging to detect unless proper monitoring and auditing mechanisms are in place (Gotthardt et al., 2020). Therefore, organizations need to set up stern security systems such as guaranteeing integrity checks on code, continued monitoring, and regular audits of bot activity. However, beyond all this, access to the bot configuration files should be limited and only authorized personnel should be allowed to alter bots.

Lack of Proper Governance and Monitoring

Lack of governance and monitoring frameworks is another big security risk in the deployment of RPA. Many organizations deploy bots very fast, without proper creation of governance models that define protocols for security, audit trails, and monitoring capabilities. Such deployments are then not monitored well, and any anomalies or unauthorized activities may never be easily identified. A misconfiguration of a bot can lead to data corruption because the bot might malfunction or act out of character, causing partial transactions and security gaps.

Real-time observation of bot activities and keeping an extensive log of their operations is an essential necessity to ensure the RPA system's integrity. In addition to governance guidelines that should include periodic review of bot performance, security audits, and compliance checks to ensure that bots are working as designed, and quick identification and resolution of potential security risks.

Integration Risks with Third-Party Systems

The RPA systems typically leverage connections with several third-party applications, cloud services, and external APIs to power end-to-end processes. Although integration may enhance the usability and scope of an RPA, integration does pose significant security risks against the process. Third-party systems are hardly operating at the same security protocol as the internal; thus, third-party connections may introduce vulnerabilities into the RPA environment. For example, if the RPA bot communicates with a poorly secured third-party API, attackers can abuse a weak spot to access sensitive data or break automated workflows.

Third-party systems integrated into the organizations' RPA platforms are rather an Achilles' heel. Regular updates and patches for third-party software, use of secure API connections, as well as thorough security assessments of external partners will help markedly reduce risks associated with integration.

Insider Threats

Insider threats are one of the biggest security risks for deployments of RPA, where an employee or contractor with access to bot configurations or sensitive data will misuse their privileges. Malicious insiders will control bots in executing unauthorized action, data leak, or sabotage automated processes for personal gain or to hurt the organization. Besides, careless inside people may inadvertently open RPA systems to risk, for example, by sharing the credentials of a bot or forgetting to log out from a system.

To counter this, organizations should ensure adoption of stringent access controls, roles-based access, and observation of all interactions between employees and bots. Organizations can also help reduce the instances of insider attacks by ensuring comprehensive security training and awareness programs are in place to inform workers on best practices about dealing with sensitive data as well as using RPA systems effectively and safely.

OPERATIONAL VULNERABILITIES IN RPA SYSTEMS

It is the place where RPA makes business productivity more efficient by automating repetitive business processes, but the operational risk that springs from its implementation will bring about huge disruptions and performance issues. Those vulnerabilities are most usually related to inherent RPA system limitations as well as the way they interact with all other business processes. One of the effects it may show is on scalability, reliability, and overall system performance. This is important because it is the only possible way to ensure the operation of the RPA system at its maximum potential and to continue delivering value without jeopardizing business continuity.

Scalability and Maintenance Challenges

Scalability of automation is one of the leading operational vulnerabilities in RPA systems (Hofmann et al., 2020). RPA bots are programmed to automate rule-based, repetitive tasks, but while organizations try to scale up these systems toward more sophisticated or bigger workflows, they encounter great challenges. RPA systems are normally not designed for large, dynamic settings with a mix of different tasks and greater decision-making complexity (Syed et al.,2020). Consequently, the load of data or the unwarranted changes in workflows are challenging to handle by bots and have led to bottlenecks and inefficiency in systems.

The second complexity involved in maintenance in scaling an RPA system continues. Due to more numbers of deployment of bots in multiple departments and workflows, the demand for monitoring, updating, and troubleshooting such a system becomes heavier. Keeping so many bots requires constant manual intervention as they usually don't function as one would prefer them to in the event of changes to processes or when bots hit unforeseen errors. This maintenance burden increases the operational cost of RPA as organizations might use plenty of resources ensuring that everything operates well.

Bot Fragility and Process Changes

Bot fragility is also another operational vulnerability in RPA systems; they break easily due to changes in processes. Most RPA bots operate according to a predefined set of steps mimicking real human behavior based on specific conditions. This may be okay for static, repetitive processes, but problems occur when business processes even remotely change. Changes to the application UI or systems upgrades and variations in the input of data may freeze the operation of an RPA bot. Since humans do not possess the adaptational cognition power of humans, alterations cannot be made in a bot without recourse to either redesign or hand-programming (Leno et al., 2021).

These are risky and may lead to a high rate of bot failure and also require frequent adjustments and updates. For instance, if a bot is designed to input some data in an application interface, then if the 'button' or a form layout is changed slightly, it may break the bot; this means time lost due to downtime with productivity reduced and a decline in business processes.

Error Handling and Exception Management

Error handling and exception management are among the most important constituents of operational resilience in any automated system (Bygstad, 2017). Error and exception mechanisms in RPA systems, unfortunately, are quite weak in many cases for handling errors and exceptions during the automation process. Bots may fail due to missing data, system outage, or variations in the input being processed; however, most RPA implementations are not prepared to manage situations like these properly. Otherwise, without establishing proper monitoring systems, bots may well stop processes altogether or return wrong answers possibly without giving any form of notice.

For example, an accounts payable bot may stumble upon a bad or incomplete entry in the data-entry process and have nowhere to mark the error or forward it up for human intervention, thus waiting around. On the other hand, the bot can continue processing the wrong data, and the errors passed down the line will lead to erroneous outputs that affect other processes. A disorganized method of handling errors can severely disrupt business activities by wasting one's time and increasing the operational cost.

Dependency on Legacy Systems

Many organizations are using RPA as their workflow modernization solution while deploying it on top of existing legacy systems in most cases. One of the best advantages of RPA is that it can work cross-platform and interface with legacy software through its UI. Still, this poses huge operational vulnerabilities. Legacy systems are usually not streamlined for automation and probably contain outdated or inconsistent software; hence, operation by bots will hardly be possible there with any reliability.

Bots operating within a legacy system are also prone to errors due to either the lack of stability or performance provided by such systems in comparison to modern applications. Moreover, legacy systems can be more challenging to upgrade in some scenarios, which complicates integration with RPA when these legacy systems pose problems of incompatibility or performance bottlenecks to an automation process. The operational risks will increase if legacy systems are not documented properly, making it challenging for the developers of RPA to develop or sustain stable automation processes.

Monitoring and Oversight Issues

Most RPA systems are implemented with a presumption of very high levels of autonomy and infrequent oversight. This creates operational vulnerabilities to be highly likely because most RPA systems do not fully support underpinnings of monitoring and oversight mechanisms. Bots may continue running even when faults happen, or transactions cease to achieve the expected outcomes providing the amassing of problems before their critical nature is detected. If real-time monitoring is not provided, many business operations, data inaccuracies, and disruption may result.

Furthermore, distribution of monitoring and oversight across multiple bots or RPA systems can result in inconsistencies and fragmentation in an organizational automation strategy. If there is no integrated platform that monitors the performance of the bots and solves the issues that may emerge, organizations are to stumble in managing and scaling their automation activities. This could eliminate process optimization opportunities and may increase the possibility of operational failure.

Compliance and Regulatory Challenges

It is operational vulnerabilities that one finds in RPA systems due to the complexity of meeting the compliance of certain industry standards. For instance, industries relating to health, finance, and legal services normally have strict regulatory compliance standards about how the data is processed, stored, and secured. Nonchalant configuration and oversight of such systems might leak holes that breach the compliance requirements. For instance, machines that handle personal information should be subject to all applicable data protection regulations, including the General Data Protection Regulation, and an unsupervised bot system can easily violate these rules in the first place, especially if such bots lack an appropriate procedure for processing or securing data properly.

However, regulation frameworks should be updated or monitored and bot workflows refreshed as the process must ensure that RPA follows all the regulations placed in industries. Non-compliance may attract legal penalties of different kinds, reputational damage, and customers' mistrust, effects that majorly affect businesses.

DATA PRIVACY AND COMPLIANCE RISKS IN RPA SYSTEMS

The risks associated with data privacy and compliance have come into sharper focus as organizations increasingly rely on RPA for its efficiency-enhancing and operation-cost-reducing value. Sensitive data, consisting of personal information, financial records, and confidential business data, are very frequently associated with RPA systems, and thus, raise the question of how that data is handled, stored, and secured. Organizations should conduct the appropriate treatment of sensitive data for critical reasons: to avoid legal, financial, and reputational risks; assure compliance with regulations; and avoid breaches, penalties for non-compliance, and loss of customer trust and stakeholder relationships.

Handling of Sensitive Data

One of the major concerns of data privacy in RPA systems is sensitive and personally identifiable information (PII). Several organizations use large volumes of sensitive data within RPA systems for automating payroll for their employees, customer account information, or even patient health records. The risk of breaches increases with high-class breaches if the RPA system is not securely developed or implemented.

Perhaps the most fundamental issue is that bots are processing data that is either unencrypted or not strongly secured. For instance, the bot may temporarily retain sensitive information in logs or files for processing purposes, which then places them at risk of either internal or external threats if proper encryption measures are not in place. It is thus important that RPA uses encryption, data masking, and other protective measures when processing sensitive information so that the information is not inadvertently exposed.

Data Retention and Access Control Issues

Data retention policies, or the lack thereof, can also raise significant privacy concerns in RPA systems. For instance, by default, bots are set to process and store data for a long period, depending on the nature of the automation task. In absence of a data retention policy, RPA systems may retain more sensitive information for longer than it is needed, thereby increasing the risk of data misuse or breaches. Over-retention of sensitive data, particularly PII, may lead to violations of the law regarding privacy and could go against the GDPR or CCPA, which prescribe erasure and data anonymization when an organization has successfully fulfilled its purpose for collecting data.

Another problem with RPA systems is access control. Bots need to have access to databases and other systems to do their work quickly and efficiently. However, access controls if misconfigured can lead to bots entering information, they are not granted access to and should not process. Furthermore, poor segregation of duties or absence of audit trails will eventually point out who accessed what. Organizations remain susceptible to insider threat or data leakage. Strong access control policies and role-based access along with the provision of audit mechanisms are the essentials ensuring that bots have data access only to those they need for their specific functions.

Compliance with Data Protection Regulations

Most challenging aspect when deploying RPA is following data protection regulation. These have included GDPR, CCPA, the Health Insurance Portability and Accountability Act (HIPAA), and the Payment Card Industry Data Security Standard (PCI DSS). These standards obligate organizations to handle data with a significant level of care in terms of taking care of data privacy, security, and proper processing practices (Anagnoste, 2018). The implication of implementation of RPA systems to automate processes involving regulated data indicates that there is an element of compliance risks if the organization fails to meet standards set.

For instance, the General Data Protection Regulation demands that an organization must get explicit consent from a person before processing his or her personal data and that his or her data need to be capable of being deleted or anonymized on demand. It thus forces Personal data-being-handled RPA bots to be developed in such a way that adheres to these obligations which at large makes sure data is processed, stored, and deleted the proper way. Besides, the rights of the data subjects such as right to be forgotten or right to portability will be utilized by the bots. When the corresponding regulatory requirements are absent while designing RPA processes, it would affect the big fines and legal wrangles in some processes.

Finally, systems of RPA will have to prepare for data handling according to audit and compliance standards. It involves maintaining records of all interactions with data, being transparent about how data is processed, and allowing for regular audits of RPA processes to comply with constantly changing regulations. Where such regulatory scrutiny is high, like in health or finance, failure to abide by privacy laws incurs catastrophic consequences, including loss of accreditation and reputational damage.

Cross-Border Data Transfers and Compliance

Deploying RPA is common with multinational organizations. The big issue usually involves crossing the borders to transfer data. Other risks are related to privacy and compliance. Regulations do vary from country to country on how data should be treated and transferred across borders (Syed et al., 2020). For example, GDPR requires tight control over the transfer of personal data outside the EEA, which means that institutions have to be able to assure adequate measures for protecting data in the destination country.

RPA may unintentionally violate cross border data transfer rules by, for example, if bots are trained to extract and process data from the source whose privacy laws are different. It requires clear policies when dealing with cross-border data, including the use of standard contractual clauses, binding corporate rules, or the existence of localization measures in respect of applicable law.

Third-party risks in an RPA environment

In most RPA deployments, many organizations end up buying into third-party vendors for RPA platforms, cloud services, or integration with third-party systems. Such deployments do increase the complexity of data privacy and compliance risks since the third-party vendors may have varying security practices compared to an organization and may store data in jurisdictions with lesser stringent data protection laws. Organizations may be exposed to additional risks because the vendors do not conform to similar standards on privacy and regulation.

Organizations, therefore, must vet carefully the third-party vendors selected and ensure that they have rich data-protection policies. The data privacy and security obligations in the contract should be specific and clearly delineated, and audits of the third-party systems must take place regularly to confirm compliance. Active management of third-party risks can, therefore, reduce the potential for data breaches or compliance violations originating from external partnerships.

THIRD-PARTY SOFTWARE AND INTEGRATION VULNERABILITIES IN THE RPA SYSTEM

The third-party software used by organizations is very normal, where it allows extension of RPA bot capabilities by connecting them to various enterprise applications, cloud services, and external tools. However, this reliance on third-party software leads to a myriad of vulnerabilities that would further compromise the security, stability, and effectiveness of RPA systems. These are because integrating different systems may be challenging, security lacks in third-party solutions, and potential mishandling and noncompliance with data. The risks that need to be understood and mitigated in this case are very crucial so that organizations can successfully and safely implement RPA deployment over time(Osmundsen et al., 2019).

Reliance on Untested Third-Party Application

Most RPA systems rely on third-party applications for specific functionalities-in many cases, document processing, data extraction, or system integration (Leno et al., 2021). However, organizations rarely inspect their third-party solutions and how they're integrated into their overall RPA architecture (Sobczak, 2021). Unchecked third-party software has vulnerabilities or security flaws that a hacker can use against it and create a weak link in the entire RPA architecture. For example, outdated software security patches or weak encryption protocols malware has the potential to unlock vulnerabilities into data breach or illicit access to sensitive information.

It may not have been developed with a focus on the specific needs and risks of an RPA system. Hence, such integration into an RPA environment may be incompatible and slow and unsafe due to unnoticed weaknesses. In this regard, the organization has to adequately evaluate and assess the security and performance requirements of any kind of third-party software intended to be integrated.

Vulnerabilities in APIs and Integration Points

Most RPA bots interact with third-party software in the form of APIs. Using APIs, bots can effectively send and receive data between systems to result in seamless integration with other platforms. However, if the APIs are not properly secured or even poorly designed, there is a possibility that deep vulnerabilities may easily find their way into the RPA environment. In most cases, attackers tend to target the easy entry points-the APIs that offer a weakness in authentication, encryption, or a lack of rate limiting.

If an RPA system interacts with a thirdparty API that suffers from SQL injection vulnerabilities or cross-site scripting (XSS) attacks, or other types of data manipulation, sensitive data might be exposed, or attackers could exploit the system for unauthorized access to internal systems. Therefore, misconfigured APIs, weak API keys, or malicious users who find an open and illegal way into unauthorized areas of the system may compromise the whole RPA environment. Strong controls like encryption, tokens for authentication, and monitoring in APIs mitigate such risks and validate RPA integrations.

Data Transmission and Privacy Risks

The installation of third-party software used to integrate with RPA generally involves transferring sensitive information across systems. It transmits the data, which would prove privacy risks and breaches if it is not secured. Many third-party solutions handle sensitive information related to customers, financial transactions, or health records that have to be protected by different data protection laws, including GDPR, HIPAA, or CCPA. Sensitive information may leak when transmitted through if third party systems do not have appropriate encryption or data masking protocols.

Moreover, the third-party providers might have a history of maintaining data in areas that are not aligned with an organization's policies about personal data safety. Especially in the case of cross-border transfer, companies may not be compliant with the policies of their third-party providers. For example, integrating a cloud-based RPA tool with a third-party software would mean all the data is stored in regions where data protection laws are not as strict as in other regions. Therefore, such integrations increase the risk of non-compliance to global regulations. Data should be protected by having organizations maintain strict encryption standards for all data transmissions and by requiring third-party providers to honor applicable data privacy laws.

Software Updates and Patch Management

Another major vulnerability related to third-party software integration is the issue of staying updated. Some third-party software vendors regularly go to develop and release patches and updates that fix bugs, enhance functionality, or fix newly discovered security vulnerabilities. If these patches are not applied in time, they would leave an entire system open to vulnerabilities in regard to known threats.

Updates of several third-party solutions may become difficult to manage for the organizations, if not streamlined using a centralized approach to manage the patches. Additionally, software updates may have compatibility issues that may impact the existing RPA workflows by making the bots less functional or about to cause process interruptions which do not align to the expectations of the owner. Comprehensive Patch management approach with timely updates and compatibility tests is vital to ensure security and functionality of integrated systems along with RPA.

Third-Party Vendor Security Risks

Apart from the technical weaknesses the third-party software may have on its own merit, there are other security issues in involving third-party vendors. Most organizations do not know the security culture of whom they outsource to, given that the said third-party vendors must have adequate security measures in their midst. Probably third-party vendors are relatively less strict on security measures, and thus most likely to serve as a point of entry for hackers in case their systems are compromised. If hackers succeed to infiltrate the systems of a vendor, either through software or services provided, they are pretty sure to access the RPA environment of the organization.

Third-party service providers subcontract some areas of the services to other companies. The complexity of the security environment increases and raises the risk of exposure to vulnerabilities. An organization must conduct an in-depth security analysis on third-party vendors and sign strict contractual agreements stating the obligations of data protection and security. Organisations must also regularly audit and assess the third-party vendors to ensure that they meet the requirements

Lack of Control Over Source Code

Organizations that depend on third-party software for their RPA integrations have no control over the source code for any of the solutions that they are using. This considerably limits the effectiveness of controlling the underlying code of solutions when discoveries of vulnerabilities take place on third-party software. Without having access to that source code, an organization has little choice other than to wait for the vendor to release updated patches to correct security vulnerabilities. That is unless the vendor takes such a long time before noticing and fixing vulnerabilities, the RPA system by the organization remains open to attack.

Another pertinent issue is that the third-party software is closed source, meaning the codes cannot be reviewed for backdoors and undetectable vulnerabilities. This circumstance adds to the difficulty of carrying out complete security assessments on such software and ensuring that the organization's security requirements are met. Organizations should only engage with vendors offering access to their source code for auditing or use open-source alternatives with great community support for continuous security updates.

The integration of third-party software can revolutionize the functionality of RPA systems but automatically throws open a whole range of vulnerabilities that need to be handled properly. From unchecked software risks and insecure API threats to risks of data transmission and threats from vendors, organizations have to gear up ahead of the play in securing their RPA ecosystems. This would include rigorous security reviews, strict policies on encryption and access, as well as robust patch management practices. Dealing with these vulnerabilities enables organizations to be able to ensure that their RPA systems are secure, compliant, and effective.

HUMAN ERROR AND GOVERNANCE VULNERABILITIES IN RPA SYSTEMS

Even though RPA systems are designed to minimize the human intervention aspects of workflow processes for streamlined smooth operation, one of the most crucial vulnerabilities surrounding the deployment of the systems remains that of human error. Human mistakes in the configuration, handling, and maintenance of RPA bots bring various problems that vary from inefficiencies in the processes to risks to security. Furthermore, poor governance models that govern the development and deployment of RPA systems exacerbate these vulnerabilities, leading to disruptions in operations, non-compliance with norms, and the heightened vulnerability to cyber attacks. Human mistakes and governance weaknesses like the above call for appropriate mixes of good training, monitoring, and positive governance to achieve true long-term success and security for RPA systems.

Errors in Bot Configuration and Development

Humans err when they configure and develop bots, which happens to be a very common area of mistake in RPA systems (Cooper et al., 2019). Setup and programming are actually much needed for RPA systems to work. Even minor configuration errors of a bot may cause highly significant errors within an automated workflow-incorrect data mapping, for instance, or logical flaws and inaccurate rules that may significantly cause error within the system (Madakam et al., 2019). Such misconfigurations could lead to processing incorrect data, failure in accomplishing tasks, and also by unintentional leaks of data when over-access is granted to these bots handling sensitive data.

For example, a bot that is written very poorly to handle a financial transaction can even process a duplicate payment or maybe fails to process the most critical transactions, leaving one to financially lose or damage the reputation. Another example might be where developers failed to include error-handling mechanisms in the bots, hence the system may not recover when something goes as expected, thus leading to operational down time or processing interruption. To mitigate these risks, the developers and users of RPA within organizations must be adequately trained and tested to confirm the accuracy and functionality of bot configurations before deployment.

Lack of Standardized Governance Frameworks

Most organizations implement RPA without clear governance frameworks. Such open pathways become vulnerable to inconsistent practices and enable a higher likelihood of errors and security threats. Governance is essential in governing the entire lifecycle of the bots, from design and implementation to monitoring and decommissioning. Without standardized governance protocols, various teams in an organization will design and deploy bots with disjointed methods, further fragmenting workflows and creating inefficiency (Lacity & Willcocks, 2018).

This is why the lack of governance increases the likelihood of compliance violation. For example, RPA systems handling confidential information may lack regulations on uniform security and compliance checks. This puts the organization under risk factors, especially by high-risk industries like finance and healthcare. Organizations have to ensure that they provide proper governance policies with defined responsibilities, security standards, and continuous monitoring of bot activities.

Insufficient Oversight and Monitoring

Even once the RPA bots are deployed, oversight by humans is typically required for ongoing proper operation and security, but under-monitored or accountability-less mistakes are often not reported, leaving vulnerabilities or errors that can unlock the whole system. The bot might run into an unexpected change in the environment: perhaps some system updates or changes in sources of data-it will fail or produce incorrect outputs. Without adequate surveillance, these issues may go unnoticed until they cause enough operational commotion.

In addition, a lack of monitoring of the bots leaves it vulnerable to malicious attacks. Cybercriminals can exploit the vulnerabilities of unmonitored RPA systems in order to inject malware, change data, or to gain improper access to internal systems. Continuous monitoring systems, coupled with regular audits, are very critical to making sure that RPA bots operate securely and efficiently. Organizations must also clearly define accountability for bot performance. Any problems must be resolved promptly by responsible individuals.

Lack of Role-Based Access Control

One more human-error-predisposing weakness results from inappropriate access control to the RPA system implementation. Organizations may fail to implement role-based access controls (RBAC) strictly limiting which users could create, modify, or execute RPA bots. This may result in wrong staff members who have full, unbridled access to bot configurations and who may change them for erroneous reasons or propagate security risks. A user with poor training or education may inadvertently open a sensitive system or information to access by a bot and, thereby, increase the potential for data breaches or system failures.

In the absence of strict access controls, internal actors—malicious or careless may misuse the RPA system for unintended purposes. The risk can therefore be mitigated by enforcing strict role-based access controls wherein only authorized personnel are allowed to manage or change RPA bots. This should include limiting access to sensitive functions and requiring multi-factor authentication for users with high-level privileges.

Failure to Update or Decommission Bots

Bots developed using RPA are likely to get outdated overtime because business processes change, software environments change, and compliance requires may change. Yet, human error may prevent organizations from decommissioning or updating obsolete bots, thereby exposing them to errors or security breaches. Outdated bots may be dependent on legacy systems that have lost their security or may not be in accordance with the new regulations that are enforced, and by so doing, expose an organization to risks if not corrected.

Moreover, dormant or unfollowed bots that continue to be connected to live systems can eventually put an organization on the danger list as the hackers will seize any unmonitored resources to enter the internal network. Organizations should, therefore, have a lifecycle management policy where periodic reviews on the performance and relevance of the bots are conducted. The processes of decommissioning unnecessary bots should be secure with their access cut off and system dependencies terminated.

Lack of Thorough Documentation

Further, poor documentation will exaggerate the weaknesses caused by human errors and bad governance. It is tricky to debug and identify fault lines or expose root causes of security failures if the processes regarding the development, configuration, and maintenance of bots are not well documented. Further, it makes it difficult to transfer knowledge in an organization if new members join and makes it tough for them to manage or even update the systems correctly.

There should also be documentation of all the lifecycle stages of the bot such as the design specifications, security protocols, error-handling procedures, and ongoing monitoring requirements. This will ensure proper documentation of RPA systems for management even in cases where personnel may have changed or systems may be upgraded.

Even though RPA systems are aimed at reducing the intervention of humans into routine tasks, human error and failures in governance also pose a major risk in the effective deployment and operation of an RPA system. These may be configuration mistakes and failing governance structures or incomplete monitoring and role-based access control, among others. These can have far-reaching effects on the security, efficiency, and compliance of the RPA systems. This would enable organizations to identify these vulnerabilities and mitigate all the intrinsic risks they face, thus ensuring their RPA ecosystems are safe.

MITIGATING INTRINSIC VULNERABILITIES IN RPA SYSTEMS

Organizations should ensure that their Robotic Process Automation (RPA) systems are used safely and efficiently by having an understanding of intrinsic vulnerabilities. The threats include errors in configuration to governance gaps and therefore organizations need to have multiple strategies to deal with the issues arising. Proactive planning, continuous monitoring, and appropriate governance frameworks can protect RPA deployments from devastating risks to applications' performance, security, and compliance. A holistic approach to RPA management ensures that the automation effort brings the benefits it is intended to provide without bringing in unnecessary risk.

Building Strong Governance Frameworks

A sound governance framework to deal with the vulnerabilities brought about by RPA is one of the best ways to achieve easy addressing. Governance plays a critical role in defining the policies, procedures, and controls required to manage RPA systems from development through deployment and on-going operations (Kirchmer, 2017). Ensuring good governance structures mean that bots are developed according to the best set practices and industry standards while at the same time reducing the risk of security lapses and operational inefficiencies.

Governance frameworks should entail standards in bot development, testing and monitoring and decommissioning. Organizations should also ensure compliance with their rules in the fields of security and privacy by having explicit policies(Kirchmer, 2017). For example, organizations can put up automated controls that initiate bot updates upon change from a regulatory requirement or security patch. Another approach would be putting up cross-functional governance that includes IT, operations and compliance stakeholders; this would increase accountability further, for RPA systems are administered holistically.

Implementing Rigorous Bot Development and Testing Processes

The risks that come as a consequence of bot development must therefore be abated by the organizations; in this case, through very strict development and testing processes. This all starts with strict configuration standards for bot setup so all workflows, logics, and access points are correct at setup. Developers should develop according to safe coding practices, and all bots should be subjected to a rigorous test, including security testing, to at least identify and correct such potential vulnerabilities before using them (Jimenez-Ramirez et al., 2019).

Testing should mimic reality with respect to simulating real-world scenarios to test that bots will behave in the way expected under various conditions. This includes stress tests aimed at identifying how bots will respond in light of changing the system environment or to unexpected inputs. By catching such configuration errors early in the development process, organizations will avoid costly disruptions and potential security breaches down the line.

Continuous Monitoring and Auditing

Once deployed, continuous monitoring and auditing would be essential to keep the deployed RPA system at a high level of security and operational efficiency. Real-time monitoring tools can monitor how the bots are performing, and anomalies or irregularities might indicate errors, failures, or even security threats(Santos et al., 2020). Regrade activities to monitor bot activities ensure that RPA systems are operated correctly and stay abreast of up-to-date security measures.

In addition to that, monitoring should include the health of the RPA ecosystem in its integration with other systems and third-party software, not only the performance of each bot. Any instance, which doesn't follow the expected behavior, should raise alarm. With this, possible weaknesses can be addressed right away (Geyer-Klingeberg et al., 2018). Automated logging and reporting systems provide more ease for the monitoring process; problems are documented early.

Enhancing Role-Based Access Control (RBAC)

Access control should be performed in a role-based model in an organization. This would ensure that only authorized people can create, modify, or execute bots while minimizing accidental changes on an RPA system. This way of access control also prevents inappropriate changes on workflows that contain sensitive data (Geyer-Klingeberg et al., 2018).

Besides RBAC, MFA should also be implemented for users with high privileges. This provides a second layer of security so that in case the credentials are compromised, unauthorized access to sensitive RPA functions is not easy. The user roles and rights of the RPA system must be reviewed periodically. In this respect, this process enables only necessary personnel to retain access to the RPA systems.

Ensuring Comprehensive Training and Documentation

Appropriate training programs can minimize human errors since employees are equipped through proper skills and knowledge required to operate RPA systems effectively. In addition to the above, developers, operations teams, and other stakeholders should receive regular training on bot configuration, security best practices, and error handling. Training programs should be amended according to any new developments in RPA technologies so that staff are conversant with the latest abilities and risks.

Training aside, proper documentation for all RPA systems must also be in place. Such documentation covers every aspect of the bot's lifecycle, ranging from development specifications to error-handling procedures and system dependencies.

Systems that are well-documented are easier to troubleshoot and can be upgraded or secured, especially when personnel or changes to the systems occur.

Establishing a Lifecyle Management Process for Bots

Another crucial process to minimize attacks through vulnerabilities is the establishment of a well-defined lifecycle management process of RPA bots. Defunct or obsolete bots should be safely and securely removed from access to systems and data sources while their dependencies are resolved. If inactive bots are not properly decommissioned, security gaps would often remain open to exploitation by attackers.

Lifelong management should also include upgrading of RPA bots where business processes, system environments, or regulatory requirements change. Reviewing and updating bots is a way through which organizations ensure automation systems' relevance, safety, and compliance.

Strengthening Third-Party Integration Security

Many of these RPA systems rely on third-party software and integrations, so such tools add to the vulnerabilities based on the software as well. Organizations need to look much more deeply at third-party tools before their deployment in the RPA environment, understand the security posture and patching processes for such software, as well as industry standard compliance.

Once adopted and integrated into a system, third-party tools must be monitored for the security vulnerability potential. Third-party access can also be limited to software scope: allowing third-party tools only access to data and systems they require to perform their operations.

FUTURE TRENDS AND CHALLENGES IN SECURING RPA SYSTEMS

As more sectors embrace Robotic Process Automation (RPA) and the concept continues to grow and adapt, its security landscape changes. The most prominent trend appears to be the incorporation of artificial intelligence and machine learning into RPA frameworks in support of more intelligent and adaptive automation. Of course, this new advancement calls for emerging security challenges (Siderska, 2020). With AI-driven bots becoming more autonomous and decision-capable, the risk of AI-driven bots acting based on flawed data or unwanted scenarios escalates. While data fed into AI-driven RPA systems calls for accuracy and security to avoid the cascading failure causing operational disruptions or compliance violation(Ivančić et al., 2019).

The more the organization extends the scope of RPA deployments, it becomes intricate in complexity while handling those systems. More bots means more integration points with other systems and applications, creating a larger attack surface for cyber-attacks (Ivančić et al., 2019). That means organizations will have to adapt to more advanced forms of monitoring tools relying on AI to identify anomalies and have breaches in real time. Continuous training for employees will also be needed to adequately recognize emerging threats (Siderska & Jadaan, 2018).

It challenges the securing of automated processes as more shift to working from anywhere and cloud-based RPA solutions. Organizations have many employees working remotely from their RPA systems and must, therefore, employ effective security, including VPN, MFA, and strong access controls, to prevent unauthorized access and exposure of sensitive data (Hallikainen et al., 2018). Additionally, the situation regarding security becomes worse because organizations depend on third-party vendors for software and services, which implies that the responsibility is put squarely on the organization to ensure that all these third-party vendors are meeting the same intense security standards.

Another dimension where the regulatory landscape is expected to evolve is in data privacy and security. Organizations will face the challenge of becoming compliant with this evolving regulatory landscape in safeguarding sensitive data when using RPA systems. The potential implication could be more robust data handling practices, with the addition of transparency and auditability of the RPA process.

Such great promises the future holds for RPA in more efficiency and productivity, and concerning the incident of security, organizations must be aware of it and address related security concerns. Businesses can mitigate risks while successfully deploying RPA systems in an increasingly complex digital environment through investments in sophisticated security technologies, a culture of security awareness, and keeping pace with changes in the regulatory environment.

CONCLUSION

In conclusion, securing RPA systems represents an ongoing challenge for which a multi-faceted approach must be developed to overcome the inherent vulnerabilities and evolving threats in today's dynamic technological landscape. As organizations increasingly adopt RPA as one of their means for streamlining operations and enhancing efficiency, security at every stage of the automation lifecycle ought to be adopted to ensure appropriateness. This includes robust governance frameworks, rigorous development and testing procedures, and continuous monitoring and improvement through individual and organizational culture development.

With AI and ML integrated into RPA systems, it does come with certain opportunities and challenges that are bound to arise (Enríquez et al., 2020). Though meant to be leveraged to enhance automation capabilities, they come with new risks that an organization has to face. Knowing the risks helps protect sensitive data as well as remain complaint with all regulatory requirements. The complexity of the deployments of RPA increases significantly, thereby demanding more advance security measures on the organizations' part that can effectively manage this much broader attack surface, especially when third-party integrations and remote work are considered.

Over the coming years, organizations must take proactive steps into dealing with all the security challenges RPA systems pose. The window in this case would be for discovery of new threats, education of employees, and changes in the law. By prioritizing security and resilience, the full unlocking of the benefits from RPA will be limited to the risks. Only a holistic and proactive approach to securing RPA systems will safeguard organizations from vulnerabilities while, at the same time, ensure that automation efforts are contributing positively toward operational excellence and business continuity. Only through exercising diligence and foresight can organizations now chart their course through the confusing play of RPA security and thrive in an increasingly automated world.

REFERENCES

Agostinelli, S., Marrella, A., & Mecella, M. (2019). Research challenges for intelligent robotic process automation. In *Business Process Management Workshops: BPM 2019 International Workshops, Vienna, Austria, September 1–6, 2019, Revised Selected Papers 17* (pp. 12-18). Springer International Publishing. DOI: 10.1007/978-3-030-37453-2_2

Aguirre, S., & Rodriguez, A. (2017). Automation of a business process using robotic process automation (RPA): A case study. In *Applied Computer Sciences in Engineering: 4th Workshop on Engineering Applications, WEA 2017, Cartagena, Colombia, September 27-29, 2017* [Springer International Publishing.]. *Proceedings*, *4*, 65–71.

Anagnoste, S. (2018, May). Robotic Automation Process–The operating system for the digital enterprise. In *Proceedings of the International Conference on Business Excellence* (Vol. 12, No. 1, pp. 54-69). DOI: 10.2478/picbe-2018-0007

Beuerle, M. (2019). *Development of low-scaling methods to calculate ground state energies and analytical gradients based on the adiabatic-connection fluctuation-dissipation theorem* (Doctoral dissertation, lmu).

Bygstad, B. (2017). Generative innovation: A comparison of lightweight and heavyweight IT. *Journal of Information Technology*, *32*(2), 180–193. DOI: 10.1057/jit.2016.15

Cooper, L. A., Holderness, D. K.Jr, Sorensen, T. L., & Wood, D. A. (2019). Robotic process automation in public accounting. *Accounting Horizons*, *33*(4), 15–35. DOI: 10.2308/acch-52466

Enríquez, J. G., Jiménez-Ramírez, A., Domínguez-Mayo, F. J., & García-García, J. A. (2020). Robotic process automation: A scientific and industrial systematic mapping study. *IEEE Access : Practical Innovations, Open Solutions*, *8*, 39113–39129. DOI: 10.1109/ACCESS.2020.2974934

Geyer-Klingeberg, J., Nakladal, J., Baldauf, F., & Veit, F. (2018). Process Mining and Robotic Process Automation: A Perfect Match. *BPM (Dissertation/Demos/Industry)*, *2196*, 124-131.

Gotthardt, M., Koivulaakso, D., Paksoy, O., Saramo, C., Martikainen, M., & Lehner, O. (2020). *Current state and challenges in the implementation of smart robotic process automation in accounting and auditing*. ACRN Journal of Finance and Risk Perspectives. DOI: 10.35944/jofrp.2020.9.1.007

Hallikainen, P., Bekkhus, R., & Pan, S. L. (2018). How OpusCapita Used Internal RPA Capabilities to Offer Services to Clients. *MIS Quarterly Executive, 17*(1).

Hofmann, P., Samp, C., & Urbach, N. (2020). Robotic process automation. *Electronic Markets, 30*(1), 99–106. DOI: 10.1007/s12525-019-00365-8

Ivančić, L., Suša Vugec, D., & Bosilj Vukšić, V. (2019). Robotic process automation: systematic literature review. In *Business Process Management: Blockchain and Central and Eastern Europe Forum: BPM 2019 Blockchain and CEE Forum, Vienna, Austria, September 1–6, 2019* [Springer International Publishing.]. *Proceedings, 17*, 280–295.

Jaunat, J., Garel, E., Huneau, F., Erostate, M., Santoni, S., Robert, S., Fox, D., & Pasqualini, V. (2019). Combinations of geoenvironmental data underline coastal aquifer anthropogenic nitrate legacy through groundwater vulnerability mapping methods. *The Science of the Total Environment, 658*, 1390–1403. DOI: 10.1016/j.scitotenv.2018.12.249 PMID: 30677999

Jimenez-Ramirez, A., Reijers, H. A., Barba, I., & Del Valle, C. (2019). A method to improve the early stages of the robotic process automation lifecycle. In *Advanced Information Systems Engineering: 31st International Conference, CAiSE 2019, Rome, Italy, June 3–7, 2019* [Springer International Publishing.]. *Proceedings, 31*, 446–461.

Kirchmer, M. (2017). Robotic process automation–pragmatic solution or dangerous illusion? *BTOES Insights*, (June), 17.

Kokina, J., & Blanchette, S. (2019). Early evidence of digital labor in accounting: Innovation with Robotic Process Automation. *International Journal of Accounting Information Systems, 35*, 100431. DOI: 10.1016/j.accinf.2019.100431

Kreuzer, M. P. (2014). *Remotely piloted aircraft: Evolution, diffusion, and the future of air warfare* (Doctoral dissertation, Princeton University).

Lacity, M., & Willcocks, L. P. (2018). *Robotic process and cognitive automation: The next phase*. SB Publishing.

Leno, V., Polyvyanyy, A., Dumas, M., La Rosa, M., & Maggi, F. M. (2021). Robotic process mining: Vision and challenges. *Business & Information Systems Engineering, 63*(3), 301–314. DOI: 10.1007/s12599-020-00641-4

Leno, V., Polyvyanyy, A., La Rosa, M., Dumas, M., & Maggi, F. M. (2021). Action-oriented process mining: Unveiling the control flow of actions in process models. *Information Systems, 102*, 101782.

Madakam, S., Holmukhe, R. M., & Jaiswal, D. K. (2019). The future digital work force: Robotic process automation (RPA). *Journal of Information Systems and Technology Management*, *16*, e201916001. DOI: 10.4301/S1807-1775201916001

Menolfi, D., Jiang, W., Lee, B. J., Moiseeva, T., Shao, Z., Estes, V., Frattini, M. G., Bakkenist, C. J., & Zha, S. (2018). Kinase-dead ATR differs from ATR loss by limiting the dynamic exchange of ATR and RPA. *Nature Communications*, *9*(1), 5351. DOI: 10.1038/s41467-018-07798-3 PMID: 30559436

Moffitt, K. C., Rozario, A. M., & Vasarhelyi, M. A. (2018). Robotic process automation for auditing. *Journal of Emerging Technologies in Accounting*, *15*(1), 1–10. DOI: 10.2308/jeta-10589

Osmundsen, K., Iden, J., & Bygstad, B. (2019). Organizing robotic process automation: balancing loose and tight coupling.

Santos, F., Pereira, R., & Vasconcelos, J. B. (2020). Toward robotic process automation implementation: An end-to-end perspective. *Business Process Management Journal*, *26*(2), 405–420. DOI: 10.1108/BPMJ-12-2018-0380

Siderska, J. (2020). Robotic Process Automation—A driver of digital transformation? *Engineering Management in Production and Services*, *12*(2), 21–31. DOI: 10.2478/emj-2020-0009

Siderska, J., & Jadaan, K. S. (2018). Cloud manufacturing: A service-oriented manufacturing paradigm. A review paper. *Engineering Management in Production and Services*, *10*(1), 22–31. DOI: 10.1515/emj-2018-0002

Sobczak, A. (2021). Robotic Process Automation implementation, deployment approaches and success factors–an empirical study. *Entrepreneurship and Sustainability Issues*, *8*(4), 122–147. DOI: 10.9770/jesi.2021.8.4(7)

Syed, R., Suriadi, S., Adams, M., Bandara, W., Leemans, S. J., Ouyang, C., ter Hofstede, A. H. M., van de Weerd, I., Wynn, M. T., & Reijers, H. A. (2020). Robotic Process Automation: Contemporary themes and challenges. *Computers in Industry*, *115*, 103162. DOI: 10.1016/j.compind.2019.103162

Willcocks, L. P., Lacity, M., & Craig, A. (2015). The IT function and robotic process automation.

Chapter 5
Intelligent Robotic Process Automation (RPA) Development

Mohit Yadav
https://orcid.org/0000-0002-9341-2527
O.P. Jindal Global University, India

Tanupriya Choudhury
University of Petroleum and Energy Studies, India

Shivani Dhand
https://orcid.org/0000-0002-4809-1365
Lovely Professional University, India

Xuan-Hoa Nghiem
https://orcid.org/0000-0003-2292-0257
International School, Vietnam National University, Hanoi, Vietnam

ABSTRACT

This chapter deals with growth and deployment, changing business operations with Intelligent RPA. It brings together the traditional automation with AI, machine learning, and cognitive technologies in handling complex processes to enhance the efficiency of operations and reduce costs. The chapter outlines the critical enablers of Intelligent RPA, the benefits attached to it during implementation, the challenges that are met during its implementation, and finally the future trends defining its direction. As such, it examines the management, operational, social, and ethical effects of Intelligent RPA adoption for organizations, emphasizing transparency, workforce transformation, and ethical governance. Further, as enterprises transition to Intelligent RPA, they will be forced to balance the celebration of technological

DOI: 10.4018/979-8-3693-4365-4.ch005

accomplishment with people-oriented means to sustainable value creation. Lastly, the paper concludes that Intelligent RPA can eventually alter the work nature in almost every industry while factoring consideration of the ethical and social implications of automation.

INTRODUCTION TO INTELLIGENT ROBOTIC PROCESS AUTOMATION (RPA)

Conventional wisdom in managing businesses is brought about by Robotic Process Automation (RPA). These have been exercising repetitive and rule-based processes automation that brings high efficiency, reduced error rates, and lower operational costs. Traditional RPA applications cover a broad spectrum of industries such as finance, healthcare, supply chain management (Aguirre & Rodriguez, 2017). The tasks performed in such operations include data entry invoice processing and report generation. However, the search of organizations for a higher level of efficiency and adaptability triggered the next generation of RPA known as Intelligent RPA (Syed et al., 2020). While integrating AI, ML, and Cognitive technologies, it has further extended the possibility of offering traditional RPA to systems with capabilities that are much more advanced, making the system handle complex tasks and even more intricate decision-making processes.

The difference of Intelligent RPA is that it introduces AI and cognitive technologies like NLP, OCR, and machine learning to distinguish from traditional RPA, which is great in automating structured, rule-based tasks but labors in unstructured data or decision-making tasks (Chakraborti et al., 2019). Intelligent RPA can analyze, interpret, and make decisions by using unstructured data available as emails, scanned documents, or voice commands. This capability allows organizations to automate more comprehensive processes and, consequently, improve their overall business performance and agility (Hoffmann et al., 2020). For instance, in smart RPA, NLP can guarantee that customer service interactions are automated; thus, chatbots can be depended on to answer more complex questions by understanding the nuances of human language and, hence, giving contextual accuracy responses.

By integrating AI and machine learning, intelligent RPA systems learn from the data patterns and thus make autonomous decisions. Intelligent RPA does not face these disadvantages of traditional RPA, whereby changes in a process necessitate manual updates continuously. Intelligent RPA can amend the workflows depending on new inputs of data and can learn from previous experiences, rather than the case with traditional RPA. As such self-learning may improve the efficiency in the process and even facilitate the automation of more dynamic as well as other unpredictable tasks like fraud detection or predictive maintenance, businesses are able to enhance their adaptability and robustness to fast changes of the market or any unexpectable disturbance (Willcocks et al., 2024).

Further, the smart RPA is quite importantly instrumental to workforce transformation. Although job displacement on account of automation is a very significant concern, smart RPA also enables employees to focus even more on higher-value tasks that involve creativity, critical thinking, and emotional intelligence. Through the reduction of redundant work, which includes tasks such as problem-solving, innovation, and decision-making, the workforce is freed up for strategic activities: these are paramount in today's knowledge economy, respectively (Lacity & Willcocks, 2021). With intelligent RPA, the expertise that a business has can be maintained by shifting employees into positions that involve managing, maintaining, and enhancing automation.

Another reason that fuels the uptake of intelligent RPA is the prospect of tremendous cost savings and operational efficiencies. Intelligent RPA will automate complex and previously impossible-to-automate tasks, thereby streamlining business processes, reducing errors, and ensuring higher compliance. As this technology continues to move in its evolutionary track, there is every hope that it will unleash its full impact in shaping the world of work and business, catalyzing innovation, and keeping organizations in check as companies become part of an increasingly digital world.

Intelligent RPA represents a step change in the evolution of automation technology and is an extension of traditional RPA where AI and cognitive technologies are enabled to support automation. Intelligent RPA will have the potential to automate more decision-based complex tasks, which will bring the much-needed disruptive changes into industries, improve agility in businesses, and unleash innovative pathways toward workforce transformation.

KEY TECHNOLOGIES IN INTELLIGENT RPA

The development of Intelligent RPA has been underscored by a set of advanced technologies that take the basic and classic capabilities of automation forward (Syed et al., 2020). The prime objective of bringing together AI, ML, and cognitive systems together is to help businesses become able to automate those tasks that are more complex and dynamic in character. Some of the key technologies that drive the functionality and effectiveness of Intelligent RPA are as follows.

1. Artificial Intelligence and Machine Learning

The basis of intelligent RPA is artificial intelligence and machine learning, transforming static automation into adaptive self-learning systems. Through AI, RPA bots learn to mimic human thinking, process information, identify patterns, and draw on learned knowledge to make decisions (Chakraborti et al., 2020). Machine learning takes it a little further by learning from inputs and perfecting its performance through experience. Unlike the conventional RPA, which depends on the operation being based upon the pre-existing rules, AI and ML make Intelligent RPA capable of following variability, such as changed data or workflows, and can optimize processes without any human involvement (Osmundsen et al., 2019).

2. Natural Language Processing (NLP)

NLP is extremely useful for Intelligent RPA systems so they can read and interact in human language (Leno et al., 2021). This will allow an RPA bot to read unstructured text data, like emails, chat messages, or voice commands, extracting meaning from the same, which is a prerequisite for automating customer service and high-frequency communication tasks like document processing. Real-time linguistic interpretation of NLP-enabled RPA bots means that the bots can answer queries with an accuracy level paralleled only by human employees; they can extract relevant information from documents or even initiate processes based on voice commands (Enríquez et al., 2020).

3. Optical Character Recognition (OCR)

This is where OCR happens, and actually does most of the work that would allow Intelligent RPA to work with unstructured data. Here, RPA bots could accept scannned copies of documents, images, or PDFs for easy reading and subsequent transformation into machine-readable format for easy extraction. This is especially very useful in financial services, healthcare services, and legal services where many processes predominantly rely mainly on information incorporated into a document (Osmundsen et al., 2019). With the integration of OCR, Intelligent RPA even automates processes such as invoice processing and form filling, manages contracts, and reduces even the chances of human intervention and errors (Agostinelli et al., 2019).

4. Cognitive Decision Engines

Cognitive decision engines are used to improve the capabilities of decision making in Intelligent RPA systems for processing a complex scenario, thereby providing context-aware solutions (Suri et al., 2017). These are based on the holistic application of AI, ML, and rule-based algorithms for the evaluation of multiple variables to analyze and estimate outcomes and come up with autonomous decisions. Cognitive decision engines will help the intelligent RPA to extend beyond mundane automations alone. They can perform the tasks where some judgment is required such as fraud detection, compliance checks, or risk assessment. They help organizations to generate more accurate and efficient results by simulating human-like capabilities for decision-making while generating automated processes (Syed et al., 2020; Agostinelli et al., 2019).

5. Integration with Cloud Computing and APIs

Cloud infrastructure and Application Programming Interfaces (APIs) provide Intelligent RPA with the scalability to grow with the flexibility of scale, cutting across multiple platforms as well as various applications. With cloud-based infrastructure, Intelligent RPA can be deployed quickly, handling large quantities of data in real-time across different systems (Osmundsen et al., 2019). APIs enable seamless connectivity between the RPA bot and the enterprise system, allowing the bots to interact with other software applications, databases, and external services. That enables Intelligent RPA to operate across a diverse IT ecosystem towards achieving more efficient and agile automation.

6. Advanced Analytics and Insights

Advanced analytics monitoring and optimization of performance is another important technology in Intelligent RPA: the integration of analytics into the RPA system allows organizations to know how automation is working, to be able to identify bottlenecks, and to have real-time insights into the performance of operations (Hofmann et al., 2020). Advanced analytics abilities also include predictive abilities, which help the Intelligent RPA system predict future trends or issues-for instance, potential process failures or probable customer needs. It thus leads to constant improvement in all the automated workflows, thereby further improving a decision and operational efficiency.

7. Robotic Vision

This is one of the emerging technologies of Intelligent RPA, which is robotic vision; it makes bots be able to "see" and interpret visual data. Computer vision integrated with RPA enables the bot to decode images, videos, or any other form of visual input. It, therefore, enables them to do quality control or product inspection and even visual data analysis. This is particularly helpful to the industries in manufacturing and retail regarding monitoring and tracking critical visual inspections.

These advanced technologies combined allow Intelligent RPA to assume any form of simple automations as well as complex processes requiring decision making, thus giving businesses a power tool intended to drive digital transformation and foster operational efficiency (Anagnoste, 2017).

BENEFITS OF INTELLIGENT RPA

The adoption of Intelligent RPA brings in many advantages to businesses. This encompasses the incorporation of traditional automation capabilities with AI, ML, and cognitive technologies. The advantages are definitely beyond improvement in efficiency. It transforms the very nature of how organizations operate, innovate, and compete in a digital-first economy. Below are key benefits of Intelligent RPA:

1. Enhanced Efficiency and Accuracy

One of the strongest strengths of Intelligent RPA is to enhance operational efficiency and accuracy. This is because automated routine, repetitive tasks minimize the chances for human error and execute the same processes without failing in accuracy. With AI and ML, the bots could take more complex tasks such as decision-making, pattern recognition, and analysis (Enríquez et al., 2020). Further, productivity and processing speed will increase, and manual intervention during processing will be reduced.

2. Scalability and Flexibility

Intelligent RPA systems scale easily while granting the potential of automating any process within an organization across various departments. When businesses' needs expand, these systems can be scaled up to handle a broader volume of data and tasks with little increased investment. Intelligent RPA is also flexible; it can adapt to changing workflows, business environments, or regulations (Osmundsen et al.,

2019). Whether it's automating financial reporting or customer service interactions, Intelligent RPA provides the agility needed to meet evolving business demands.

3. Cost Savings and Return on Investment (ROI)

Using Intelligent RPA to automate tasks yields considerable cost savings (Lacity & Willcocks, 2016). Because this approach eliminates unnecessary manual labor on repeating tasks and reduces errors that may subsequently lead to costly rectifications, it is possible for organizations to decrease the operational costs. Furthermore, Intelligent RPA has a higher processing speed, which means more throughput in the same number of resources. This effectiveness translates into massive ROI since organisations can achieve much with a reduced number of human capital resources, devoted to strategic planning and innovation.

4. Improved Compliance and Risk Management

Intelligent RPA ensures compliance and risk management as it carries out activities as intended, following rule-based and regulation-driven mandates (Anagnoste, 2018). The high command of automation systems easily logs every action taken, thereby creating detailed audit trails that guide organizations on their role in the compliance to the given industry standards and regulatory measures. Intelligent bots are thus capable of sifting through data and potentially identifying risks, such as fraud, as well as risk data breaches, which may improve the strength of the organization's risk management capabilities. The real-time identification of risks helps companies mitigate their chances of experiencing some form of compliance violation and, therefore, having financial penalties due to such misgivings.

5. Optimized Customer Experience

In customer-facing roles, for instance, Intelligent RPA increases the prospects of bettering the customer experience through automation and personalization in interactions. For instance, customer service is where bots powered by NLP respond to enquiries fairly rapidly and correctly to offer personalized support with no human agents. Robotic process automation systems may also reduce the processing of the time consumption in onboarding, resolve issues faster, and ensure that customers get a response within their most sought time to fulfill their needs. This helps in increasing customer satisfaction, retention, and loyalty as the human workforce is left free to cater to more complex interaction (Agostinelli et al., 2019).

6. Continuous Learning and Improvement

One of the features that constitute Intelligent RPA differs from standard RPA is that intelligent RPA can learn as time progresses. On the basis of machine learning algorithms, intelligent RPA systems are capable of analyzing the prior performance data in a bid to find opportunities for processes' optimization. This self-learning ability makes the bots more efficient and effective as they can respond to changes in business environment or changes in process requirements without needing to repeatedly be programmed for updates. Gradually, this results in processes being more efficient and less prone to errors or inefficiencies (Enríquez et al., 2020).

7. Better Workforce Productivity and Engagement

The intelligent RPA automates the mundane, boring work and frees human talent for more creative, strategic, and decision-based work that makes more sense to be done by humans. When the workforce is allowed to contribute to higher-value-creating endeavors, there is increased employee engagement and productivity. Employee skill sets can also be changed to include management and optimization of RPA systems as new career paths open.

8. Data-Driven Decision Making

Intelligent RPA systems are usually combined with advanced analytics, which helps in feedback mechanism on performance in terms of output in real time. By analyzing massive amounts of data about processes, these systems create insights that can lead to wiser decision making. Organizations can utilize them to identify trends in their workflows, to enhance the workflow process, and to predict future demands. A more data-driven approach aids businesses become even better decision-makers toward improving competitiveness and market positioning.

Although the traditional version of RPA automation brings benefits to the efficiency, cost reduction, compliance, and customer satisfaction of business operations, Intelligent RPA introduces greater possibilities for business augmentation in the most comprehensive sense (van der Aalst et al., 2018). In fact, it not only automates repetitive tasks but also promotes agility, scalability, and continuous improvement through integration with AI, ML, and cognitive technologies. As organizations continue to embrace digital transformation, Intelligent RPA will increasingly become important to drive innovation and operational success.

DESIGN AND DEVELOPMENT OF INTELLIGENT RPA SYSTEMS

In designing and developing Intelligent RPA systems, the development process requires the incorporation of an advanced technology plan, cooperation with key stakeholders, and business objectives. Intelligent RPA systems take the automation concept a step beyond traditional rules-based automation by including AI, ML, and other cognitive tools in the application of more advanced automation of complex activities. This section analyzes the critical stages that make up the development and design phases of Intelligent RPA systems. The stages are: process identification, architecture design, system integration, and continuous optimization.

1. Process Identification and Analysis

Identifying suitable processes to be automated is the first step towards an Intelligent RPA system. While traditional RPA is typically applied to repetitive, rule-based tasks, Intelligent RPA is applied towards more complex processes that have to do with unstructured data, decision-making, or engagements related to human language (Bédard et al., 2024). Generally, organizations have to analyze their workflows in identifying which ones may be automated-those are time-consuming and usually error-prone, with abundant manual input.

At this stage, a comprehensive process analysis is conducted to understand the workflow and present state, inputs, outputs, exceptions, and bottlenecks. Business analysts and process owners work together with creating the process map as well as defining the rules and decision-making points that will be addressed by the Intelligent RPA System (Lievano-Martínez et al., 2022). Such an assessment enables businesses to identify potential benefit areas for auto-categorization: potential savings in cost, efficiency benefits, or better servicing a customer.

2. Architecture and Technology Selection

After identifying the processes to be automated, design system architecture and select the technologies. The Intelligent RPA architecture must support integration of AI/ML algorithms, NLP, and other cognitive technologies. For that purpose, a cloud-based architecture will provide scalability, flexibility, and real-time processing (Jha et al., 2021).

Technology choice is of highest significance during the development stage. Smarter RPA platforms provide a range of tools and libraries, that support different use cases ranging from document processing to predictive analytics. Technology choice is indeed determined by the level of complexity in automation and the existing infrastructure of the organization (Chanopas et al., 2006). For instance, OCR might be needed for applications such as scanning documents and NLP to allow for automated customer interaction. The solution should also be developed with APIs along with other integrations that make it possible for the system to interface with existing enterprise software including a CRM system or ERP system.

3. Development and Bot Design

Once the architecture and technologies have been chosen, actual development of the RPA bots begins. In the case of Intelligent RPA, the bots are constructed for undertaking predefined tasks under static conditions along with demonstrating their ability to adapt to changing conditions and learn from data. For such a goal, it becomes essential for developers, AI specialists, and business stakeholders to work effectively with each other so that the desired processes can be effectively managed by the bots.

During the development phase, machine learning models learn to read data for patterns and will then make decisions. For example, bots can be programed to classify emails from customers, check for aberrations in financial transactions, or offer solutions based on historical data. AI-driven RPA bots are created in such a manner that they learn over time. Hence, continuous training and refinement of the machine learning models are major components of success in the long run (Tatikonda et al., 2022).

Of course, a part of the process of bot development must deal with UI design. The bots have to fit in with the applications and systems. They must handle user interactions without interrupting them. UI design must pass the test, wherein its usability, as well as its reliability, are tested.

4. Testing and Validation

Before the Intelligent RPA system is installed, it must be extensively tested and validated as a holistic performance testing simulating real-world scenarios, designed to accommodate a healthy number of bugs and badly spiked performance or process failures (Leno et al., 2021). Functional and non-functional testing ensure that the bots can execute an entire range of tasks-from simple data processing all the way up to complicated decision-making.

Some of the essential testing parameters for Intelligent RPA include guaranteeing the correctness and dependency of a machine learning model. This could be done by validating multiple models on various datasets, fine-tuning those models depending upon value returned as well as error generated. Tests for scalability, security, and compliance are very important, especially in industries like finance and healthcare, which rely much on these regulatory guidelines.

5. Deployment and System Integration

At the end of this testing and validation of an Intelligent RPA system, it becomes ready for deployment. In this step, the RPA bots are integrated with the existing systems, applications, and databases of an organization. There can be some configuration of APIs and secure connections between the bot and other kinds of software tools. Cloud deployment is very common, ensuring that the system can be scaled up according to demand and huge volumes of data.

Deployments also act as a time for change management. Those affected employees need to be trained on how to interact with bots, how to handle exceptions when they occur, and how to keep the RPA system running (Lacity & Willcocks, 2016).

Stakeholder communication and involvement needs to be a part of the process of the deployment so that everything goes as smoothly as possible, and the organization adapts the system without much restriction in place.

6. Continuous Monitoring and Improvement

Development of Intelligent RPA systems does not end with deployment, and it needs to be constantly monitored and improved to ensure maintenance of optimal performance. Since the Machine learning models need continuous training as well as fine-tuning to evolve with the arrival of new data inputs or change in business conditions, their performance is regularly audited and performance reviews to ensure that the bots operate efficiently and are generating the ROI one would expect.

Organisations can then also use the output of Intelligent RPA systems to find more automation opportunities. Advanced analytics will allow organisations to find insights from the current performance of processes, thereby identifying where there is a scope for further optimisation or automation. That process of continuous improvement thus ensures that the Intelligent RPA system continues delivering value in relation to business needs changing over time (Suri et al., 2018).

The strategic approach follows the Intelligent RPA system design and development. There are multiple phases within which AI, machine learning, and cognitive tools are combined to deliver such a system: identification to continuous improvements, with the support of technical teams and business stakeholders at every stage to improve the system, making it more efficient in operational terms, scalable, and suitable for better decision-making (Anagnoste, 2018). Advanced technologies can transform the workflows of organizations and help enhance agility, cost savings, and innovation.

CHALLENGES IN INTELLIGENT RPA IMPLEMENTATION

Intelligent Robotic Process Automation indeed offers several benefits, but the implementation thereof is quite challenging for an organization. Organizational change management and integration of advanced technologies are key sources of these challenges besides ensuring time consistency in the efficiency and accuracy of automated systems. The major challenges encountered during the introduction of Intelligent RPA are the following:

1. Integration with Existing Systems

The biggest challenge with Intelligent RPA is integrating these newer forms of automation with the existing systems and other applications in an enterprise. Most firms would rely on hundreds of software packages, comprising CRM systems, ERP systems, and solution-specific applications. For such intelligent RPA bots to fetch data, process information, or carry out work, they must communicate seamlessly with these systems.

Legacy systems will probably not have the necessary APIs or flexibility needed for easy integration, which would mean that the process of deploying RPA bots becomes more complicated (van der Aalst et al., 2018). Customizations in old systems also tend to have a negative impact on the RPAs and would mean that the solution developed will have trouble running flawlessly. IT teams often must work significantly more to establish compatibility and to build custom integrations to tackle this, and hence they tend to take much more time and incur much higher costs in the implementation phase.

2. Process Standardization and Complexity

An intelligent RPA will do its best work in an organization with well-defined and standard processes (Syed et al., 2020). Most organizations, however, have an inconsistent workflow in the department or region. Location-, department-, or business unit-based inconsistencies may make it even more confusing to automate because bots need strict and structured workflows to work at their optimal capacity.

More complex decisions, human judgment, or unstructured data are harder to automate. Intelligent bots have AI and ML capabilities. Some of this complexity can be addressed sometimes, but considerable time and resource investment goes into creating the bot, fine-tuning the bot, and so forth. Organizations have to invest in the right kind of process re-engineering and standardization before deploying Intelligent RPA so that the system works optimally in all use cases.

3. Change Management and Employee Resistance

Intelligent RPA implements significant change, especially in terms of how business activities are conducted and the way workers interact with technology (Lacity & Willcocks, 2016). The most significant challenge relates to staff resistance to the automation process. Workers will resist it because it threatens to take away jobs, reduce tasks, and lower levels of engagement by requiring new skills.

To overcome the challenge, effective change management strategies include communicating clearly on the purpose of RPA, its benefits, and how it would affect the employees in the organization. Training programs should be developed to ease employees into new roles-often bot management or more added-value tasks involving critical thinking and decision-making. Employers could engage and support employees through the journey toward the transition, which could reduce resistance to the change and increase adoption rates.

4. Data Security and Compliance Concerns

For example, some of the RPA systems are smart and frequently access sensitive data. Data security and compliance with relevant laws therefore become an issue. Those that handle data related to customers, financial matters, or personal health records must adhere to data protection regulation such as GDPR, HIPAA, or industry standards (Mendelson, 2017). Unless correctly configured, the bots will expose organizations to some risks such as losing control over information, unauthorized access, non-compliance with the law, among others.

Organizations need to ensure proper security measures are in place for the data these bots access (Elliott, 2010). This encompasses encryption, access controls, and audit trails to trace bot activity. Also, periodic updates will be needed in regard to emerging threats, and audits must be conducted periodically to ensure compliance continues.

5. Complexity in Managing and Scaling Automation

The second implementation challenge of Intelligent RPA is the management and scaling of automation initiatives across the organization. As companies expand their RPA programs to include more processes, governance, bot management, and consistency among different departments become huge challenges in this area (Syed et al., 2020). Lack of central systems to manage bots, update machine learning models, and track performance may create inefficiency and operational challenges.

Scaling Intelligent RPA requires a strong governance framework that outlines bot development, deployment, and maintenance guidelines (Yang et al., 2011). Organizations must clearly define roles and responsibilities of bot managers and developers to ensure the system works efficiently as it scales. Lack of willingness to confront the governance challenges stated above leads to an overcomplicated automation program that is difficult to handle and optimize.

6. Ongoing Maintenance and Model Training

Intelligent RPA systems, especially AI- and ML-powered ones, also demand a continuous form of maintenance and model training for it to remain effective. ML models need continuous updates with fresh data to ensure it continues to make accurate predictions. In the long run, processes may change, or business needs evolve, making the RPA bot need reconfiguration or retraining to adapt new scenarios (Kothandapan, 2021).

This type of maintenance is quite resource-intensive, as performance needs to be monitored, and algorithms changed as well as processes fine-tuned by skilled personnel. In the absence of committed resources and good-quality maintenance, performance of RPA systems may degrade, and it may lead to inefficiencies, errors, and even potential process breakdowns.

7. Cost and Resource Constraints

High initial technology, infrastructure, and human investments are involved in an intelligent RPA. The move to add AI and ML to standard RPA products complicates the price of developing more complex products compared to other RPA solutions. It may not be easy for organizations to provide adequate resources, especially when they lack in-house competencies to design and manage intelligent automation solutions.

The ROI of Intelligent RPA also may not be immediate since it demands organisations accommodate the costs of training, testing, and deploying the system, plus long-term maintenance of bots and machine learning models (Chugh et al., 2022). It would be relatively tough for SMEs to justify investment in Intelligent RPA, even though long-term benefits compensate for it.

Where intelligent RPA becomes truly transformative in enabling complex process automations, organizations would have to overcome the major challenges it has in place before its full utility can be tapped. Some of the significant impediments are system integration, complexity of processes, resistance from employees, data security, and the burden of maintenance (Gianni & Gotzamani, 2015). All these must be approached with caution, involving essential investments in right technologies, strong commitment to the aspects of change management and governance. If such challenges are successfully identified and addressed by a company, it can well ensure capturing the intrinsic potential of Intelligent RPA toward better operational efficiency and cost-saving factors with innovative zeal.

FUTURE TRENDS IN INTELLIGENT RPA

The history of Intelligent RPA is going to redefine the future of business automation with unprecedented levels of efficiency and innovation in different organizations. As the advancements in AI and automation technologies continue, new trends are underway to frame the landscape of Intelligent RPA (Gill et al., 2022). Some of the biggest trends that will be shaping the landscape of Intelligent RPA in the years to come are discussed in this section.

1. Hyper-Automation

Hyper-automation is an emerging trend that will change the trajectory for Intelligent RPA. This phrase describes the use of a bunch of Advanced Technologies, including RPA, AI, ML, and process mining, which can almost everywhere automate virtually any business process. It is going to be beyond traditional RPA, which maintains its focus on automating repetitive, rule-based, routine tasks, and pushes more complex workflows with cognitive decision-making and unstructured data into the horizon for upgrading automation (Zhao et al., 2022).

In terms of the future, companies will embrace more aspects of hyper-automation and develop end-to-end automated solutions with the ability to analyze, decide, and execute on their own, independent of human intervention. This will result in operations that attain improved levels of operational efficiency as well as agility, in addition to better decision-making through analytics and AI-driven insights.

2. AI-Enhanced Decision Making

As Intelligent RPA matures, AI-enhanced decision-making will be added to the automation solutions. Future generations of RPA systems will begin using more advanced AI capabilities, such as natural language processing, predictive analytics, and deep learning, in more complex activities of judgment, reasoning, or even making decisions in the future. All this should contribute to RPA bots doing more than just simple task automation and carrying out higher-level functions like identifying trends, making predictions, or even optimizing business strategies.

AI-driven RPA systems will also learn over time from the data that may be collected, which means that automation processes have the potential to continually self-improve and become more refined with time (Tatikonda et al., 2022). This will make much more dynamic and adaptable automation solutions available for faster responses to rapidly changing business environments and shifting requirements in customer communication.

3. Integration with IoT and Edge Computing

Another emerging trend in the future of Intelligent RPA is its integration with the Internet of Things (IoT) and edge computing. IoT devices generate quite a tremendous amount of data from various sources like sensors, machines, and connected products (Premavathi et al., 2024). With integration of RPA with IoT, organizations can automate data collection, analysis, and decision-making processes for real-time information generated from these devices.

More capabilities would be added with edge computing for improving latency and response times, allowing Intelligent RPA systems to manage more timely-sensitive tasks, such as predictive maintenance or real-time monitoring. Convergence points in this regard are RPA, IoT, and Edge Computing, the area from which new opportunities emerge for manufacturing, health service, and logistic functions.

4. Low-Code/No-Code RPA Platforms

While enterprises look for more agile means to digital transformation, low-code/no-code RPA platforms will gain significant traction. It will enable non-technical users to develop and deploy RPA bots with little to no coding, democratizing the automation process across the organization. This trend will be driven more by business users-building their own automation solutions, such as operations managers and analysts, versus looking to IT departments.

The low-code/no-code RPA platforms will become even more powerful through drag-and-drop interfaces, templated examples, and AI-driven recommendations to speed up bot development (Sufi, 2023). This will make Intelligent RPA easier to adopt faster and within broader ranges of industries and functions, so the organizations can accelerate the scale of automation much faster.

5. Intelligent Document Processing

Intelligent RPA will change vastly in one of the significant areas: intelligent document processing. While the bulk of business processes depend on unstructured data-the email, contract, invoice, and report - recent years have been real issues in automating them due to an inability to build a simple method to read and extract information from unstructured documents (Ling et al., 2020).

Future Intelligent RPA systems will contain rich cognitive technologies such as OCR, NLP, and AI-based data extraction to process and analyze unstructured data with high accuracy. All of these will enable businesses to automate much more complex tasks, like legal contracts for analysis, claims processing, customer service questions, etc., difficult to automate using traditional RPA.

6. Intelligent RPA as a Service (RPAaaS)

The advent of cloud computing has also led to Intelligent RPA as a Service (RPAaaS), which enables the ability for organizations to access and make use of RPA solutions on a cloud-based platform without the need for large on-premise infrastructure. RPAaaS offers the scope of scalability, flexibility, and lower cost, along with no reliability issues. This makes it an appealing model for all business sizes.

RPAaaS platforms are likely to get more inducted with AI in the future, thus providing out-of-the-box intelligent automation solutions tailored specifically for industries or use cases (Durairaj, 2016). The cloud-based platforms are going to give organizations access to leading-edge automation technologies, obviate large upfront investments, and permit rapid and efficient scaling of automation initiatives.

7. Focus on Human-Bot Collaboration

In the future, humans and bots will be merged-a collaboration rather than a replacement relationship. The nature of these relationships will continue to grow in the direction of a complementary relationship. RPA systems will be designed in the future with the objective of enhancing human capabilities to free up these workers' time to focus on more creative and higher-value tasks related to critical thinking and emotional intelligence. Intelligent RPA bots can be deployed to handle mundane and repetitive activities and free up the human employees for more strategic roles (Chakraborti et al., 2020).

This trend will transform the way businesses operate and manage their workforce in terms of talent development. Business will have to invest in upskilling and reskilling individuals to ensure effective work through automation technologies. For that reason, human-bot collaboration becomes a symbiotic relationship that ultimately increases productivity but ensures there is still a human element in decision-making and innovation.

8. Increased Focus on Ethical AI and Automation

The more Intelligent RPA becomes powerful, the greater the ethical considerations. Increased use of AI and automation raises such questions as data privacy, mass job displacement, and biased decision-making processes. Organizations will focus more on developing ethical frameworks in guiding the responsible application of Intelligent RPA (Khogali & Mekid, 2023).

Future trends are expected to include the design of transparent and accountable AI in RPA systems as well as implementation of regulations and guidelines to ensure equal fairness, security, and compliance. Automating these operations will command attention from companies on how their stakes are contributing positively to harmonious social interaction and ought to be ethical.

It is about to transform the nature of business operations for any establishment, with the integration of the most advanced AI technologies and automation. The other trends, such as hyper-automation, AI-factored decision making, IoT integration, and low-code/no-code platforms, are supposed to guarantee the adoption of even more complex solutions for automation (Sufi, 2023). As organisations continue to embrace Intelligent RPA, focus will be placed on human-bot collaboration, ethical considerations, and how automation will only help in making more efficient operations, innovation, and customer experiences. Intelligent as well will be the next generation of RPA empowering organizations to reach new levels of productivity and agility in the increasingly automated world.

IMPLICATIONS OF INTELLIGENT RPA DEVELOPMENT

The development and implementation of Intelligent Robotic Process Automation systems carry deep implications across managerial, practical, social, and ethical dimensions. As organisations start adopting intelligent automation technologies, these implications shape the way businesses would operate, the way the employees relate with technology, and societal management of the effects of increased automation (Koljonen, 2023). This section expands further on broader implications of Intelligent RPA along managerial, practical, social, and ethical perspectives.

1. Managerial Implications

The only managerial perspective on Intelligent RPA could come in the form of leadership and at an organizational level. Primarily, the acceptance of Intelligent RPA will alter how leaders run the organization and make decisions. That would mean managers adopting a new paradigm in which automation technologies would play an important role in business processes and strategy execution.

- Strategic Decision Making: Intelligent RPA enables managers to make data-driven decisions by automating complex tasks with data analysis, pattern recognition, and forecasting. It dislodges the manager from time-consuming detailed operational work and pushes them onto strategic decisions at the more elevated level. Situational awareness through real-time insights enhances the accuracy of decisions and improves responsiveness to ensure businesses are agile in a competitive market.
- Workforce Transformation: With Intelligent RPA assuming this kind of work, managers must reconceptualize workforce planning and the allocation of human resources. Instead of focusing on employees merely as critical thinkers, creativity nurturers, and emotive intelligencers, they need to focus on new talent development strategies to build a culture that fosters continuous learning and upskilling in preparation for collaboration with automation technologies.
- Governance and Oversight: Intelligent automation necessitates the existence of strong governance frameworks. This means that managers have to align all RPA implementations with the aims of organizations, compliance to regulations, and scalable. Proper oversight mechanisms must exist to monitor the performance and effectiveness of bots in ensuring that they yield consistent results.

2. Practical Implications

The effects of Intelligent RPA are massive, touching the core of changing the way organizations do things, becoming more efficient, and offering more value to customers (AlbertElijah, 2023). It affects the optimization of costs and agile operations.

- Operational efficiency: The number of mistakes made on laborious processes decreases and the speed at which work is completed increases, as such labor-intensive processes are automated. Processes that were only feasible with manual input can be taken over by bots, and human resources may thus be freed up for more strategic use. Cumulatively, the organization may reap significant advantages in terms of time saved, productivity, and cost efficiency at scale.

- Cost Savings: The implementation of intelligent RPA reduces operational costs, primarily in this category, by conserving human resources to be used in routine and repetitive operations. The high capital investment in developing intelligent bots is very costly; however, the long-term savings can be extremely more value-for-money. This applies well in high volumes of transactional industries such as banking, customer service, and supply chain management, among others.
- Scalability and Flexibility: Intelligent RPA provides theoretical advantages in the form of scalability. Once a bot is designed and trained, the enterprise can deploy it across different processes and departments with only minimal incremental costs. This allows the business to scale its automation program rapidly as and when required to accommodate changes in workloads or shifts in market demand with minimum disruption.

3. Social Implications

This trend is likely to have deep social effects in the sphere of employment, the dynamics of labor, and roles of society (Madakam et al., 2019). As increasingly more organizations inevitably come to rely on Intelligent RPA for solving business problems, the nature of work will change as will the way in which people participate in the workforce.

- Workforce Displacement and Job Creation: As Intelligent RPA replaces some jobs that have quite routine and repetitive working patterns, at the same time, it will also give birth to new ones requiring cognitive skills, creativity, and emotional intelligence. Workers thus have to be ready for various new jobs or jobs pertaining to the management of automation systems, developing algorithms, and undertaking higher-level tasks that cannot be automated. The adjustment to these new jobs will, however, be bumpy at best and short-term job displacement and challenges to being social, particularly among low-skilled employees.
- Upskilling and Lifelong Learning: Therefore, the social consequences of Intelligent RPA suggest a shift of focus toward the acquisition of lifelong learning and upskilling. Governments, the educational institutions, as well as businesses, must make it easier to reskill so that workers are seen to respond to new challenges created by the shifting demand in the labor market. Without adequate upskilling, those who are benefiting from an automated world will continue excluding those who aren't.

- Social Equity: Intelligent RPA will further estrange the social systems if the benefits of automation do not reach equitable levels within the society. When automation brings economies of scale to businesses, the loss of low-wage laborers would create a far bigger economic difference among people. Both policymakers and business leadership need to come together to work towards this transition to the automated economy without widening the gap between the advantaging and disadvantaging group of people.

4. Ethical Implications

The intelligent implications of RPA are critical for two main reasons: automation technologies become an integral part of decision-making processes and affect human-to-human communications. Transparency, fairness, accountability, and privacy are some of the most prominent considerations that are thought to be indispensable in such a context (Ahmad, 2024).

- Bias and Fairness: AI and machine learning-based smart RPA systems may perpetuate biases unintentionally especially where it comes to hiring, lending, or decisions about customers. This becomes an ethical dilemma for the organizations involved about the fairness of the procedures, non-discrimination in decision-making, and need for unbiased algorithms, auditing of the decisions, and monitoring for unintended biases (Chakraborti et al., 2019).
- Transparency and Accountability: The more Intelligent RPA systems move toward decisions, the more an organization needs to assure transparency in the way in which decisions are being made by bots. Translucency to stakeholders about how bots function and the mechanism to the decision-arriving process; who is accountable in case of a mistake, failure, etc should be accounted for. Building transparent systems is very important to maintain trust in their customers, employees, and regulators.
- Data Privacy and Security: Intelligent RPA systems handle sensitive data, including personal and financial details. Ensuring that these systems protect data privacy and adhere to the regulation of GDPR becomes a 'moral imperative'. Organizations must have elaborate security protocols, encryption, and access controls to protect data against unauthorized access or breach. Overlooking this privacy may lead to serious legal and reputation risks.
- Human-Centric Automation: Other ethical considerations would be the balance automation brings between itself and human experience. As RPA systems take over routine tasks, organizations must ensure that there remains a continued centrality of human dignity and value within the automation strategy. The principles should thus focus on symmetry so that bots supplement

the human workforce rather than replace, leading to collaborative workplaces instead.

Development and deployment of smart RPA creates managerial, practical, social, and ethical implications on a massive scale. It has opportunities to promote improvement in efficiency, reduction in cost, and changing forms of industries. However, the challenges associated with it are workforce displacement, biased data, unwanted social equity, and issues related to data privacy. As organizations embrace and scale Intelligent RPA, it is important that such a comprehensive approach considers such implications, with technical soundness and social and ethical responsibility in automation initiatives. This will allow all benefits of Intelligent RPA to be fully unlocked while generating positive outcomes for businesses, employees, and society at large.

CONCLUSION

It is the new advancements beyond only automation that have brought intelligent RPA, set to rapidly alter the business landscape today, together with AI, machine learning, and other cognitive technologies. Thereby, as organizations continue with intelligent automation, the point is that efficiency, scalability, and reductions in cost lie to their advantage. Intelligent RPA systems, for example, do more than just routine work automation and will automate more difficult decision-driven procedures, thereby freeing human workers to concentrate on higher-value activities, such as innovation, strategic planning, and customer engagements.

However, the mammoth challenges of developing and implementing Intelligent RPA call for movement and should be approached with great caution. To unleash the complete potential of automation, the organizations need to work on thudding technical complexities, disruption in the workforce, and integration. Along with this, they need to setup a governance framework that would ensure transparency, accountability, and a business alignment. Equally relevant to this is the focus put on upskilling and preparing employees for this shift towards a more automated future where interaction between human workers and intelligent bots are part of life.

With this in place, intelligent RPA has tons of promise for future development, including natural language processing, advanced analytics, and deeper learning capabilities. These trends shall lead to more intuitive, autonomous, and adaptive systems in which problems could be foreseen and solved within real-time environments. Hence, with evolution in automation technologies, ethical matters involving bias, data privacy, and social equity must forever be at the top of implementation strategies.

While Intelligent RPA is a transformative opportunity, such adoption needs to be managed carefully so that it brings the corresponding benefits. Adoption of automation requires a holistic approach to address the technological as well as human aspects of automation to bring sustainable value in a competitive position in the world that will increasingly become automated. A balance will be required in the future direction between engineering technological innovations and a human-centric approach to automation, ensuring Intelligent RPA contributes to an overall augmentation of human potential rather than diminish it.

REFERENCES

Agostinelli, S., Marrella, A., & Mecella, M. (2019). Research challenges for intelligent robotic process automation. In *Business Process Management Workshops: BPM 2019 International Workshops, Vienna, Austria, September 1–6, 2019, Revised Selected Papers 17* (pp. 12-18). Springer International Publishing. DOI: 10.1007/978-3-030-37453-2_2

Aguirre, S., & Rodriguez, A. (2017). Automation of a business process using robotic process automation (RPA): A case study. In *Applied Computer Sciences in Engineering: 4th Workshop on Engineering Applications, WEA 2017, Cartagena, Colombia, September 27-29, 2017* [Springer International Publishing.]. *Proceedings*, *4*, 65–71.

Ahmad, A. Y. A. B. (2024). Ethical implications of artificial intelligence in accounting: A framework for responsible ai adoption in multinational corporations in Jordan. *International Journal of Data and Network Science*, *8*(1), 401–414. DOI: 10.5267/j.ijdns.2023.9.014

AlbertElijah, R. R. (2023). RPA: Transforming Organizational Processes with Intelligent Automation. *International Journal of Advanced Engineering Technologies and Innovations*, *1*(02), 201–218.

Anagnoste, S. (2017, July). Robotic Automation Process-The next major revolution in terms of back office operations improvement. In *Proceedings of the International Conference on Business Excellence* (Vol. 11, No. 1, pp. 676-686). DOI: 10.1515/picbe-2017-0072

Bédard, M., Leshob, A., Benzarti, I., Mili, H., Rab, R., & Hussain, O. (2024). A rule-based method to effectively adopt robotic process automation. *Journal of Software (Malden, MA)*, *36*(11), 2709. DOI: 10.1002/smr.2709

Chakraborti, T., Sreedharan, S., Grover, S., & Kambhampati, S. (2019, March). Plan explanations as model reconciliation--an empirical study. In *2019 14th ACM/IEEE International Conference on Human-Robot Interaction (HRI)* (pp. 258-266). Ieee.

Chanopas, A., Krairit, D., & Ba Khang, D. (2006). Managing information technology infrastructure: A new flexibility framework. *Management Research News*, *29*(10), 632–651. DOI: 10.1108/01409170610712335

Chugh, R., Macht, S., & Hossain, R. (2022). Robotic Process Automation: A review of organizational grey literature. *International Journal of Information Systems and Project Management*, *10*(1), 5–26. DOI: 10.12821/ijispm100101

. Durairaj, S. (2016). A Comprehensive Study on the Business Application of Robotic Process Automation: Exploring Scalability, Integration, Cognitive Capabilities, and Governance. *Global journal of Business and Integral Security.*

Elliott, C. (2010). Botnets: To what extent are they a threat to information security? *Information Security Technical Report, 15*(3), 79–103. DOI: 10.1016/j.istr.2010.11.003

Enríquez, J. G., Jiménez-Ramírez, A., Domínguez-Mayo, F. J., & García-García, J. A. (2020). Robotic process automation: A scientific and industrial systematic mapping study. *IEEE Access : Practical Innovations, Open Solutions, 8*, 39113–39129. DOI: 10.1109/ACCESS.2020.2974934

Gianni, M., & Gotzamani, K. (2015). Management systems integration: Lessons from an abandonment case. *Journal of Cleaner Production, 86*, 265–276. DOI: 10.1016/j.jclepro.2014.08.023

Gill, S. S., Xu, M., Ottaviani, C., Patros, P., Bahsoon, R., Shaghaghi, A., Golec, M., Stankovski, V., Wu, H., Abraham, A., Singh, M., Mehta, H., Ghosh, S. K., Baker, T., Parlikad, A. K., Lutfiyya, H., Kanhere, S. S., Sakellariou, R., Dustdar, S., & Uhlig, S. (2022). AI for next generation computing: Emerging trends and future directions. *Internet of Things : Engineering Cyber Physical Human Systems, 19*, 100514. DOI: 10.1016/j.iot.2022.100514

Hofmann, P., Samp, C., & Urbach, N. (2020). Robotic process automation. *Electronic Markets, 30*(1), 99–106. DOI: 10.1007/s12525-019-00365-8

. Jha, N., Prashar, D., & Nagpal, A. (2021). Combining artificial intelligence with robotic process automation—an intelligent automation approach. *Deep learning and big data for intelligent transportation: enabling technologies and future trends*, 245-264.

Khogali, H. O., & Mekid, S. (2023). The blended future of automation and AI: Examining some long-term societal and ethical impact features. *Technology in Society, 73*, 102232. DOI: 10.1016/j.techsoc.2023.102232

. Koljonen, J. (2023). Benefits and Risks Measurement Model: In different managerial positions during RPA implementation.

Kothandapani, H. P. (2021). Integrating robotic process automation and machine learning in data lakes for automated model deployment, retraining, and data-driven decision making. *Sage Science Review of Applied Machine Learning, 4*(2), 16–30.

Lacity, M. C., & Willcocks, L. P. (2021). Becoming strategic with intelligent automation. *MIS Quarterly Executive, 20*(2), 7.

Leno, V., Polyvyanyy, A., Dumas, M., La Rosa, M., & Maggi, F. M. (2021). Robotic process mining: Vision and challenges. *Business & Information Systems Engineering*, *63*(3), 301–314. DOI: 10.1007/s12599-020-00641-4

Lievano-Martínez, F. A., Fernández-Ledesma, J. D., Burgos, D., Branch-Bedoya, J. W., & Jimenez-Builes, J. A. (2022). Intelligent process automation: An application in manufacturing industry. *Sustainability (Basel)*, *14*(14), 8804. DOI: 10.3390/su14148804

Ling, X., Gao, M., & Wang, D. (2020, November). Intelligent document processing based on RPA and machine learning. In *2020 Chinese Automation Congress (CAC)* (pp. 1349-1353). IEEE. DOI: 10.1109/CAC51589.2020.9326579

Madakam, S., Holmukhe, R. M., & Jaiswal, D. K. (2019). The future digital work force: Robotic process automation (RPA). *Journal of Information Systems and Technology Management*, *16*, e201916001. DOI: 10.4301/S1807-1775201916001

Mendelson, D., & Mendelson, D. (2017). Legal protections for personal health information in the age of Big Data–a proposal for regulatory framework. *Ethics, Medicine, and Public Health*, *3*(1), 37–55. DOI: 10.1016/j.jemep.2017.02.005

. Osmundsen, K., Iden, J., & Bygstad, B. (2019). Organizing robotic process automation: balancing loose and tight coupling.

Premavathi, T., Shekhar, A., Raj, A., Mohan, K., Palaniappan, D., & Shukla, M. (2024). The Utilization of Internet of Things (IoT) in the Field of Robotics Process Automation. In *Applications of New Technology in Operations and Supply Chain Management* (pp. 337–359). IGI Global. DOI: 10.4018/979-8-3693-1578-1.ch018

Sufi, F. (2023). Algorithms in low-code-no-code for research applications: A practical review. *Algorithms*, *16*(2), 108. DOI: 10.3390/a16020108

Suri, V. K., Elia, M., & van Hillegersberg, J. (2017). Software bots-the next frontier for shared services and functional excellence. In *Global Sourcing of Digital Services: Micro and Macro Perspectives: 11th Global Sourcing Workshop 2017, La Thuile, Italy, February 22-25, 2017, Revised Selected Papers 11* (pp. 81-94). Springer International Publishing. DOI: 10.1007/978-3-319-70305-3_5

Syed, R., Suriadi, S., Adams, M., Bandara, W., Leemans, S. J., Ouyang, C., ter Hofstede, A. H. M., van de Weerd, I., Wynn, M. T., & Reijers, H. A. (2020). Robotic process automation: Contemporary themes and challenges. *Computers in Industry*, *115*, 103162. DOI: 10.1016/j.compind.2019.103162

Tatikonda, V., Venigandla, K., & Vemuri, N. (2022). Transforming customer banking experiences: Ai-driven rpa for customized service delivery. *International Journal of Developmental Research, 12*, 60674–60677.

Van der Aalst, W. M., Bichler, M., & Heinzl, A. (2018). Robotic process automation. *Business & Information Systems Engineering, 60*(4), 269–272. DOI: 10.1007/s12599-018-0542-4

Willcocks, L. P., Hindle, J., Stanton, M., & Smith, J. (2024). *Maximizing Value with Automation and Digital Transformation: A Realist's Guide*. Springer Nature. DOI: 10.1007/978-3-031-46569-7

Yang, J., Yu, T., Jian, L. R., Qiu, J., & Li, Y. (2011). An extreme automation framework for scaling cloud applications. *IBM Journal of Research and Development, 55*(6), 8–1. DOI: 10.1147/JRD.2011.2170929

Zhao, X., Oseni, T., & Medishetty, B. T. (2022, October). Overview of Business Hyper-automation. In *2022 IEEE International Conference on e-Business Engineering (ICEBE)* (pp. 100-105). IEEE. DOI: 10.1109/ICEBE55470.2022.00026

Chapter 6
Generative AI as a Tool for RPA

Shashank Mittal
https://orcid.org/0000-0002-8666-5323
O.P. JIndal Global University, India

Rohit
https://orcid.org/0009-0004-1796-5252
Gulf Medical University, UAE

ABSTRACT

This chapter delves deeper into the inclusion of generative AI in RPA and the transformative possibility it brings regarding business operation. Issues of companies pertaining to the operational dimension and their evolution from being rule-based to AI-powered systems will also be discussed. This part of the chapter will focus on how generative AI can transform decision-making processes through a better quality of decision-making support and generate content, which would prove to be an instrumental aspect for application in future scenarios. Additionally, the chapter elaborates on other significant advantages of this integration - efficiency and adaptability - as well as describes some challenges and limitations for organizations, including ethical concerns and those workers displaced by this new trend. The future RPA trends driven by generative AI - namely, hyper-automation and cognitive automation - are also explored in the book.

DOI: 10.4018/979-8-3693-4365-4.ch006

INTRODUCTION TO GENERATIVE AI AND RPA

Generative AI and Robotic Process Automation, or RPA, are two of the most transformative technologies working in today's digital world. Generative AI refers to the use of artificial intelligence algorithms to create new content, whether it's text, images, code, or even processes. Unlike the traditional AI system, which focuses on pattern recognition and decidability based on rules or data, generative AI can produce results that are novel through learning from the large dataset as well as by generating results that simulate human-like creativity and problem-solving (Balaguru, 2024).

RPA is technology that aims at automating repetitive, rule-based tasks normally carried out by humans within business processes. Such tasks usually involve data entry, the generation of reports, and transaction processing. RPA mimics the movement of humans while emulating interaction with a digital system, performing predefined workflows intended to make an operation efficient, minimize the volume of human errors, and save costs (Beerbaum, 2023a). Due to its usual rule-based premise, it has until now been limited to performing those jobs that entail decision-making as well as flexibility in the face of unstructured data.

This is where generative AI and RPA converge at a new frontier in the world of automated technology. Business now can automate tasks that were believed to lie outside the realm of automation by the combination of creativity and cognition abilities of the generative AI with RPA efficiency. The capabilities of RPA are furthered by generative AI in processing unstructured data, making real-time decisions, and even generating new workflows because of the patterns that it has learned from past activities. This integration makes RPA, a purely automating tool, into an intelligent system that can orchestrate a broad spectrum of business operations (Beerbaum, 2023b).

The core of generative AI has an inhumane capability of learning from data and also modifying the behaviors on new tasks. It will help in the enhancement of processes related to RPA automatically by analyzing the data collected from the automation process with suggestions for optimization. For example, while using generative AI in the automation of customer services, auto-generated response to customer queries can be adjusted in terms of service offerings and, therefore interaction facilitation (Bi, 2023). Through the previous justifications, the use of AI-generated responses also provides quicker and personalized responses to customers than the traditional human responses, which are likely to improve customer experience and satisfaction.

Another potential benefit of generative AI is its ability to automate tasks that demand creativity and complex decision-making, areas most conventional RPA solutions are challenged in. For instance, the generation of content for reports, contracts, or even marketing material can be automated with the aid of AI-driven tools. Such tools can draft outlines and give recommendations in keeping with what it has learned through user feedback over time to further enhance the quality and accuracy of the output.

Integrating these two technologies, therefore, presents challenges associated with generative AI in RPA, some of which include data privacy and security as well as ethical issues related to the use of AI, including when it comes to deployment of generative models whose outputs can be produced without direct human oversight. Opportunities exceed challenges because, with generative AI-enabled RPA, a business can only scale its operations with cost control in an increasingly automated world while retaining competitive advantage (Feuerriegel et al., 2024).

Summary The advent of generative AI in RPA marks one significant step in the automation journey. Here, with RPA process efficiency and the flexibility and intelligence of generative AI, organizations can automate more complex tasks, improve decisions, and unlock new opportunities for growth and innovation.

EVOLUTION OF RPA: RULE-BASED TO AI-POWERED AUTOMATION

Early Stage: Rule-Based Automation

RPA started from the solution of handling repetitive, rule-based tasks that require human intervention but do not demand better cognitive skills. Actually, in its initial form, RPA was designed to mimic human actions in interacting with software systems, such as clicking through menus, entering data, or copying and pasting information from one application to another. This was a highly effective form of automation because it could reduce human errors dramatically in the structured environment, where tasks kept following predefined, predictable paths (Gotthardt et al., 2020).

However, traditional RPA was a rather limited technology. It worked well when the process in question was clearly defined and was only too eager to fail when some human judgment or creativity was needed to perform the task, or when unstructured data was part of the process involved. If an RPA system was good only as good as the rules followed by it, it could not be robust in the environments where processes shift regularly or where complex decision trees are employed (Lievano-Martínez et al., 2022). The requirement for more intelligent systems came forward as organizations looked for more advanced automation solutions.

Rise of AI-Augmented Automation

Advancements in AI technologies have made them complement RPA systems and helped them to execute rules that are not even easy. AI alone brought upon machine learning, natural language processing, and computer vision to the automation capabilities of RPA. It made it possible for RPA to handle larger, more complex tasks by learning from patterns in data and slowly improving over time (Lo, 2024).

It can, for example process unstructured data such as emails, documents, and images, which most of the traditional RPA systems cannot. Artificial intelligence now understands and responds based on text-based queries or instructions through NLP, allowing automation to solve problems such as answering customer support questions or by analyzing large datasets with more accuracy. Similar to this, computer vision allows an RPA bot to "see" and interpret information in a scanned document or an image/PDF, which can easily automate tasks like processing documents or extracting data (Lo et al., 2024).

Adaptability has also accompanied AI into RPA systems. Apart from strict rule-based operations, AI-powered RPA has the flexibility to adopt changes in the workflow or system environment. Even if there is a variation in the process, such as a new data format being introduced or an interface changing, AI may learn how to change its approach without necessarily requiring extensive reprogramming. This evolution seriously reduces downtime and increases the agility of automated processes.

Generative AI: The Next Stage of RPA Evolution

Generative AI represents the latest development in creating an RPA as part of an automation system that would leap from merely performing predefined tasks into new processes, workflows, and content. Generative AI is different from other systems because it generates novel outputs, and therefore, results in creating automated responses, reports, drafting emails, or even proposing improvements to processes. This makes RPA significantly more flexible as generative AI can now perform creative work activities that require decision-making and intricate problem-solving (Mandvikar & Achanta, 2023).

For example, a generative AI-enhanced RPA bot could send out the invoices but also generate customizable client communication messages from historical interactions, report writing, and even optimize workflows based on learning curves from historical data. This has actually transformed RPA from a tool for repeated, routine tasks into a system that could handle dynamic, knowledge-intensive processes, which many people deemed impossible to automate.

RPA has made this gargantuan leap in evolution from rule-based systems to becoming an AI-powered solution. Adding the layers of machine learning, NLP, computer vision, and generative AI makes RPA more than a simple automation tool. In that way, RPA changes from being a simplistic tool into becoming an intelligent mechanism capable of automating complex, data-driven processes within the execution of businesses by enhancing operational efficiency and leveraging new business capabilities. This is the beginning of something new in the age of automation where the collaboration between humans and machines can bring unprecedented productivity and innovation (Negoiţă & Borangiu, 2023).

GENERATIVE AI APPLICATIONS IN RPA

1. Automated Content Generation

One of the most impactful applications of generative AI in Robotic Process Automation (RPA) is automatic content generation. It includes anything from the drafting of reports, emails, and contracts to the generation of marketing material and customer communications. Generative AI is learning how to generate quality, written content that can be indistinguishable from its human source using existing data. An example of this would be that RPA bots in business environments can produce meeting minutes, generate performance reports, or even write customer follow-up emails based on existing interactions with customers (Pahune & Rewatkar, 2024).

Beyond time savings, this kind of automation ensures consistency and accuracy in communications-automatically ironing out wrinkles and problems. Generative AI tools can also tailor content to specific target audiences; therefore, they make for ideal tools for general customer service or marketing automation with personalized communications. As the AI learns from feedback and improves with time, businesses can continue to enhance the quality of automated content to more closely meet their needs.

2. Improving Decision Making and Predictive Analytics

With generative AI, on the other hand, RPA can facilitate decision-making and predictive analytics processes. This is because the algorithms of AI can be trained and analyzed against historical data, and then the information obtained from such analysis can be fed into business processes by RPA bots (Pedretti et al., 2021). For example, in the supply chain management space, AI-infused RPA will analyze past data trends for demand, levels of stock, and shipping times to predict needs that will ascertain future demands and auto-order processes to be taken in light of the projections.

For example, in financial services, generative AI may assist RPA bots in generating financial forecasts, risk assessments, and fraud detection reports through the real-time analysis of large datasets. The insights derived through AI may be much more accurate than those obtained using traditional rule-based approaches, as generative models keep learning with new data. It means that business decisions will be faster and with much greater confidence, as they are more reliant on data.

3. Workflow Optimization and Process Automation

This role of Generative AI in workflow efficiency is one thing, and taking a process further toward automation in the running of organizations is quite another. Generative AI can help identify bottlenecks and propose improvements for the process, and then even generate an entirely new workflow based on historical data analysis. RPA systems, in traditional fashion, automate predefined tasks, but with generative AI, they can evolve to new situations by dynamically reconfiguring workflows (Siderska, 2023).

Generative AI, for instance, in customer service, can screen the types of inquiries that customers make and automatically generate scripts or workflows guiding the RPA bots to deal with complex queries more efficiently. AI-enabled RPA can also begin creating new task sequences whenever business processes change without necessarily having to be reprogrammed manually. Such flexibility makes generative AI an essential part of the more fluid and knowledge-intensive business process automation environments.

4. Structuring Unstructured Data

Traditional RPA systems are essentially incapable with regards to unstructured data, such as text, images, and audio files. Generative AI, on the other hand, makes use of natural language processing and computer vision to 'read' and act on unstructured information. For example, within a legal or compliance department, generative AI will be able to read and interpret contracts, and results summed up or key terms extracted for automated processing (Siderska et al., 2024).

For example, in healthcare, generative AI can read through medical records and extract relevant information to automatically generate summaries that RPA bots can process for claims or schedule appointments (Venigandla, 2024). The capacity to handle unstructured data opens up an incredibly expanded area of tasks that RPA can automate, freeing businesses from operations that contain various forms of data and sources.

5. Customer Support and Virtual Assistants

Another important application of generative AI is for customer service and support. Equipped with generative AI, virtual assistants and chatbots can really engage with customers in much more complex ways to generate responses that are contextual, or even personal. Different from chatbots that merely repeat lines of script, generative AI bots can even change their lines on the fly upon the inputs given by customers, making the interaction much more natural and human-like (Sowjanya & Vijaya Chamundeeswari, 2024).

For example, RPA bots empowered by AI would be able to process customer-related queries with the help of user-specific information for formulating customized replies. This would also be possible through generative AI to predict the future needs of the customer based on its previous behaviors. In such a scenario, customer satisfaction is enhanced while human agents are guided toward more complicated or sensitive matters, and the overall efficiency of customer service operations would be increased (Söylemez, 2024).

Generative AI applications in RPA would revolutionize business process automation and thereby decision-making. With its integration, it extends the capabilities of RPA significantly-from creating content to improving workflows, to handling unstructured data, and to helping customers with complex matters. The superpower combination enables the automation of more complex tasks and more personalized services with much more efficiency, at lower operational costs and with improved scalability of the automation systems (Toxtli, 2024).

KEY BENEFITS OF USING GENERATIVE AI WITH RPA

The integration of generative AI with Robotic Process Automation brings a host of transformative benefits to the organizations looking to enhance their automation capabilities. Advanced cognitive capabilities of AI lend more flexibility, efficiency, and capability to handle complex tasks to the systems of RPA. Given below are the key benefits to using generative AI in RPA:

1. Increased Ability to Deal with Unstructured Data

The most significant limitation of traditional RPA is its structured data model that leaves many processes containing unstructured information out of the scope of automation. Generative AI solves this problem by allowing RPA systems to read and understand unstructured data, including text and images, and even voice inputs (Ghaisas & Singhal, 2024). Using NLP and machine learning, AI processing can read documents, summarize data, and give answers to queries immediately. This capability makes possible wider scopes of business work being automated from looking through documents to reaching customers and others in enhancing the general utility of automation systems (Liu et al., 2023).

2. Optimized Efficiency and Productivity

Generative AI-driven RPA offers better efficiency by having higher ranges and complexities of creative tasks with less human intervention. AI RPA can generate reports, compose drafts of emails, compare large datasets, and even make some decisions based on learned patterns. This increases productivity because it frees employees from repetitive work, allowing them to focus on more strategic tasks that require human intellect. Also, the RPA bots can function 24/7 without feeling tired, giving excellent and continuous results and letting businesses scale their processes logically (Noy & Zhang, 2023).

3. Real-Time Decision Making and Responsiveness

Traditional RPA systems are dependent upon predefined rules and workflows. A Generative AI brings a sense of decision making and adaptation to the RPA. An AI-driven bot can parse data in real time, identify trends or anomalies, and then make decisions based upon that. In finance, an AI-enhanced bot in RPA can highlight suspicious transactions, predict future trends, or suggest actions based on real-time insights. The technology allows companies to adjust their RPA to developments in the marketplace or about change in processes without having to be structurally reprogrammed (Brynjolfsson et al., 2023).

4. Increased Efficiency and Reduced Errors

Generative AI ensures the process output is more accurate and freer from errors mainly on repetitive, high-volume processes due to a lesser prospect of human error. As AI will learn from the data and hone the outputs over the life cycle, the RPA systems become more reliable to output more accuracy in certain task areas including data entry, report generation, and customer service interactions (Mulukuntla, 2022). Improvement of accuracy also reduces the costs that an error in those tasks would have incurred otherwise it would ensure the fulfillment of regulations and addition to efficiency in operational quality.

5. Cost Savings and Scalability

The hybridization of generative AI with RPA would bring in the big cost-cutting benefits to the businesses. Complex tasks that are usually supervised by a human can be automated for cost reduction, but simultaneously the quality of output shall improve. AI-driven systems can scale more efficiently with an unproportionate increase in volumes of work without needing human resources to grow in the same proportion (Chavan & Chavan, 2024). Here lies the opportunity of continuous cost-cutting and the expansion of businesses' automation capabilities while further refining the processes through AI

6. Customized Customer Interface

To do that, the generative AI helps the RPA bots deliver highly personalized customer experiences in being able to analyze past interactions and generate real-time responses. For example, in customer service, AI-enhanced RPA can draft custom emails or chat responses so that every customer receives a response that feels unique and relevant to the inquiry they submitted to receive. This approach often increases customer satisfaction and loyalty while saving time and effort in dealing with each interaction (Banala, 2024).

7. Continuous Learning and Process Improvement

In the meantime, in contrast to classical RPA systems, which function based on static rules, generative AI will learn new things from fresh data; and then, based on that knowledge, it will adjust its own processes correspondingly. With this continuous learning capability, business processes can improve continuously over time. For instance, generative AI could analyze data from some automated process, identify inefficiencies in that process, and suggest corresponding optimizations (Robert, 2024).

The integration of generative AI and RPA opens up new possibilities for businesses: automating many more tasks, making better decisions, and increasing the overall efficiency of their system. RPA systems empowered by AI can also provide the competitive edge needed to handle unstructured data and deliver personalized customer experiences. By investing in these benefits, organizations may streamline operations, reduce costs, and scale their automation strategies more effectively and get prepared for long-term prosperity in an increasingly automated world (Banala, 2024).

CHALLENGES AND LIMITATIONS OF USING GENERATIVE AI IN RPA

Although integration of generative AI into RPA presents many benefits to a business, it also encompasses some issues and limitations that the business will have to address. Whereas automation through AI heightens the demand for businesses to pay attention to technical, ethical, as well as operational obstacles having a negative impact on the effectiveness of their automation system via RPA, some of the key challenges and limitations include:

1. Data Privacy and Security Concerns

Probably the biggest challenge in using generative AI in RPA operations relates to how data are going to be managed, regarding data privacy and security. Typically, generative AI requires wide volumes of data for model training and output generation, often including sensitive information like details from customers or financial records and intellectual property. Thus, improper handling of such data may bring about a breach and put organizations facing serious legal and reputational risks (Balaguru, 2024).

In addition, as AI-run automation spreads widely, the compliance to GDPR and other data protection regulations becomes increasingly difficult. Therefore, the organizations must have clear frameworks about protecting their data with AI models so that these computers do not inadvertently break laws on privacy or misuses of sensitive information inadvertently (Beerbaum, 2023a).

2. Ethical Considerations and Bias

Such biases, if present in the training data, are inherited by the generative AI systems. For example, an AI model might be biased on gender or race and produce biased output on sensitive attributes. This would spread widely to areas such as hiring, customer service, and finance, where fairness and objectivity are important aspects (Beerbaum, 2023b).

Another ethical problem with generative AI is transparency and accountability: because AI models can generate content or decisions completely on their own, tracing rationale for specific outputs may prove quite challenging, thus raising questions concerning accountability. Organizations should therefore ensure AI systems are transparent, interpretable, and fair, hence avoiding unintended consequences as well as retention of stakeholder trust.

3. Technical Complexity and Integration Issues

It is quite technical and complex, and so calls for specialist skills in order to integrate generative AI into existing RPA systems. While classic RPA platforms operate on pre-defined rules, generative AI models use more sophisticated algorithms and machine learning techniques which, in turn, require ongoing training and maintenance. Business organizations thus necessitate highly specialized skill sets to configure, monitor, and optimize these systems (Bi, 2023).

It also makes the current systems incompatible with legacy systems. There are many organizations that are still using legacy software and infrastructure, which might not be compatible fully with AI, or the extent to which older software and infrastructure have been used might not support AI, so lots of time and human effort is wasted to make integration possible, upgrade infrastructure or build new solutions to make up for the shortcomings of integration between AI and current legacy systems (Feuerriegel et al., 2024).

4. High Initial Investment and Maintenance Costs

While RPA's generative AI can eventually bring long-term cost savings, the initial investment in AI technology, infrastructure, and talent is massive. Financial resources are needed for training AI models, deploying them at scale, and maintaining performance over time. By comparison, small businesses or those with more limited budgets find it hard to justify these on-premise costs. And where ROI is not immediately apparent (Gotthardt et al., 2020).

Moreover, AI models should be updated and retrained constantly for them to continue maintaining precision and performance. Maintenance and monitoring of AI systems are resource-intensive, especially for organizations that do not have a core expertise in developing or running AI.

5. Limited Capability to Manage AI Outputs and Quality Control

Generative AI models are supposed to create new outputs. Not always would those outputs be of perfect quality or specifically required to perform the intended task. Examples include AI-generated content requiring review and fine-tuning for it to meet quality standards. The list might include legal documentation, marketing materials, communications with customers, among others; these requirements somewhat reduce the efficiency RPA so readily offers (Lievano-Martínez et al., 2022).

Furthermore, it is also hard to handle the uncertainty of the output of the AI. Generative AI systems are often "black boxes," which means it is not easy to understand the way in which they arrive at their conclusions or recommendations. This lack of transparency makes it difficult for businesses to accept the AI's outputs without proper monitoring and quality control processes being embedded.

6. Issues of Scalability

While RPA systems are, by definition, scale-friendly, adding generative AI to the mix complicates scaling operations. The more advanced these AI models become, the more complex tasks they are able to take on, meaning computational resources expand exponentially in training and in running them as well. That means it can be quite challenging to manage the infrastructure requirements involved, especially when trying to scale operations across multiple departments or geographic regions (Lo, 2024).

Another aspect of scalability is striking a balance between automation and human oversight in AI-driven RPA systems. As increasingly large decisions are left to generative AI systems, organizations need to make sure they have mechanisms to audit the output produced by AI and prevent errors from propagating at scale.

7. Change Management and Workforce Resistance

Introduction of generative AI into RPA can create resistance on the part of employees, who may be afraid of losing their jobs or cannot cope with new workflows that demand automation with AI power. Thus, proper change management strategies are needed to make the transition smooth and drive away the fears related to job losses. Organizations must train or upskill and reskill their workforce so that employees can work with AI systems and use them to boost their productivity (Lo et al., 2024).

While RPA integrated with generative AI has a lot of promise towards the automation of complex tasks and making business operations more complex, it also creates quite a number of challenges in front of the organizations that need to be addressed. Data privacy and ethical concerns, technical complexity, and high costs are some of the challenges that a company needs to overcome in its pursuit of successfully implementing AI-driven automation. If organizations address all these concerns smartly and foster a culture of transparency and collaboration between humans and machines, they can harness the full transformative power of generative AI in RPA (Mandvikar & Achanta, 2023).

FUTURE TRENDS: IMPACT OF GENERATIVE AI ON RPA EVOLUTION.

It symbolizes the beginning of the generation age of RPA technology. At the same time, as these technologies continue to advance, business will look towards some newness, upon which more intelligent, adaptive, and scalable solutions will remake whole industries. Some of the main future directions-the impact of generative AI on RPA evolution-will be summarized here below:

1. Hyper-Automation

In the next couple of years, it is expected that hyper-automation - multiple advanced technologies such as generative AI and RPA, machine learning, and natural language processing, working together to automate an entire business process end-to-end-is a topic that would likely unravel. Hyper-automation is well beyond traditional RPA task-specific automation, incorporating AI-powered decision-making, predictive analytics, and continuous optimization of processes (Negoiţă & Borangiu, 2023).

Business processes, hence, would be automated not only of the routine kind but of the dynamic, knowledge-intensive kind. For example, AI-enhanced RPA could analyze market trends, prepare reports, and autonomously change strategy in real time, which will significantly position businesses for strong competitive advantage. Hyper-automation will nudge organizations toward being totally autonomic, with human intervention needed only when acting as oversight and strategic decision making.

2. Cognitive Automation

Cognitive automation uses AI technologies to mimic the human cognitive functions of understanding, reasoning, and problem-solving into automated processes. In this regard, what generative AI would be making possible is supporting cognitive automation much more in generating more complex, non-linear tasks requiring creativity and decision-making capabilities for RPA systems. For instance, it can automatically write contracts, design marketing campaigns, or even innovate solutions to the business problems that arise from analyzing historical data integrated with real-time insights (Pahune & Rewatkar, 2024).

Businesses will automate tasks considered to require human intelligence with the advancement of cognitive automation, thus reducing reliance on manual labor and enhancing the pace of operational efficiency.

3. Process Discovery and Optimization by AI

Generative AI is poised to change discovery and optimization of business processes in organizations. Traditional RPA systems rely on pre-defined workflows, while AI-based RPA systems will discover inefficiencies and bottlenecks in business processes from historical data autonomously. AI-based RPA not only executes tasks but also builds new workflows or adjusts existing ones according to insights they gain from real-time data. This ability will make the systems more adaptable and responsive to ongoing changes in business operations (Pedretti et al., 2021).

Businesses will increasingly rely in the future on AI-driven process discovery tools automatically mapping and suggesting the optimum workflows without involving human cognitance. This will hasten the digital transformation agenda and support continuous improvement in operations.

4. Personalized and Contextual Automation

Therefore, with generative AI, RPA systems will be able to create even more personalized and context-aware automation. Generally, the better the AI system understands context, the more that individualized responses and actions can be generated by the RPA bot for individual users or in individual situations. For example, in customer service, an AI-powered bot in RPA could create specific suggestions for a customer based on the interaction history, preferences, or real-time needs of the customer (Siderska, 2023).

This will be a trend even for automating customer service with its further extension into areas like employee training, marketing, and sales, hence generating outputs unique to user profiles and contextual data. This will eventually bring enhanced user experience and more meaningful engagement with customers and employees in general.

5. Better Collaboration Between AI and Humans

Now, with generative AI becoming a part of RPA, the trend is away from replacing the human and towards human enhancement. In the future, AI-powered RPA systems will accompany and support humans to better enhance their decision-making capabilities, creativity, and production. For example, generative AI might assist professionals in creating an initial version of a report, a creative design, or a technical document, but the final act of review and strategic decision shall be that of humans (Siderska et al., 2024).

This is when there would be more collaboration between humans and AI to come up with a new model of augmented work where human expertise and AI-driven automation will serve to deliver higher levels of innovation and efficiency. Thus, the future of RPA will be defined by this partnership, where AI takes care of routine and creative tasks, and humans take care of more strategic and high-impact activities.

6. Scalability and Democratization of RPA

Generative AI will enable real scalability and thus make such accessibility possible for organizations of any size. Lowered cost and installation will be the progress of AI-driven automation, bringing it to even SMEs across the industry so they can take advantage of advanced RPA technologies, no longer having an extensive team of developers needed to implement automation (Sowjanya & Vijaya Chamundeeswari, 2024).

This democratization of RPA will see it adopted almost in every industry and help further the efficiency and innovation processes at all business levels. Because AI-powered RPA will scale further in more complex processes within the organization, widespread digital transformation across the global economy is anticipated.

7. Ethical and Governance Frameworks for AI-Driven RPA

With generative AI increasingly incepted in RPA, ethical implications for AI-powered automation are likely to grow on the horizon. Data privacy, bias, accountability, and transparency will be placed to be debated. Long into the future, organizations will have to develop strong governance frameworks so that AI-driven RPA would be well-positioned for ethically compliant and managed use.

This is likely to create an industry standard and best practices for responsible AI usage in automation. As companies start proactively dealing with these challenges, they will begin to outpace their competition by establishing trust with their customers, employees, and regulators (Söylemez, 2024).

This is going to redefine the future of automation by making it both intelligent and more adaptive and scalable once generative AI is integrated into RPA. RPA systems, with generative AI, will increasingly be expected to solve complex problems while offering superior value across industries-from hyper-automation to cognitive capabilities, personalized workflows to ethical considerations. Artificial intelligence in RPA becomes a magnet for innovation, efficiency, and growth as this core nature of how organizations operate shifts in the digital era (Toxtli, 2024).

IMPLICATIONS

Integrating generative AI into RPA holds deep implications in most of the areas, involving not only the practical and managerial implications but also ethical and social implications. The adoption of these technologies in AI-driven automation creates a need for an understanding of its very extensive effects in making work operations more efficient, affecting the nature of leadership, workforce, and the society at large.

1. Practical Implications

From a pragmatic perspective, generative AI in RPA really delivers real value for complex, creative, and knowledge-based tasks involving material not previously known to be directly achievable with traditional RPA. In sum, AI-powered RPA supports the automation of complex and creative content generation, data analysis, and decision-making processes that significantly improve business process efficiency, cost of operations, and service levels (Venigandla, 2024). For example, an RPA system driven by AI can be used by organizations to automatically generate custom marketing documents, draft legal papers, or draft predictive financial reports for enhanced productivity and minimized effort on the human side.

While these practical advantages come with their own costs, such as data infrastructure and AI model management, businesses need to be investing in scalable technology and proficient people to maintain and improve these systems over time. This makes the implementation processes a little more complex, but long-term operational gains can make many of these investments worthwhile (Ghaisas & Singhal, 2024).

2. Managerial Implications

An example would be the addition of generative AI to RPA; this will need changes in management approaches regarding workforce management, change management, and leadership. Managers will have to adapt to such changing work situations from the paradigm of manual task execution through the one dominated by AI-driven automation as they motivate a culture of collaboration between AI systems and human employees. Leadership needs to shift its emphasis on upskilling and reskilling as it equips employees to work effectively with AI, in addition to using automation to enhance their productivity (Liu et al., 2023).

From then onwards, managerial focus will shift from routine and task-oriented management to supervision of strategic decision-making activities. With routine and creative processes automated, managers will have the bandwidth to focus on innovation and business growth and long-term planning. There are challenges ahead in terms of how to manage employee resistance when they may fear job displacement as a result of automation. Effective management of change must be the course through these concerns, where employees in a future with humans and AI work together in mutual respect and value (Noy & Zhang, 2023).

3. Ethical Implications

RPA development brings generative AI with it and significant ethical issues related to transparency, accountability, and bias. In the matter of emerging autonomy in AI model-generated content or even the generation of decisions, understanding how such systems work is paramount to transparency. Organizations need to create frameworks for explaining AI output and decision making, especially in regulated industries like finance, healthcare, and law (Brynjolfsson et al., 2023).

Moreover, AI systems are biased themselves. They learn from history data, which could often contain societal prejudices. This can result in a bias against certain groups of people because of automated processes like hiring, credit scoring, or customer interactions. Businesses will need to take appropriate measures-making that kind of robust safeguarding focus on detection tools and diverse data sets-to ensure that AI-driven RPA systems are fair and equitable. Another significant concern will be the ethical governance of AI, along with guidelines that will set out who uses the data, what a company's decisions about the use of automation should be on, and on what lines of ethics (Mulukuntla, 2022).

4. Social Implications

As RPA system capabilities develop the capacity to undertake mundane as well as creative pieces of work, certain jobs will fall redundant because people rely more and more on such functions, mainly in traditional areas such as data entry, customer service, and content generation. Strip away a couple of jobs in turn, but AI would end up opening new ones where specific skills are needed in the management of AI, data science, and automation oversight (Banala, 2024).

Furthermore, democratizing automation would further alienate the haves from those who can afford to invest in AI-driven RPA and the have nots. Large enterprises will benefit from increased productivity and innovation, while smaller business enterprises do not have a chance to keep pace with the AI train. This will only continue social inequalities unless AI technology becomes more accessible and economical.

To be more positive, generative AI in RPA can indeed improve quality of life since mundane, routine, and repetitive tasks can be eliminated to allow productive hours for employees in meaningful, creative, and strategically designed tasks. All this will be reflected in greater satisfaction at work and a better balance between work and life, thus taking care of employee welfare in the broader sense (Chavan & Chavan, 2024).

Practical implications of this type of AI in RPA would include efficiency, managerial challenges, ethical considerations, and societal impacts. Its potential to transform does have challenges that must be carefully managed from tremendous opportunity to automate complex processes enhance productivity, and drive innovation to really require careful management about the benefits and pains of this transformative technology. This would perhaps work to the organization's advantage by showing responsible practices in helping AI and human workers collaborate, hence unlocking the full potential of generative AI in RPA with a value addition on business and society.

CONCLUSION

Thus, with generative AI attached to RPA, now changes can be made in businesses and insights gained in how to implement automation and then maintain operational efficiency. Therefore, they can move beyond rule-based automation with AI-enhanced capabilities to automate repetitive, creative, and decision-intensive processes. It brings forth greater productivity levels, smooth workflow, and a more flexible approach towards market demands by companies. Innovation, higher scalability, and hyper-automation open new avenues for digital transformations in a world of today and tomorrow, helping firms remain innovative.

But at the same time, transformational changes do come with a suite of challenges that will be managed very carefully. While generative AI may turn super effective for RPA, these include high technological complexity, workforce adaptation, ethical concerns, and data governance - work that will require careful leadership effort. Leadership must therefore upskill, facilitate human collaboration with AI systems, and set transparent frameworks for safe AI engagement. These efforts are crucial not only to excavate the true value from generative AI but also to instill trust, which might otherwise reduce these potential risks of automation.

On a more macro level, the sociological effects of generative AI in RPA should not be overlooked. It will lead to unemployment and social imbalances, but it may also open avenues for innovation as well as creation of new, high-skilled jobs. Being socially aware and ensuring equal access to this AI technology, organizations can work towards a fairer future while considering the benefits of automation maximally.

As such, generative AI for RPA holds the promise of revolutionizing the future of work and business processes. Creating practical managerial ethical, and social consequences would require effort in harnessing technical benefits of AI-driven automation. Responsibility in adoption and visionary leadership are hoped to ensure sustainable growth and a positive impact for society while that will unleash the transformative power of generative AI for organizations.

REFERENCES

Balaguru, S. (n.d.). *Beyond automation: Redefining healthcare revenue cycles through RPA*. NLP and Gen AI.

Banala, S. (2024). The future of IT operations: Harnessing cloud automation for enhanced efficiency and the role of generative AI operational excellence. *International Journal of Machine Learning and Artificial Intelligence*, *5*(5), 1–15.

Beerbaum, D. (2023). Generative artificial intelligence (GAI) ethics taxonomy-applying ChatGPT for robotic process automation (GAI-RPA) as business case. *Available atSSRN* 4385025.

Beerbaum, D. (2023). Generative artificial intelligence (GAI) with ChatGPT for accounting–a business case. *Available atSSRN* 4385651.

Bi, Q. (2023). Analysis of the application of generative AI in business management. *Advances in Economics and Management Research*, *6*(1), 36–36. DOI: 10.56028/aemr.6.1.36.2023

Brynjolfsson, E., Li, D., & Raymond, L. R. (2023). *Generative AI at work (No. w31161)*. National Bureau of Economic Research. DOI: 10.3386/w31161

Chavan, P., & Chavan, P. (2024, June). Data migration between on-prem to cloud using generative AI to reduce costing and overheads. In *2024 International Conference on Innovations and Challenges in Emerging Technologies (ICICET)* (pp. 1-7). IEEE.

Feuerriegel, S., Hartmann, J., Janiesch, C., & Zschech, P. (2024). Generative AI. *Business & Information Systems Engineering*, *66*(1), 111–126. DOI: 10.1007/s12599-023-00834-7

Ghaisas, S., & Singhal, A. (2024). Dealing with data for RE: Mitigating challenges using NLP and generative AI. *arXiv preprint arXiv:2402.16977.*

Gotthardt, M., Koivulaakso, D., Paksoy, O., Saramo, C., Martikainen, M., & Lehner, O. (2020). Current state and challenges in the implementation of smart robotic process automation in accounting and auditing. *ACRN Journal of Finance and Risk Perspectives*, *9*(1), 90–102. DOI: 10.35944/jofrp.2020.9.1.007

Lievano-Martínez, F., Fernández-Ledesma, J., Burgos, D., Bedoya, J., & Builes, J. (2022). Intelligent process automation: An application in manufacturing industry. *Sustainability (Basel)*, *14*(14), 8804. DOI: 10.3390/su14148804

Liu, Y., Yang, Z., Yu, Z., Liu, Z., Liu, D., Lin, H., Li, M., Ma, S., Avdeev, M., & Shi, S. (2023). Generative artificial intelligence and its applications in materials science: Current situation and future perspectives. *Journal of Materiomics*, *9*(4), 798–816. DOI: 10.1016/j.jmat.2023.05.001

Lo, W., Yang, C.-M., Zhang, Q., & Li, M. (2024). Increased productivity and reduced waste with robotic process automation and generative AI-powered IoE services. *Journal of Web Engineering*, •••, 53–88. DOI: 10.13052/jwe1540-9589.2313

Lo, W., Yang, C. M., Zhang, Q., & Li, M. (2024). Increased productivity and reduced waste with robotic process automation and generative AI-powered IoE services. *Journal of Web Engineering*, *23*(1), 53–87. DOI: 10.13052/jwe1540-9589.2313

Mandvikar, S., & Achanta, A. (2023). Process automation 2.0 with generative AI framework. *International Journal of Science and Research (Raipur, India)*, *12*(10), 1614–1619. DOI: 10.21275/SR231021200626

Mulukuntla, S. (2022). Generative AI: Benefits, limitations, potential risks, and challenges in healthcare industry. *EPH-International Journal of Medical and Health Science*, *8*(4), 1–9.

Negoiţă, R., & Borangiu, T. (2023). Robotic process automation of inventory demand with intelligent reservation. *Studies in Informatics and Control*, *32*(2), 5–14. DOI: 10.24846/v32i2y202301

Noy, S., & Zhang, W. (2023). Experimental evidence on the productivity effects of generative artificial intelligence. *Science*, *381*(6654), 187–192. DOI: 10.1126/science.adh2586 PMID: 37440646

Pahune, S., & Rewatkar, N. (2024). Cognitive automation in the supply chain: Unleashing the power of RPA vs. GEN AI.

Pedretti, A., Santini, M., Scolimoski, J., Queiroz, M., Toshioka, F., Rocha, E., & Ramos, M. (2021). Robotic process automation extended with artificial intelligence techniques in power distribution utilities. *Brazilian Archives of Biology and Technology, 64*(spe). https://doi.org/DOI: 10.1590/1678-4324-75years-2021210217

Siderska, J., Aini, S. N. B. M., & Kedziora, D. (2024, March). Complementing robotic process automation with generative artificial intelligence (ChatGPT), case of Robocorp. In *Future of Information and Communication Conference* (pp. 37-53). Cham: Springer Nature Switzerland.

Siderska, J., Aunimo, L., Süße, T., Stamm, J., Kedziora, D., & Aini, S. N. B. M. (2023). Towards intelligent automation (IA): Literature review on the evolution of robotic process automation (RPA), its challenges, and future trends. *Engineering Management in Production and Services, 15*(4), 90–103. DOI: 10.2478/emj-2023-0030

Sowjanya, S., & Vijaya Chamundeeswari, V. (2024, February). Information extraction using RPA and generative AI from unstructured documents: A case of invoices. In *International Conference on Computational Intelligence in Data Science* (pp. 250-264). Cham: Springer Nature Switzerland. DOI: 10.1007/978-3-031-69986-3_19

Söylemez, I. (2024). Accelerating low-code automation development with generative artificial intelligence.

Toxtli, C. (2024). Human-centered automation. *arXiv preprint arXiv:2405.15960.*

Venigandla, K., Vemuri, N., & Nnamere Aneke, E. (2024). Empowering smart cities with AI and RPA: Strategies for intelligent urban management and sustainable development. *International Journal of Scientific Research and Management, 12*(4), 1117–1125. DOI: 10.18535/ijsrm/v12i04.ec02

Chapter 7
Exploiting Cloud Computing in RPA

Sandeep Kumar Singh
https://orcid.org/0000-0002-1741-7254
O.P. Jindal Global University, India

Rohit
https://orcid.org/0009-0004-1796-5252
College of Healthcare, Management, and Economics, UAE

Ajay Chandel
https://orcid.org/0000-0002-4585-6406
Lovely Professional University, India

Tanupriya Choudhury
University of Petroleum and Energy Studies (UPES), India

ABSTRACT

The integration of cloud computing with Robotic Process Automation (RPA) changes the very course of the way business is conducted, bringing about improvements in business operations efficiency, scalability, and adaptability. This chapter introduces the reader to the architecture of cloud-based RPA and discusses its benefits: that is, a lowered cost of operations and rapid deployments. Discussions include the rise of cloud-native RPA platforms and tools, as well as the importance placed on security and compliance in the cloud environment. On the other side, the chapter also analyses other issues and limitations organizations may face when implementing cloud-enabled RPA, including issues with data privacy and integration complexity among others. Outlook on cloud-enabled RPA is also reviewed in terms of implications for business and the workforce. Based on this, this chapter is specifically looking at various considerations in enabling insight into the means through which

DOI: 10.4018/979-8-3693-4365-4.ch007

organizations can transform their processes through cloud-enabled RPA while propelling innovation into the future.

AN INTRODUCTION TO CLOUD AND RPA INTEGRATION

Amongst two major technologies that are reshaping efficiency, scalability, and innovation in the ever-changing digital transformation landscape across industries are Robotic Process Automation (RPA) and cloud computing. RPA is designed to automate repetitive rule-based tasks by modeling human interactions with digital systems. This facilitates making opportunities for the business to streamline operations, cut human errors, and optimize resource utilization. On the other hand, cloud computing offers on-demand access to computing resources-technically, any kind of server, storage, databases, software, application, and other related facilities over the internet. By unifying these two technologies, organizations can achieve enhanced flexibility, scalability, and cost-effectiveness, really taking automation to new heights (Kunduru, 2023).

The association of RPA with cloud platforms is revolutionizing the method by which businesses are now approaching automation. Traditionally, RPA installations were mostly on-premise and therefore presented their share of difficulties such as involving a huge upfront cost to implement the infrastructure, scalability issues in terms of achieving substantial scale, and requirements for constant maintenance and update. The entrance of cloud computing, however has reoriented the face of RPA, providing an alternative to on-premise deployment (Jha et al., 2021). In a cloud-based RPA, the organization makes available automation tools and resources through cloud infrastructure to dispense with the need for significant capital investment in hardware and software. This allows companies to step up their automation activities rapidly according to changing needs. This is without being restricted by the physical infrastructure.

In other words, the first benefit of using cloud-integrated RPA is scalability. Traditionally, scaling the RPA environment was expensive and time-consuming because additional hardware, along with software licenses, had to be deployed in an on-premises environment. Near limitless scalability means businesses can scale up or down automation initiatives in a flash, whether automation of a few processes or RPA across departments and regions, with agility born from cloud-based RPA as business needs evolve. Apart from that, cloud-enabled RPA provides access to all the latest updates and innovations in RPA technology so the organization can stay ahead on the curve of automation capabilities.

Another pertinent advantage of cloud-enabled RPA is cost efficiency as cloud platforms engage with a pay-as-you-go model, thus an organization incurs costs only for the resources it has utilized. One of the key advantages is that it allows organizations to optimize their cost for the budget and reduces operating cost because they no longer have to upkeep pricey on-premise infrastructure (Madakam et al., 2019). More importantly, cloud-based RPA solutions improve centralized management and maintenance, free up the burden of internal IT teams, which enables them to deploy automation initiatives faster and faster implementation. Collaboration support by the cloud enables distributed teams to access and manage RPA processes from any part of the world, highly valuable in today's increasingly distributed and globalized workforce.

Incorporating RPA with cloud architecture further enables organizations to embrace advanced technologies, like AI and ML, that can benefit their automation processes. Integration would provide access to AI and tools that enable it to analyze huge volumes of data, such that the RPA system makes intelligent decisions, learns patterns, and improves over time. With convergence of cloud, RPA, and AI, organizations can leave the boundaries of simple task automation for more complex, cognitive types that need automation on unstructured data and dynamic decision-making approaches (Douglas & Roger, 2024).

Cloud computing with RPA thus forms a great business enabler to facilitate operational optimization and add depth to the capabilities of automation in a business. Scalability, cost efficiency, and the advanced resource base of the cloud will enable businesses to harness the full potential of RPA for further competitive advantage in an increasingly automated world.

ARCHITECTURE OF CLOUD-BASED RPA

Cloud-based architecture for Robotic Process Automation differs from typical solutions found in on-premise; software offers users more scalability, flexibility, and accessibility. With infrastructure usage via cloud deployment, the RPA systems can run very well with reduced overhead burden in the management of equipment and related software in-house. Below is an in-depth elaboration of the central components that explain the cloud-based RPA architecture.

1. Cloud Infrastructure Layer

Underlying this scenario lies the cloud infrastructure. Involving all the computing resources one needs to have in order to run automation processes, it involves servers, storage, and networking all delivered by Cloud Service Providers, including AWS,

Microsoft Azure, or GCP. It provides scalability-on-demand, so an organization can expand or contract its own computing needs based on real-time demand. Hosting RPA solutions in the cloud provides business with the mitigation of the high cost that exists when a business buys and makes initial and ongoing investments in a physical hardware structure (Koppisetti, 2024).

Cloud infrastructure can also enable multi-tenancy; in which a number of organizations can share the same computing resources without their data or operations interfering with the others. It allows smaller businesses to get huge enterprise-grade automation tools at one-tenth of the price, while big enterprises are using RPA to run huge global operations in completely minimal IT-infrastructure setups.

2. RPA Orchestration Layer

The orchestration layer is part of the architecture of cloud-based RPA responsible for governing and orchestrating the execution of processes by multiple bots and workflows. It determines when the right bot should be triggered at the right time, whether by predefined or real-time triggers, while monitoring the automation task's execution-hence ensuring that the activity may execute smoothly and efficiently (Lo et al., 2024).

Most cloud-based RPA platforms utilize centralized orchestration tools which offer a single entry point for handling all the bots and processes of an organization. These tools allow customers to monitor the status of workflows, pool resources in real-time and ensure high bot availability. On the cloud layer, the orchestration layer enjoys full global accessibility, which allows the organizations to manage their RPA processes irrespective of where they are located in the world. This enables an organization to have distributed teams working together and sharing their resources (Douglas & Roger, 2024).

3. RPA Bot Layer

The RPA bot layer is the actual work done through the software bots that form the automation engines. In cloud-based RPA, these bots can be deployed and executed within the cloud infrastructure or in a hybrid environment where the execution of bots both happens within the cloud as well as on local machines. Cloud-native bots can scale elastically, which means that according to the workload, the required number of bots can be spun up at peak times for handling increased tasks and scaled down at times of low activity.

In cloud-based architecture, bots are sometimes categorized into attended and unattended bots. Attended bots are utilized in parallel with human operators for tasks that require human input, and unattended bots are utilized with no human when available for repeated processes with no human intervention (Lo et al., 2024). Both the bots can take advantage of cloud-based deployment as both bots can work from anywhere and can view applications and data hosted in cloud environments in real-time.

4. Cloud-Based Data Management Layer

This massive data, generated and used by the automation process, will have to be dealt with at the data management layer of the cloud-based RPA architecture. Data can be safely stored, managed, and accessed from anywhere in cloud architecture. As a result, an easy interface between RPA bots and enterprise systems such as ERP, CRM, or legacy databases is created. Cloud storage would be doing this, with data all ready to be pushed up to bots for extraction, transformation, and reporting (Manukonda, 2024).

In addition to these, cloud-based data management provides robust security and compliance features to ensure that sensitive data is encrypted and protected in transit and at rest. Nearly all the cloud RPA platforms support compliance with industry standards such as GDPR, HIPAA, and PCI-DSS, allowing organizations to safely automate business processes involving sensitive information without risking breaches or non-compliance.

5. AI and Machine Learning Integration Layer

The benefit of this architecture being cloud-based is that it permits seamless integration of machine and artificial intelligence technologies. Cloud platforms provide access to leading AI services, such as NLP, image recognition, and predictive analytics, with which it is possible to enhance RPA capabilities. With the inclusion of AI, the functionality of cloud-based RPA systems can now perform complex tasks, including processing unstructured data, make intelligent decisions, and learn from historical data on how to optimize workflows (Moderno et al., 2024).

This layer allows an organization to move beyond simple rule-based automation and cognitive automation, in which the bots can adapt to changing conditions and make decisions based on patterns while even predicting what the future may bring. Cloud-based AI services are extremely scalable, and an organization can experiment with AI-driven RPA without investing in dedicated hardware or infrastructure.

6. Security and Governance Layer

Security and governance are of great concern in the cloud-based architecture of RPA. Cloud environment is equipped with integrated security features, such as encryption, identity management, and access controls; hence, RPA processes are safeguarded from unauthorized access and breaches. Governance tools on cloud RPA platforms enable organizations to monitor and control who accesses automation, what data can be used, and how bots interact with sensitive information (Tiron-Tudor et al., 2024).

Governance features are necessary for ensuring that the automated processes thus developed meet the regulatory needs of such industries as finance, healthcare, and even government. Cloud-based RPA architecture ensures that security protocols established are followed consistently, regardless of whether bots or workflows are being executed somewhere else in the cloud.

A cloud-based RPA architecture will allow a powerful framework to be utilised in support and management for scale-based automation solutions. Advanced orchestration, data management, and AI integration enrich capabilities, whereas scalability and adaptability of cloud infrastructure reshape how businesses think about automation. Using the power of the cloud, organizations can deploy their RPA solution faster at lower costs and greater flexibilities than traditional on-premises systems, keeping them ahead in this rapidly changing digital landscape (Kurylets & Goranin, 2024).

THE KEY BENEFITS OF THE CLOUD FOR RPA

The integration of RPA and cloud computing brings so much benefit to an organization seeking to increase the efficiency of operations, lower costs, and scale automation quickly. Cloud-based RPA presents unique advantages that are not possible with traditional on-premise automation solutions. Businesses can thus streamline their processes based on the demands presented by the fast-paced digital age, among other things. Below are some of the main benefits of using the cloud for RPA:

1. Scalability and Flexibility

An important advantage of cloud-based RPA is scalability in automation. Organisations don't need to make the same kind of big-time investment in infrastructure as they would with on-premise RPA systems to scale up. In fact, organizations can expand their automation capabilities based on a 'need-to-use' basis in cloud RPA. This elasticity helps organisations quickly add or remove RPA bots based on workload fluctuations. Deploying more bots during peak loads like the end of a financial quarter or holidays, where resources are scaled down during low activity periods to minimize cost repercussions (Nakić & Boban, 2024).

There is also the feature of accessibility with cloud-based RPA. It offers more flexibility when it comes to accessibility. Bots may be spread across several departments or geographies, not bounded by physical infrastructure. This means that global teams collaborate as efficiently as possible on automation projects, with central management and control provided by cloud-based orchestration tools.

2. Cost Efficiency

Secondly, cloud RPA saves cost significantly more than on-premise RPA implementations. Cloud uses the pay-as-you-go pricing model; an organization only pays for its resources used and, thus, completely avoids the pre-investment of big servers, software licenses, and IT professionals. This approach is very friendly to small and medium-sized enterprises that cannot invest in expensive infrastructure (Alirezazadeh, 2020).

In addition, the cost of maintenance and updating on-premise systems is not incurred by the cloud. Routine maintenance, updating security, and hardware upgrades are managed by cloud service providers, freeing internal IT teams to focus on more strategic initiatives. The decreasing cost of operations makes cloud-based RPA an attractive option for businesses wanting to optimize budget but expand their ability to automate.

3. Rapid Deployment and Implementation

The infrastructure of the cloud is already built, so already in place, which makes it ready for use, expediting the deployment of RPA solutions. Organizations will deploy and configure RPA bots without the often complex installation and integration processes usually associated with on-premise systems. This can be leveraged by business to speed up task automation and derive RPA benefits in hours or days rather than weeks or months (Alirezazadeh & Alexandre, 2020).

Moreover, cloud-based RPA platforms come with prebuilt templates and libraries for general automation use cases, ensuring that organizations can deploy bots with minimal inconvenience. These out-of-the-box solutions prove most valuable to businesses interested in automating repetitive tasks such as data entry or invoice processing-in other words, processing support inquiries from customers-by adopting customer support-all without having to create custom bots from scratch.

4. Better Collaborations and Accessibility

Cloud-based RPA allows for distributed teams to collaborate. Distributed teams are widely dispersed across locations; they can access the cloud by using an internet connection from anywhere around the world, easily accessing and controlling the flow of automation. This capability is particularly valuable in the increasingly remote workplace-where businesses increasingly rely on distributed teams to manage global operations.

The centralized nature of cloud-based RPA platforms also enables teams to work collaboratively toward bot development and optimization. Access to shared data and workflows in real-time helps enable team members to collaborate, irrespective of the office or time difference. Cloud platforms also allow access to RPA dashboards and analytics from anywhere. This helps users gain valuable insights into process performance and bottlenecks (Kehoe et al., 2015).

5. Continuous innovation and upgrades

The cloud-based RPA platforms are always updated with the latest features and innovations. This means that there is no need for manual upgrades and maintenance, unlike in an on-premise solution, providing automatic updates along with new functionalities. This automatically ensures organizations can stay on the cutting edge of automation technology without placing a single additional resource into system upgrades (Kunduru, 2023).

Cloud-based RPA platforms also accommodate advanced technologies, such as AI and ML. These tools can help businesses extend rule-based automation and build cognitive automation in their workflows. The AI-powered bot can process unstructured data while making decisions based on patterns and learn from the previous experiences to optimize further workflows.

6. Enhanced Security and Compliance

These cloud service providers have a strong security measure toward the protection of data and compliance with all sorts of industry rules. In fact, most of the RPA platforms that lie in the cloud come with inbuilt security features like encryption identity management, and access controls to ensure sensitive information security. Such measures enable organizations to hold a secure environment for their automation processes and comply with data privacy regulations like GDPR, HIPAA among others (Jha et al., 2021).

Also, cloud RPA platforms greatly enhance disaster recovery and business continuity capabilities. As the data is in the cloud, organizations can easily recover from disturbances such as system failures or cyber-attacks without much downtimes and un-interrupted automation operations.

7. Global Accessibility and Real-Time Processing

Cloud-based RPA systems achieve global accessibility, allowing organizations to deploy bots in multiple regions and across different time zones. This is also very significant for companies with headquarters operating across the globe since such companies could standardize automation processes in all the locations without wavering performance. More importantly, cloud RPA provides for real-time processing where bots could potentially function at any time of day and night with no delay or restriction based on location. This sure enables faster processing of vital transactions, ultimately enhancing the overall efficiency (Madakam et al., 2019).

Now, with the core benefits of using the cloud for RPA, including scalability, cost efficiency, rapid deployment, enhanced collaboration, continuous innovation, better security, and accessibility across the globe, this is a superpower for organizations to optimize their automation strategies. Locking into the potential offered by cloud technology to fully harness the capability of RPA encourages businesses to be more agile, efficient, and innovative.

CLOUD-NATIVE RPA PLATFORMS AND TOOLS

Being cloud native, RPA platforms are designed to scale. They are flexible and powered with the full capability of cloud computing, thus making them a great environment for any scale automation that organizations may require. These add more tools that empower automation through cloud infrastructure, advanced AI services, and real-time data processing. Here is a closer look at some of the main cloud-native RPA platforms and tools available in the market today.

1. UiPath Automation Cloud

UiPath Automation Cloud is an end-to-end, cloud-native platform for building, managing, and deploying RPA solutions across various environments without any requirements of on-premise infrastructures. A fully scalable solution is offered wherein the users can build, orchestrate, and monitor the automation from a single unified platform (Douglas & Roger, 2024).

The main advantage of UiPath Automation Cloud is the simplified orchestration feature, which makes deployment rather easy. This allows users to go ahead and start automating with no necessity for any server setups, along with their maintenance hurdles. It integrates AI tools like natural language processing (NLP), machine learning (ML), and more that enables bots to perform tasks in addition to simple rule-based automation.

The flexibility of UiPath's cloud offering is there because it can support both attended and unattended bots. It supports the most perfect usage range, from the most back-office back-to-end customer-facing applications and puts in place a global reach and perfect security that would make businesses feel confident and comfortable in across-border automation while ensuring regional regulations would be respected in data privacy (Koppisetti, 2024).

2. Automation Anywhere A360

Automation Anywhere A360 is another cloud-native RPA leader that offers a variety of automation capabilities fully delivered through the cloud. It is well known mainly because of its easy-to-use interface, making it very easy for both technical and non-technical users to design and deploy automation workflows.

Automation Anywhere A360: Provides ready-built templates and bot designs for the most frequently repeated tasks. This makes it easier to automate a task in less time. The AI and ML services also support automation, enabling bots to execute even complex tasks such as the processing of documents, extraction of data, and even decision-making. The bot store available here in the platform is a marketplace where the users can buy pre-built bots, and even components accelerate the activities of automation (Lo et al., 2024).

One of the crucial features of Automation Anywhere A360 is its cloud-native tool, Bot Insights, which now presents real-time analytics on performance from a bot, process efficiency, and ROI. Thus, organizations are able to gain actionable insights into the effectiveness of their RPA initiatives and can continuously improve things.

3. Microsoft Power Automate (formerly Microsoft Flow)

It is cloud-based and fully compatible with the Microsoft toolset, among them Office 365, Dynamics 365, and Azure. This is why this solution is particularly effective for organizations that already rely on Microsoft products and integrate easily into the existing work flows of those organizations (Manukonda, 2024).

Power Automate allows users to automate simple and complex processes with a mix of low-code and drag-and-drop tools and AI-powered features. It offers cloud-native AI Builder services where users can integrate intelligence into the flows of their automation workflows, such as sentiments, form processing, and object detection (Moderno et al., 2024).

However, one of the distinctive features of Power Automate is its connectors library, which can integrate with hundreds of third-party applications and services. In other words, users could work to automate workflows across any number of applications-from CRM systems to social media platforms-and term it the kind of versatile tool that any sized business would need.

4. Blue Prism Cloud

Blue Prism Cloud, which was previously known as Thoughtonomy, provides enterprise-class intelligent automation as a service, with a completely cloud-native RPA platform that brings together traditional RPA, cloud computing, AI and cognitive services for the most scalable automation solution available today (Tiron-Tudor et al., 2024).

The offering of Blue Prism Cloud is further marked by the concept of Digital Workforce. Here, a software bot is used as the virtual workforce, and it will work in conjunction with human employees. The platform offers powerful orchestration tools that help organizations manage and deploy bots efficiently across their operations. It also offers multiple AI services; therefore, it can allow bots to process unstructured data to create conversations and make intelligent decisions based on real-time information (Kurylets & Goranin, 2024).

Blue Prism Cloud offers an integrated analytics dashboard that provides insights into bot performance, effectiveness of automation, and process optimization. There is also cloud-native architecture, allowing bots to be easily deployed across organization locations, and strong security and compliance features.

5. Google Cloud RPA Solutions

Not an independent RPA platform, the Google Cloud RPA solution offers powerful tools to be used in combination with other RPA systems to strengthen automation capabilities. Google Cloud's AI and Machine Learning Services: The platform offers a wide array of AI and machine learning services. Included here are Google Cloud AI, which empowers a bot to process natural language, images, or even extract data from documents.

Integration with platforms like UiPath and Automation Anywhere will allow organizations to have more and better ways to operate with AI capabilities in line with automation workflows. For instance, an RPA bot may utilize Google's NLP services for the automation of customer service processes or apply Google's document AI for working with invoices and contracts much faster (Nakić & Boban, 2024).

Google Cloud further offers solid tools for data storage and management, which enables bots to access large volumes of data on the fly for processing and analyses. Integration of cloud infrastructure with RPA systems provides business enterprises with robust scalable solutions for more complex automation issues with complex data-driven requirements.

6. IBM Robotic Process Automation on Cloud

IBM Cloud-based Robotic Process Automation integrates IBM Watson and other advanced AI tools to the RPA workflows, offering an automation platform from the cloud. In this manner, organizations can automate activity across applications, including legacy systems, with cognitive capabilities for more sophisticated automation (Alirezazadeh, 2020).

IBM RPA on Cloud has hybrid cloud-native architecture with the ease of fast deployment flexibility and scalability. It provides capabilities for end-to-end automation-from bot creation and management to advanced analytics and reporting. It also integrates with IBM's greater set of cloud and AI services and, thus, appropriate for organizations looking to include AI-powered decision-making and predictive analytics in their automation plan.

Flexibilities and scalability can be achieved by organizations in regard to high-level capabilities in achieving successful automation at scale through cloud-native RPA platforms and tools. UiPath, Automation Anywhere, Microsoft Power Automate, Blue Prism Cloud, and IBM RPA on Cloud enable businesses to automate a wide range of processes-from simple repetitive tasks to complex, AI-driven workflows-with AI integrations from Google Cloud (Alirezazadeh & Alexandre, 2020). These tools prove to be the cornerstone for an organization that aspires to leave an indelible mark in today's digital landscape by optimizing to the maximum potential of RPA and harnessing the power of the cloud.

SECURITY AND COMPLIANCE FOR CLOUD RPA

When organizations have started using more and more cloud-based Robotic Process Automation, security and compliance have taken on a much greater role. While cloud-based RPA offers scalability, flexibility, and cost savings, it exposes challenges to data privacy, regulatory compliance, and system vulnerability. Thus, organizations must ensure that their cloud RPA platforms are secure and meet minimum regulatory standards in terms of protection of sensitive data and stakeholder trust. This chapter discusses the salient features of security and compliance related to cloud RPA.

1. Data Encryption and Access Control

Some of the crucial aspects of security regarding cloud RPA environments concern the confidentiality of data being stored and then transacted. Major cloud RPA service providers are offering secure mechanisms for encrypting the data and protect the information, thus. Data encryption at rest, stored on cloud servers, employs the state-of-the-art encryption protocols; if there is unauthorized access, the data will remain unreadable. Encryption is also essential in protecting data in transit. For example, communication amongst bots and the cloud servers, as well as amongst bots and enterprise systems, can be protected using techniques like TLS (Transport Layer Security) (Kehoe et al., 2015).

Encryption aside, effective access control needs to be provided in securing any cloud RPA environment. Role-based access control (RBAC) and multi-factor authentication (MFA) are among the most widely used methods for preventing unauthorized access to sensitive data and automation workflows. These controls would ensure only that the authorized users for instance administrators or bot developers will have access to, or a possibility to alter critical processes. Also, generally the cloud-based RPA platforms provide very useful centralized dashboards through which the administrators are able to control the user roles, and permissions, and thus greatly improve the security in the automation environment (Kunduru, 2023).

2. Data Privacy Compliance

With today's gaining priority on data privacy, the demands being placed on cloud RPA platforms even climb for regional and international data protection legislations. We're talking about the European Union's GDPR, U.S. HIPAA, and local state-specific laws that go highly stringent in the handling, processing, and storage of personal data.

Cloud RPA vendors ensure data localization is implemented; an organization may choose a location where data will be located depending on its regulatory needs. This means personal data will be processed in jurisdictions known to have robust data privacy legislations. Cloud RPA platforms also need to provide audit trails showing a clear view on how data was accessed, processed, and transferred. The audit logs enhance the openness, but at the same time are evidence in case the same needs to be presented in a regulatory audit (Mohanarajah et al., 2015).

Organizations using cloud RPA should check the compliance certificates offered by their chosen cloud provider, whether ISO 27001 information security management or SOC 2 system and organization controls. Compliance certifications indicate a cloud provider has implemented specific reasonable security controls with the help of appropriate policies and regulations in accordance with industry standards.

3. Identity Management and User Authentication

For example, the proper authentication and authorization of bots and users accessing platforms that conduct automation business are important to cloud RPA security. Cloud RPA platforms integrate with enterprise IAM solutions, such as Microsoft Active Directory or SS0-based systems, to streamline user authentication and manage credentials securely (Singh, 2023).

MFA integrates an additional security layer as users have to obtain one or more verification factors that can be both a password and a security token to access the cloud RPA system. This renders the probability of unauthorized access highly negligible, especially in sensitive operations or where high-privilege accounts tend to dominate. Additionally, the frequency of rotation of these credentials coupled with the use of encryption keys enhances the identity management activities so that both users and the bot uphold the highest standards of authentication protocols (Tang, 2023).

4. Bot Development and Operations Security

In a cloud RPA environment, secure bot development is very critical so that the doors to security vulnerabilities are not opened. Cloud RPA has guidelines and tools for building bots according to security best practices, which involve rules for secure coding standards, vulnerability scanning, and automated testing. From the development lifecycle of the bot, organizations can ensure that the common security risks are not present there, such as injection attacks, data leaks, or unauthorized access.

Moreover, the use of bots in the cloud has to be continually monitored and audited to identify and act on potential security threats. Most cloud-based RPA platforms have in-built monitoring tools, which track bot activities and performances in real-time and anomalies as well. Organizations can also configure alerting for suspicious activities such as suspicious bot activity or anomalies in data access patterns to rapidly respond to a security breach (Maalla, 2019).

5. Disaster Recovery and Business Continuity

Disaster Recovery and Business Continuity. Cloud RPA security is critical to involve disaster recovery and business continuity. Most of the cloud RPA-based platforms offer full backup and recovery capabilities that ensure even after the failure of the system due to natural disasters and cyberattacks, automation workflows can continue running. Automating the backup in most of the related systems stores data and automation settings in multi geographically distributed data centers that significantly reduce the risk of losing data (Axmann & Harmoko, 2020).

During any disruptions, cloud RPA platforms rapidly get the bots up and running again, thus minimizing downtime and ensuring business continuity. Organisations heavily dependent on automation for mission-critical processes such as financial transactions, customer service, or supply chain management need such a level of resilience.

6. Continuous Security Updates and Patch Management

Continuous updates coupled with security patches are hence received automatically from cloud service providers for Cloud RPA solutions, meaning the patch applying manual intervention on the part of the local IT team is not required for on-premise systems. Continuous enhancement in security measures to combat changing threats is made on cloud-based RPA solutions. This makes sure that the environment remains secure, and vulnerabilities are addressed in due course of time (Sisodiya et al., 2024).

Such providers are likely to adopt the shared responsibility model, since cloud infrastructure security would be the responsibility of the provider, while automation processes and data security belong to the organization. There is an added advantage in terms of risk mitigation against cyberattacks like ransomware, phishing, or DoS attacks by being at the latest security patches on the platform (Jha et al., 2021).

Naturally, the security and compliance aspect would figure high in cloud RPA implementations, due to sensitive data and processes involved. Cloud RPA platforms ensure a secure environment through robust encryption, access controls, data privacy regulation compliances, identity management, secure bot development, disaster recovery besides continuous updates. This allows organizations to scale up RPA solutions with confidence while having protected data so that organizations do not face compliance regulatory issues by using security best practices and the security features offered by cloud service providers (Kunduru, 2023).

CHALLENGES AND LIMITATIONS OF CLOUD IN RPA

Though cloud RPA offers flexibility, scalability, and cost-effectiveness, it has challenges and limitations. Organizations choosing cloud RPA need to navigate through technical, operational, and regulatory issues that could impinge on the effectiveness and security of automation initiatives. The present section discusses key challenges faced with cloud RPA and how these affect the implementation and operation processes.

1. Data Privacy and Compliance Concerns

Depending on the nature of data that a company generally deals with, such data may be more sensitive and thus more secure. Organisations that deal with sensitive and confidential information in finance and health services and government sectors are much in need. Cloud environments are generally considered a target for hackers and data breaches as data is generally stored and transmitted (Madakam et al., 2019).

Organizations spread across the jurisdictions add complexity as data is associated with different regional laws-for example, GDPR in the EU or the CCPA in California, U.S. It adds additional measures in terms of storage, access, and handling of personal data. Proper compliance in the cloud-based RPA environment entails monitoring, localization of data, and safe manipulation of data throughout the automation lifecycle.

Another concern that organizations have is data sovereignty, where data needs to be stored and processed only in specific geographic locations. Cloud providers have to provide region-specific data centers as well as given assurances about the location of storage, thereby making it a complex compliance issue for businesses with a global presence (Douglas & Roger, 2024).

2. Dependency on Cloud Providers

Another big limitation of cloud RPA is lock-in with the cloud service providers, both in terms of infrastructure and support. When these platforms have scalable and maintenance capabilities pretty much automatically, reliance on third-party vendors for critical business operations becomes a challenge when a cloud provider has a downtime, technical problems, or operational issues that directly affect the availability of its product (Lo et al., 2024).

Cloud service contracts typically lock an organization to one supplier and tend not to easily migrate to a different provider, much less, switch from a cloud based to on premise. Vendor lock-in can also limit the flexibility of an organization if services prove too expensive or for any other reason are not appropriate for future business. An organization must consider what services the selected cloud service provider offers, uptime guarantees, and future long-term viability to limit its dependence on a particular provider (Koppisetti, 2024).

3. Latency in Performance and Connectivity Problems

Latency in performance and connectivity problems are some of the main issues that cloud-based RPA systems are likely to encounter. This is because performance largely depends on stable internet connectivity, which may not always be the case. Therefore, poor or unstable internet connections can result in latency in performance by causing the bots to delay their ability to complete the assigned tasks. In most cases, where there is a real-time or high-speed process, slight delays can cause bottlenecks or stop the associated business operations (Manukonda, 2024).

Apps requiring great automation performance-such as finance or e-commerce-will suffer significantly at the bar of performance latency and will lose efficiency in the bot from handling big quantities of transactions or queries from customers. In others, internet infrastructure may be unreliable and not very fast enough to support cloud-based automation, which will present some challenges to maintain bot performance in some locations (Moderno et al., 2024).

Though the cloud providers are investing in upgrading network infrastructures as well as providing solutions to reduce latency, the organizations need to have robust internet connectivity that can resist the challenges posed by cloud RPA. In fact, where applicable, organisations might even consider hybrid approaches to mitigate the performance-related limitations (Kurylets & Goranin, 2024).

4. Security Vulnerabilities

Despite the very robust security measures carried out by leading cloud providers, cloud environments remain vulnerable to forms of cyber threats. Open platforms of the clouds allow several users to share resources, which can make them vulnerable to automation systems due to unauthorized access, data leaks, or malicious attacks.

In cloud-based RPA, sensitive information handled by bots includes customer data, financial records, and PII. This feature, in specific, makes the bots become prime targets for hackers. The only thing to be done is implementing access control policies and encryption standards along with continuous monitoring by organizations in order to definitely ensure minimized risk breaches. However, securing a distributed cloud environment is complex in nature, which may sometimes inconvenience businesses that do not have a dedicated cybersecurity team.

Furthermore, because cloud RPA systems operate centrally, remote access for bot developers, managers, and users must be secure, requiring measures like multi-factor authentication (MFA) and a good VPN. Weakness in these aspects can easily let the system fall under external attacks, weakening the reliability of cloud-based RPA (Nakić & Boban, 2024).

5. Challenges of Integrations

Integration with legacy or pre-existing systems is quite challenging with cloud RPA platforms. Most of the organizations rely on traditional infrastructure or enterprise software like ERP and CRM systems, which are not native to the cloud; thus, they cannot readily be compatible with automation tools based on the cloud. In many cases, it demands additional developments, custom APIs, or middleware, further adding to the cost and time for implementation.

Compatibility issues could also come from the integration of third-party services, applications, or databases when cloud-based RPA has systems under constant update or change. These failures and conflicts might heighten risks to make automated workflows not reliable or consistent (Alirezazadeh, 2020).

Organizations will have to spend a lot of time and resources developing an integration strategy that builds the least amount of friction necessary to ensure smooth communication from the cloud RPA system to its technology stack. This challenge is particularly strong for large organizations with complex IT ecosystems and demands careful planning as well as continuous support if disruptions are to be avoided (Alirezazadeh & Alexandre, 2020).

6. Cost Management

As cloud RPA provides comfort in flexibility and scalability, managing the cost that comes along with it is a matter of great worry once the number of bots and automated processes increases in scale. On-premise infrastructure tends to have costs pretty steady, while cloud services adopt a pay-as-you-go model. Organizations can be faced with escalating costs associated with storage, processing power, and data transfer as automation initiatives grow.

An unexpected usage surge-like those often faced during high-demand periods-will incur significantly higher expense, further hindering budget and forecasting. It is also less apparent expenses that may cause issues for control, such as data egress pricing or moving data out of the cloud. An essential strategy to avoid such issues is close monitoring of the usage of cloud services with optimal use of workload optimization, selection of appropriate pricing models on clouds, and auto-scaling to avoid unnecessary expense (Kehoe et al., 2015).

Benefits and advantages: added flexibility and scalability of cloud-based RPA are tremendous. Organizations must, however, prepare properly for challenges and limitations associated with it. Some such issues are data privacy concerns, dependency on cloud providers, performance latency, security vulnerabilities, integration complexities, and cost management as some significant factors that affect the successful implementation of cloud RPA deployments. Proper planning combined with proper security measures and strategic partnerships with the cloud providers will address some of these challenges, thus helping organizations fully utilize the capability of cloud RPA by minimizing risks and optimizing performance (Kunduru, 2023).

EMERGING TRENDS IN CLOUD-ENABLED RPA

As cloud computing and RPA unfold, even newer trends are emerging that promise to bring efficiency, scalability, and intelligence into the realm of automation processes. Unfolding frontiers in artificial intelligence and machine learning will hence continue to avail discoveries and advancements in cloud technology, which would make cloud-enabled RPA even more integrated, intelligent, and transformational. The section elaborates on key future trends that will shape the cloud-RPA landscape in the coming years (Mohanarajah et al., 2015).

1. AI-Augmented RPA for Cognitive Automation

Perhaps most eagerly awaited is adding artificial intelligence to envelopes of cloud-enabled RPA platforms. Traditional RPA automates rule-based, repetitive processes, whereas AI-enhanced RPA—also called cognitive automation—is trying to address more complex processes that include some form of decision-making and understanding unstructured data. It will change the applications toward image recognition, natural language processing, and sentiment analysis, which means capabilities of automation will spread much more broadly (Singh, 2023).

In the future, the AI algorithms incorporated in cloud-based RPA platforms will pattern analyze, predict, and simulate new scenarios autonomously, so bots will start performing tasks that involve processing customer support tickets, legal document analysis, or advanced data analytics. In other words, human interference will be minimized for high-volume, data-driven tasks.

2. Hybrid Cloud and Multi-Cloud RPA Deployments

Hybrid cloud and multi-cloud environments are coming on stream as well, which provides organizations with a more flexible option on how they would like to deploy RPA. With a hybrid cloud environment, organizations will be able to tap both public and private infrastructures as would be needed with sensitivity of data or special performance requirements for the RPA process. Similarly, multi-cloud strategies involve the utilization of multiple cloud providers allowing organizations to optimize on costs and reduce the occurrence of vendor lock-in and also ensure system redundancy (Tang, 2023).

The real adoption of these hybrid and multi-cloud models will surface in future cloud-enabled RPA. Such a trend would enable organisations to smoothly orchestrate automation processes in various cloud environments, showing greater resiliency and customization. For example, highly sensitive processes demanding high data protection would be run on private clouds, whereas public cloud resources would be utilized for less critical tasks to come both in security as well as in cost efficiency (Maalla, 2019).

3. More Intense Focus on Security and Compliance

Cloud-enabled RPA will surge ahead. The more pervasive it becomes, the more attention security and compliance will gain as organizational efforts; with increased regulations coupled with the data privacy concerns, expectations from cloud RPA vendors in relation to security protocols and compliance will rise. Soon enough, RPA platforms will be wired with advanced security features including blockchain validation in order to safely validate data, automate compliance monitoring and advanced encryption mechanisms.

The cloud RPA platforms will also integrate artificial intelligence-driven cybersecurity systems to detect threats in real-time and mitigate them. Anomaly detection by incorporating machine learning-based user behavior analysis will make cloud RPA much stronger with cyber-attacks, as the sensitive data in industries like health care, finance, and governments will be dealt with more securely (Axmann & Harmoko, 2020).

4. Low-Code/No-Code Platforms for Citizen Developers

Low-code and no-code development platforms will alter how organizations manage automation. These are products that would enable citizen developers, or non-technical users, to make, modify, and deploy RPA bots with minimal coding skills. Moving forward, this characteristic of cloud-enabled RPA solutions will also promote low-code/no-code capabilities so that more employees in various departments of an organization can avail themselves of automation.

This way automation is democratized, and business users will be given permission to design as well as develop bots to cater to their needs without consulting the IT teams. Platforms for cloud RPA would become even more intuitive, with drag-and-drop interfaces and templates pre-built with a set of AI-powered bot suggestions that would make development easier and less clumsy (Sisodiya et al., 2024).

5. Io-T Integration

While IoT evolves, one of the leading trends would be integration with cloud-enabled RPA and IoT. Here, the benefits of these bots automatically tracking and managing connected devices through predictive maintenance, data collection, and even remote monitoring cannot be overlooked. So, those big industrial establishments like manufacturing, logistics, and healthcare, particularly those that are operating IoT in a large number, would greatly enhance their operations (Kunduru, 2023).

For instance, RPA bots can automatically send in requests for maintenance workflows according to the data from IoT sensors, thus reducing the idle time and thus increasing operational efficiency. Cloud-based RPA systems are likely going to process real-time data streaming from IoT devices, and such an approach would be dynamic and responsive to any automation needs, thus enabling organizations to optimize their operations and cut their costs.

6. Hyper-Automation and End-to-End Process Automation

This refers to the integration of multiple automation tools and technologies, across multiple applications and processes, for end-to-end process automation. The momentum here is building up, especially for future cloud-enabled RPA platforms that would be increasingly based on hyper-automation frameworks. RPA will gain mainstream popularity now of its confluence with AI, ML, process mining, and other intelligent automation technologies toward creating completely automated workflows (Jha et al., 2021).

Such a trend will enable automation not only of individual activities but also entire processes, source to destination. For example, the loan approval process within a financial services organization can be fully automated-both the collation of customer information and credit checks, loan document generation, and fund disbursement-by deploying cloud infrastructure. By way of such cloud infrastructure, end-to-end processes are scaled and optimized in real time, so there is greater efficiency as well as cost savings.

7. RPA as a Service (RPAaaS)

Another trend that will define the future developments of cloud-enabled RPA is RPA as a Service, or RPAaaS. In keeping with agile and scalable automation, organizations need more opportunities that are going to be established through RPAaaS solutions. It lets the businesses consume RPA on a subscription or on a pay-per-use basis, especially within this model of SaaS (Madakam et al., 2019).

This would obviate the initial investment that would be required in infrastructure and give the organization flexibility to scale its automation as needed. Future RPAaaS would increasingly allow more customization so that business would be allowed to choose a certain module of automation tools and capabilities rather than implementing them in batch, thus making RPA accessible to small and medium-sized enterprises.

Cloud-enabled RPA is probably the most significant thing revolutionizing the future, and its base will be formed by AI, IoT, hyperautomation, and growth in low-code/no-code platforms. While security and regulatory compliance continue to stay at the top of agendas, stronger and more intelligent solutions would be sought to protect the data as well as maintain regulatory compliance. Hybrid and multi-cloud deployments, RPAaaS, and AI-enhanced automation would be some of the trends that businesses will embrace to unlock new levels of efficiency, scalability, and flexibility in their automation journeys. Innovation that is set to redefine how organizations will automate and optimize their operations holds a bright future pipeline in cloud-enabled RPA (Douglas & Roger, 2024).

IMPLICATION OF CLOUD-ENABLED RPA

Cloud computing with RPA brings far-reaching implications practically, managerially, ethically, and socially. The following sub-section discusses how cloud-enabled RPA impacts business and society in diverse ways by detailing its opportunities and challenges.

Practical Implications

The integration of cloud computing with RPA gives organizations several practical benefits: it streamlines processes, reduces operational costs, and boosts scalability. One of the most visible practical advantages is rapid and costless roll-out without massive investments in infrastructure. Another benefit is scalability on demand with flexibility and agility among businesses because of cloud-based platforms; automation efforts can be scaled up. There is also cloud-enabled RPA support for remote work. Remote work is today extremely important because of distributed and hybrid work environments (Koppisetti, 2024).

However, practical problems come with a need for reliable connectivity of the internet and potential latency issues when working with large sets of data or in automation processes of tasks in real-time. In organizational terms, how to integrate RPA with the existing legacy systems is an issue since the smooth interoperability of cloud-based RPA tools with on-premise infrastructure may be complex, requiring more planning.

Managerial Implication

Cloud-enabled RPA, therefore, has mixed implications for managers. While it means that the more routine repetitive tasks automated would require less resource allocation and allow employees' time to be devoted to higher value work, this in turn increases productivity by adding a potential for innovations as more time will be found in productive, creative, and strategic endeavors. Cloud-based RPA also enables the monitoring of processes in real time, which allows managers to better optimize decision-making and improve openness in operations (Manukonda, 2024).

However, it also forces managers to deal with the concerns of change management. RPA can disrupt the agility and comfort of working, and not all employees may be prepared for the same. Introducing bots to an individual's system or any worker's system may also resist as workers may fear losing their jobs due to full automation. The message that the managers need to send to the workers is that of value addition through RPA as an augmentation tool rather than a replacement (Lo et al., 2024). Managers need to ensure that the workers get enough training for collaboration with the bots. The other issues that management needs to address include proper relations with the vendors of cloud services, better control over its costs, and the management of service level agreements.

Ethical Considerations

The ethical implications run deep, especially on areas related to data privacy, employment, and accountability. High up this list is the loss of jobs through automation, such as where roles become entirely unnecessary, for example, if tasks are routine. In all ways, organizations would bear moral responsibility in workforce transition with reskilling and upskilling opportunities for displaced employees, preparing them for new roles that will leverage human creativity and problem-solving capabilities (Moderno et al., 2024).

Another very important ethical issue involves data privacy, due to the reality of cloud-enabled RPA processing huge sensitive information volumes. Organizations must demonstrate compliance and maintain standards with the regulations of data protection, such as GDPR, while also putting appropriate encryption and security in the handling of customer and employee data. Another significant consideration is accountability since automated systems make decisions that could potentially drive business operations and customer experiences. Undoubtedly, the decision-making algorithms for RPA must be properly transparent to avoid such consequences and biases while remaining human in control (Tiron-Tudor et al., 2024).

Social Implication

At a broader level, cloud-enabled RPA may lead to socially positive outcomes. For example, such automation can cause an increase in efficiency in public services and healthcare. For instance, governmental agencies can utilize the tool to automate routine administrative work and provide more time to workers who can better serve the community. Similarly, in the healthcare organization, RPA can automate the management of patient data, schedules the appointments, and billing processes while improving the quality and accessibility of service (Kurylets & Goranin, 2024).

However, the social consequences of automation have also been disconcerting and scary as this automates the economy of the business with serious worries on the widening of inequalities. For those business organizations that don't have access to the much-needed technology and skills, such businesses are going to leave them far behind with further worsening of the digital divide. This could be more crucial in developing countries or marginalized communities where even the technological infrastructure and resources can be grossly limited. To reduce these disparities, it is important to market equal opportunities in gaining access to technology so that all members of society will benefit from automation through education and training (Nakić & Boban, 2024).

With cloud-enabled RPA redefining work, the social considerations of the future of work are of vital importance. Automation will provide higher productivity but will shift employment patterns toward more freelancing and gig-based jobs. The challenge would lie in ensuring that the workers have social protections and fair labor conditions in an increasingly automated economy for policymakers as well as for businesses.

The implications of cloud-enabled RPA are very vast and go from the practical, managerial, ethical, and social to the most serious level. While cloud-RPA offers a lot in terms of scalability, efficiency, and flexibility, it also brings along serious questions about job displacement and the issue of data privacy and inequality. In order to respond to these implications wisely, organizations must take up responsible automation, bring in decision-making that is ethical, and further strive for technological advancement, looking at the well-being of employees and societies as a whole. With these in consideration, a cloud-enabled RPA can be leveraged as a transformational lever for innovation and growth while ensuring social equity and ethical accountability (Alirezazadeh, 2020).

CONCLUSION

The landscape of business automation has changed and is given unprecedented opportunities with the integration of Robotic Process Automation with Cloud Computing, upon which organizations have a chance to improve efficiency, scalability, and agility. As discussed in this chapter, cloud-enabled RPA enables businesses to automate more than they would ever imagine is possible. This promotes cost savings and frees more human time for non-repetitive and more vital tasks. Such a transformation makes the employees more productive yet enables them to focus on more valuable tasks that help to create an organization's culture of innovation and creativity.

Though cloud-enabled RPA is going to start appearing in mainstream use cases, along with its problems. The management of the infrastructure for cloud and the potential integration with other existing systems can cause some troubles for them. Moreover, org concerns related to security and compliance and ethical boundaries, such as job replacement and accountability for data privacy, will also need attention. With that said, companies have obligations toward ethical automation and settling workers well into the transition as well as being transparent about decisions made via automation.

Therefore, the opportunities look bright for cloud-enabled RPA into the future by how sophisticated changes in AI, ML, and IoT will soon become. Its hybrid and multi-cloud environment accompanied by low-code and no-code development platforms will democratize automation by its usability to various industry-specific users. However, organizations must be cautious about the social impacts of such trends to prevent the further spreading and widening of an already created gap or introducing new dilemmas among the workforce.

In all, cloud-enabled RPA is a significant evolution in the phase of business automation. With the adoption of its potential but keeping in mind challenges and ethical dilemmas it brings with itself, organizations can utilize cloud-based RPA for the transition of change. As organizations navigate these shifting waters, they must orchestrate a culture of constant learning, adaptability, and responsible automation to realize the full potential of cloud-enabled RPA and position themselves for a successful future. What will happen is that journey toward a new era of cloud-based automation has already begun, and it's the organisations prioritizing ethical practices, employee well-being, and social responsibility that will thrive in this new era of work.

REFERENCES

Alirezazadeh, S. (2020). Optimal algorithm allocation for single robot cloud systems. *arXiv*. https://doi.org//arxiv.2003.08683DOI: 10.48550

Alirezazadeh, S., & Alexandre, L. (2020). Improving makespan in dynamic task scheduling for cloud robotic systems with time window constraints. *arXiv*. https://doi.org//arxiv.2012.03555DOI: 10.48550

Axmann, B., & Harmoko, H. (2020, September). Robotic process automation: An overview and comparison to other technology in industry 4.0. In *2020 10th International Conference on Advanced Computer Information Technologies (ACIT)* (pp. 559-562). IEEE.

Douglas, R., & Roger, J. (2024). The era of intelligent automation: Cloud computing, AI, and beyond. *International Journal of Advanced Engineering Technologies and Innovations*, *1*(1), 390–414.

Jha, N., Prashar, D., & Nagpal, A. (2021). Combining artificial intelligence with robotic process automation—An intelligent automation approach. In *Deep learning and big data for intelligent transportation: Enabling technologies and future trends* (pp. 245-264).

Kehoe, B., Patil, S., Abbeel, P., & Goldberg, K. (2015). A survey of research on cloud robotics and automation. *IEEE Transactions on Automation Science and Engineering*, *12*(2), 398–409. DOI: 10.1109/TASE.2014.2376492

Koppisetti, V. S. K. (2024). Robotic process automation: Streamlining operations in the digital era. *ESP International Journal of Advancements in Computational Technology (ESP-IJACT)*, *2*(2), 74-81.

Kunduru, A. (2023). Cloud BPM application (Appian) robotic process automation capabilities. *Asian Journal of Research in Computer Science*, *16*(3), 267–280. DOI: 10.9734/ajrcos/2023/v16i3361

Kurylets, A., & Goranin, N. (2024). Design and implementation of a UMLRPAsec-extension for robotic process automation. *Electronics (Basel)*, *13*(4), 769. DOI: 10.3390/electronics13040769

Lo, W., Yang, C. M., Zhang, Q., & Li, M. (2024). Increased productivity and reduced waste with robotic process automation and generative AI-powered IoE services. *Journal of Web Engineering*, *23*(1), 53–87. DOI: 10.13052/jwe1540-9589.2313

Maalla, A. (2019, December). Development prospect and application feasibility analysis of robotic process automation. In *2019 IEEE 4th Advanced Information Technology,Electronic and Automation Control Conference (IAEAC)* (Vol. 1, pp. 2714-2717). IEEE. DOI: 10.1109/IAEAC47372.2019.8997983

Madakam, S., Holmukhe, R. M., & Jaiswal, D. K. (2019). The future digital workforce: Robotic process automation (RPA). *Journal of Information Systems and Technology Management, 16*, e201916001. DOI: 10.4301/S1807-1775201916001

Manukonda, K. R. R. (2024). Leveraging robotic process automation (RPA) for end-to-end testing in agile and DevOps environments: A comparative study. *Journal of Artificial Intelligence & Cloud Computing, SRC/JAICC-334*(3), 315, 2-5. https://doi.org/DOI: 10.47363/JAICC/2024

Moderno, O. B. D. S., Braz, A. C., & Nascimento, P. T. D. S. (2024). Robotic process automation and artificial intelligence capabilities driving digital strategy: A resource-based view. *Business Process Management Journal, 30*(1), 105–134. DOI: 10.1108/BPMJ-08-2022-0409

Mohanarajah, G., Hunziker, D., D'Andrea, R., & Waibel, M. (2015). Rapyuta: A cloud robotics platform. *IEEE Transactions on Automation Science and Engineering, 12*(2), 481–493. DOI: 10.1109/TASE.2014.2329556

Nakić, J., & Boban, D. (2024). Enhancing informal education with robotic process automation: A method for automating course search on e-learning platforms. *Journal of Communications Software and Systems, 20*(2), 215–225. DOI: 10.24138/jcomss-2023-0178

Singh, K., & Singh, P. D. (2023). Machine learning in robotics with fog/cloud computing and IoT. *EAI Endorsed Transactions on AI and Robotics, 2*. Advance online publication. DOI: 10.4108/airo.3621

Sisodiya, M., Prabhu, S., & Ashok, P. (2024, January). The transformative power of robotic process automation: Redefining business in industry 4.0. In *2024 14th International Conference on Cloud Computing, Data Science & Engineering (Confluence)* (pp. 498-503). IEEE.

Tang, H. (2023). Task scheduling strategy of logistics cloud robot based on edge computing. *Research Square*. https://doi.org/DOI: 10.21203/rs.3.rs-2423689/v1

Tiron-Tudor, A., Lacurezeanu, R., Bresfelean, V. P., & Dontu, A. N. (2024). Perspectives on how robotic process automation is transforming accounting and auditing services. *Accounting Perspectives, 23*(1), 7–38. DOI: 10.1111/1911-3838.12351

Chapter 8
Intelligence in RPA

Minakshi
https://orcid.org/0000-0001-7586-6309
King Khalid University, Saudi Arabia

Tarun Kumar
https://orcid.org/0000-0002-8507-8522
Galgotias University, India

Suman Devi
https://orcid.org/0000-0002-4038-7196
Galgotias University, India

ABSTRACT

What started as a solution to automate repetitive, rules-based tasks has expanded into RPA, incorporating artificial intelligence technologies such as machine learning, natural language processing, and computer vision, and promoting a move to intelligent process automation. This integration allows RPA to discern unstructured data, make contextual judgments, and learn and change. Conversational AI enables end-to-end process automation, ensuring human-robot cooperation, while artificial intelligence-driven analytics ensure process optimization. With the platform's increasing popularity, responsible AI and governance frameworks have become critical for ensuring circulating, fair, and ethical use. Although autonomous process estimation, self-optimizing processes, and a few other AI movements, such as automated reasoning and scheduling, are possible in the future, AI adoption in stages enables the greatest automation and excellence in action.

DOI: 10.4018/979-8-3693-4365-4.ch008

1. INTRODUCTION

Business processes are becoming more developed in light of the constantly evolving landscape. In this context, organizations are striving to come up with options to enhance their supply chains and overcome the existing challenges. RPA is a new business tool that has been widely used by companies in recent years, which has provided them with the opportunities to succeed with the help of the possibilities to improve business processes and ensure operational efficiency.

According to the modern understanding of RPA, this new business tool can perform as well as robots. Today's RPA performs tasks in a manner consistent with its programming, making it a versatile tool of great importance to users (Albus, 1994). Furthermore, RPA proves beneficial for a variety of human tasks, allowing for precise context determination. Businesses often use RPA to aid in areas such as finance, accounting, and shipping. This concept provides the appropriate tool to enhance both profits and service quality. Otherwise stated, RPAs are about to systematically overcome existing obstacles and challenges. This fact suggests that the opportunities for enhancing operational efficiency are limitless.

Simultaneously, AI develops projections and utilizes other machine learning tools to optimize operational and business processes. In other words, AI uses RPA to perform tasks, thereby automatically freeing up human resources (Pedretti et al., 2021). It opens vast opportunities for both small and large businesses to benefit from automated processes, which can significantly strengthen their standing in the business world. There are numerous reasons for business organizations to use RPAs to enhance operational efficiency and performance.

Nevertheless, as technologies evolve, simple task automation is no longer sufficient for increased competition in the data-driven, dynamic world. RPAs should integrate intelligence to automate processes, adapt to new data and conditions, and make decisions based on the data and collected insights (Chakraborti et al., 2020). To clarify, intelligence in RPA sets the limits between purely rule-based processes and endless opportunities when RPA systems are learning on their own and are able to optimize the outcomes based on the data and information. It is true to state that the integration of intelligence into RPAs makes a difference and revolutionizes business processes, making them more predictable and secure in terms of data and applied procedures (Shidaganti et al., 2021). The use of AI and machine-learning algorithms makes operations more flexible while preserving high proficiency and quality. The combination of RPAs and intelligence is beneficial for business organizations because it provides them with a greater number of opportunities to enhance effectiveness and automate the processes necessary with the least amount of human intervention.

Intelligent RPA is based on the synergy of the latest technologies and methodologies, all of which contribute to the advancement of automated process capabilities (Mohamed et al., 2022). First, natural language processing, or NLP, ensures intuitive interactions between humans and machines by enabling the bots to receive instructions through natural language. Hence, the NLP technology facilitates the delegation of tasks and enhances the overall ease of use. Second, computer vision, also known as CV, enables RPA-powered solutions to process and assess visual data. This means that tasks that were previously "unautomatizable" due to manual processing, such as photographing or shooting, transferring, and typing, now become feasible for RPA solutions (Siegert & Völker, 2021). The next technology, which attests to the substantial involvement of artificial intelligence (AI) and machine learning (ML) algorithms, is the capability of easy readjustment and learning from past instances and changes. As the data on the requirements, environment, and peculiarities of certain processes grows, RPA solutions can analyze it in terms of its experiences and history, then enrich their algorithms to refine the ratio of success.

As companies advance their digital transformation, intelligent RPA offers numerous opportunities across various industries and sectors (Ling et al., 2020). For example, in customer service, intelligent RPA powers chatbots and virtual assistants that can be available 24/7. Intelligent chatbots not only boost customer satisfaction by tailoring their services to individual consumers, but also automate a range of non-standard tasks, thereby reducing operational costs for the company (Anagnoste, 2018). Accounting and finance tools utilize intelligent RPA to automatically evaluate bank statements, identify invoices, enhance precision between actual data and the company's books and records, and detect fraud. Integration is also useful in supply chain and transportation, where algorithms based on real-time data predict the most beneficial levels of inventory, routing, and logistics. Finally, in healthcare, intelligent RPA can automate permanent administrative jobs, simplify issues related to data entry, processing, and retrieval, and, with the help of AI systems, even diagnose diseases or refine possible treatment plans (Wang et al., 2021).

As companies undergo digital transformation, intelligence in RPA is inevitable and represents a strategic advantage. The purpose of this study is to introduce and decipher intelligent RPA, revealing the ever-growing trends and technologies that make automated processes function at a completely new level. Through inspiring insights and chapters created by influencers and representatives of the industry, the book reveals the opportunities and challenges associated with RPA intelligence through inspiring insights and presents innovative ways to use new technology in practice. Thus, readers will get an opportunity to make a significant difference within a turbulent business environment and create the grounds for further progress.

Figure 1. Robotic process automation

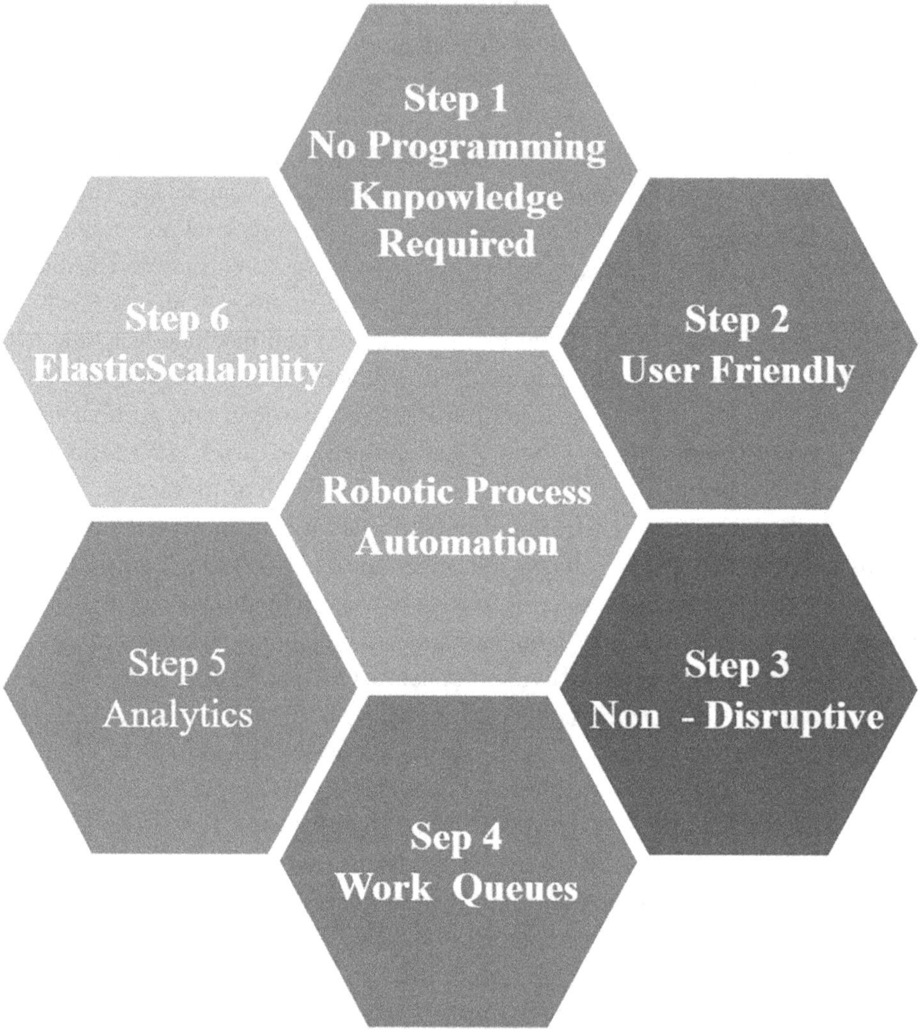

2. BACKGROUND INFORMATION AND RELATED WORK

Robotic Process Automation (RPA) is a technology that comprises the use of software robots or "bots" to automate a majority of highly repetitive, routine tasks that are usually performed by people. Such RPA technology is built to imitate human interactions with digital systems. To be more exact, "bots" can be used to perform data entries and process transactions. Moreover, this technology allows employees to respond to simple customer services questions and inquiries without having to interact with customers directly.

2.1 Evolution of RPA

While primitive robotic process automation merely executed predefined workflows in a rigid, rule-based manner, lacking adaptability, newer forms of intelligent process automation leverage machine learning and artificial intelligence. By training algorithms on immense troves of data, cognitive RPA systems can now comprehend unstructured information and constantly evolve their decision-making according to emerging patterns. This enhancement allows automated solutions to respond sensitively to atypical circumstances and efficiently handle unpredictable variances, displacing rigid constraints and enhancing organizational nimbleness.

2.2 Integration of AI with RPA

Artificial intelligence has profoundly augmented the capabilities of robotic process automation by endowing systems with human-like faculties of perception. Machine learning imports the adaptive, experiential prowess of the human brain by enabling automated platforms to perceive nuances in uncodified language and visual media. Natural language processing and computer vision furnish cognitive automation with the means to derive context and insight from amorphous sources of information, just as people intuitively comprehend uncategorized stimuli. Armed with these evolved cognitive functions, intelligent process automation possesses a limber, intuitive strength surpassing the original rigidity of rule-based automation.

2.3 Related Work

Traditional robotic process automation focused on straightforward, repetitive tasks that followed established rules and procedures. While effective for structured activities, rule-based RPA had limitations in dealing with unstructured data and adapting to changes(Aguirre & Rodriguez, 2017).

2.3.1 Advances in AI and Machine Learning

Advances in artificial intelligence and machine learning have transformed many domains by enabling adaptive systems trained on data rather than programmed instruction. In the area of RPA, AI and ML augment bots' capabilities to make sense of emails, images, and other unstructured content as well as accomplish more sophisticated objectives. Studies by (Brynjolfsson & Mcafee, 2017) underscore AI's potential to automate cognitive work.

2.3.2 Natural Language Processing (NLP)

Natural language processing allows machines to comprehend and respond to human speech and writing. Within RPA contexts, NLP can automate tasks involving textual inputs and outputs like processing customer correspondence or generating reports. Recent studies highlighted demonstrate NLP's effectiveness in improving human-machine communication, resulting in more intuitive and user-friendly RPA systems.

2.3.3 Computer Vision

Computer vision gives machines sight through computational analysis of visual inputs. In RPA, computer vision can extract data from scanned papers and pictures or detect features, as research by (Luo & Luo, 2021)illustrates. Integrating such technology with RPA significantly expands the range of inputs bots can handle and the complexity of processes they can assume.

2.3.4 Case Studies and Applications

Finance and Accounting: Research by (Villar & Khan, 2021) shows how Intelligent RPA streamlines multifaceted tasks like deceit detection, bookkeeping techniques, and budgetary disclosures, guiding to amplified exactness and competency of clerical assignments.

2.3.5 Healthcare

In medicinal services, Intelligent RPA empowers responsibilities for example patient information administration, meeting scheduling, and protection cases handling. AI-fueled RPA aids in decreasing authoritative weight and upgrading person consideration through computerized streamlining of customary assignments and innovative health assessments. .

2.3.6 Supply Chain Management

Investigations have researched the utilization of Intelligent RPA in inventory network administration, where it encourages request conjecture, stockpiling the board, and idealizing gracefully the board through computerized coordination of worldwide tasks and information examination (Ivanov, 2021).

2.3.7 Challenges and Opportunities

Indeed, the integration of AI with RPA offers multiple benefits, but it also comprises numerous challenges. The most vivid of them include:

a) Data privacy and security – it is critical to ensure the security of the sensitive data that is used and transmitted by the AI-enabled RPA.
b) Interoperability with existing systems – due to the peculiarities of some existing systems that might be outdated, the integration of AI with them can be rather complicated. Any RPA system, intelligent or not, would need to perfectly fit the technological infrastructure used in a specific organization.
c) Learning and adaptation capabilities – since it is difficult to design an intelligent RPA system that would perform tasks requiring learning and adaptation without unexpected interventions, this area of work should be significantly expanded. Further research perspectives include the development of measures to ensure data security, improvement of the RPA systems' scalability, and an increase in the interoperability of AI and RPA.

3. THE CONVERGENCE OF AI AND RPA

Artificial intelligence is a broad field of technologies and techniques formulated to allow machines to perform intelligent activities, learn from data, and solve complex issues (Szuba, 1998). As shown in Fig. 2, generative AI and robotic process automation have converged in areas such as data conversion, intelligent insights, customer-employee interactions, human-like discussions, and reshaping customer support.

Figure 2. The convergence of AI and RPA

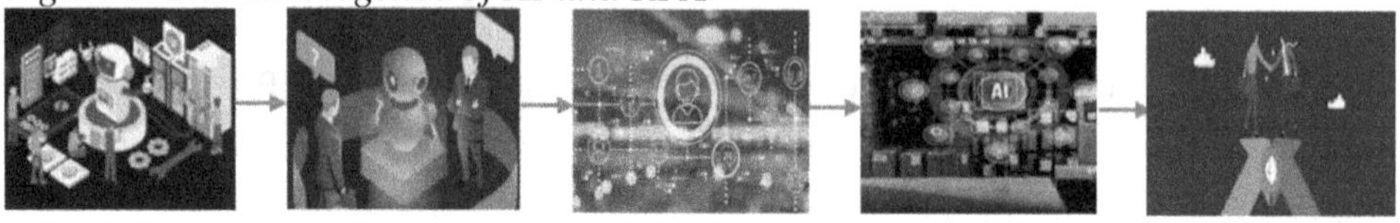

There are several key AI capabilities that contribute to the augmentation and advancement of robotic process automation, including:

1. **Machine Learning:** Developing algorithms and statistical models for machine learning enables systems to execute specific tasks without explicit programming (Paletz et al., 2018). Through AI, RPA can use techniques such as supervised, unsupervised, and reinforcement learning to identify patterns, make recommendations, and optimize a variety of processes.
2. **Natural language processing:** AI techniques can be employed for natural language processing to analyze and/or create human language. Natural language processing techniques such as text mining, sentiment analysis, and language generation allow machines to interpret and draw insights from unstructured textual data.
3. **Vision of a computer:** Computer vision technologies, thanks to artificial intelligence, can analyze, interpret, and evaluate visual image data (Szuba, 1998). This field includes technologies like object detection, image recognition, and optical character identification that RPA can use to process and interpret error-free visual data.
4. **Knowledge representation and reasoning**: It includes putting advanced reasoning into a computer so that a person can program it to behave as though it has a comprehensive understanding of a specific domain (Kant, Dixit, & Memoria, 2023).

AI and AI technologies allow RPA to process and analyze unstructured data such as spreadsheets, emails, and videos.

3.1 Making Contextual Decisions

RPA bots powered with AI techniques, such as machine learning and knowledge representation, can make decisions in context based on the patterns learned, rules, and knowledge of the domain. Therefore, AI-augmented RPA is not merely constrained to the predefined path but rather has the liberty to determine its activities based on the situation and make an "intelligent" decision.

3.2 Continuous Learning and Adaptation

Through machine learning, RPA bots can continuously enhance their performance by learning from the data they process. As they encounter more data and flow variations, the AI feature can adapt their models and make smarter decisions, leading to increased automation and accuracy ratios (Hathaway, 1982).

AI and RPA convergence have altered several business processes across many industries. Some of the most notable changes include:

1. Intelligent Document Processing: AI-driven OCR, NLP, and computer vision enable RPA to automate the extraction from unstructured documents like contracts, forms, and invoices (Hutton et al., 2001). The process resulted in end-to-end processing and the elimination of manual data capture tasks.
2. Conversational Process Automation: Conversational AI, such as chatbots and assistance, when combined with RPA, enables humans and robots to work together efficiently (Cheong & Yoon, 1999). Conversing with bots via natural language enables operational staff to execute operations, manage exemptions, and track progress.
3. Predictive Maintenance: captured data from RPA bots feeds ML models and predicts future failures (Lennon & DIRECTORATE, 2014). Organizations also use an adaptive maintenance model, whereby they decide beforehand when to change parts of the engine.
4. Fraud detection: Analytics use AI-processed data to detect and prevent fraud by training the model on the data.

These use cases are examples of the kinds of efficiency, effectiveness, and initiative to expect in almost every business operation or process with the merger of AI and RPA as intelligent automation.

4. COGNITIVE PROCESS AUTOMATION

4.1 Intelligent Process Discovery and Mining with AI Too

The integration of AI also allows automation beyond simply running predefined processes. AI can describe, analyze, and model processes on its own using novel methods to facilitate smart process automation (Hashemi, 2008).

1. **AI analyzes process data to identify opportunities for automation:** AI sorts through process-related data like system output, process analysis, and user behavior to find and understand patterns in business processes related to a-to-b operational patterns (Habibi et al., 2017). Organizations can use existing a-to-b processes to determine where the drip will feed and which chores are most suitable for automation.
2. **Automating the generation of process definitions:** By automatically establishing process characteristics and appearing to execute them on the extracted data, AI can further enhance process discovery. Applying methods like procedure mining and natural language processing, AI can extract process logic, rules, and resolution sections to create business opportunities (Witty & Lehman, 1931). Then, RPA can apply these automated process definitions to control bots, drastically simplifying traditional process mapping.

4.2 Understanding unstructured content with AI

Most of the company data is unstructured, with documents, emails, and dynamic data. In this unstructured effort, AI enables RPA to consider and assess actionable information.

1. Extracting data from text documents using OCR and computer vision. RPA can use this technique to automatically gather data for the exam, pictures, and parts (Kirkegaard & Nyborg, 2021). This enables the collection of data from every single source, allowing radical reports to be conducted without manual management or intervention.
2. Understanding Text with Natural Language Processing: In addition to structured data, AI technologies may help RPA bots understand and interpret context and meaning in unstructured text (Naglieri & Otero, 2011). Text mining, sentiment analysis, and named entity recognition are examples of NLP features that allow automation to read documents, emails, and other text-based data sources and extract useful information and insights for better decision-making and automated process execution.

4.3 AI-Powered Decision Automation

In addition to understanding unstructured content, AI technologies can also enable RPA to make intelligent decisions in the automated process.

1. Modeling Decisions from the Past: AI can develop the ability to provide actionable recommendations or outcomes by analyzing patterns in historical process data (Paik et al., 2013). This model, known in literature as a decision model, captures domain knowledge, rules, and best practices and teaches a bot how to make better decisions autonomously.
2. Turking Reasoning and Consumer Awareness: RPA now uses AI technologies like knowledge representation and reasoning, which provide reasoning or patient context to guide bot decisions more quickly and efficiently (Whitesmith, 2020). RPA bots, for example, use reasoning engines to offer firm recommendations wherever they have many options, or the context can change depending on the situation. These cognitive tools of AI help RPA pivot from a rules-based automation methodology to a cognitive automation technique, allowing them to grasp complicated data, make decisions, and then adapt and learn from the outcome (Sternberg, 1999).

5. INTELLIGENT DOCUMENT PROCESSING

Nowadays, in the digital age, organizations face an increasing number of documents of all kinds, from contracts and invoices to reports and forms. Nevertheless, document processing seems to be one of the most arduous and resource-consuming operations for all companies (Martínez-Rojas et al., 2020). On the other hand, given the rising popularity of robotic process automation, it is possible to benefit from RPA's synergy with the other two natural tools, namely, computer vision and natural language processing. Thus, by combining NLP's capabilities to interpret data with CV's capabilities to extract and process it, intelligent document processing can provide companies with numerous advantages.

First, document processing requires a variety of data operations, such as extracting, interpreting, revising, updating, and entering data into other documents. Furthermore, a single company has to deal with a plethora of document formats, including scanned images, free-form text PDFs, and even handwritten documents (Szuba, 2001). As a component of artificial intelligence, CV empowers machines to comprehend and assess visual data like images, videos, and documents, and it excels in automating the data extraction process. More specifically, we must systematically revise, update, and process the information about the company's employees, operations, clients, etc., as well as analyze the tables, forms, and other data provided in documents.

Modern CV technologies, like optical character recognition and intelligent character recognition, are also good at getting text from OCR images, which were originally used for printed documents and were later turned into images, as well as better technologies, which can read handwritten scanned documents, images with noise and distortions, and low-quality, highly compressed scans (Smith & Stanley, 1980). As such, previously laboriously processed hand-written forms, identifying objects, or completing forms using handwriting no longer require manual labor. The significance of Natural Language Processing (NLP) lies in its ability to process written language.

While CV processes content related to visual representation, how do NLP technologies interpret, understand, and process verbal information from documents and texts? NLP, or natural language processing, is the study of the computation and machine-based manipulation of written language (Katritsis, 2021). NLP is required for intelligent, efficient, and automated text processing, the extraction of important data from texts, interpretations and classifications of cryptic or non-cryptic information, and recognizing meaningful words in a word-torrent, among others. NLP extraction technologies provide start-up solutions for industries that rely on enterprise document processing solutions.

5.1 The Interaction Between CV and NLP Solutions Is Significant.

In addition to processing and understanding texts, one of NLP technologies' primary functions is entity extraction. The latter defines the task of identifying persons, organizations, locations, dates, or other entities inside the text. Accurate entity recognition would be useful for various industries, such as legal and finance. In contracts, other legal documents, risk assessments, and proper regulatory compliance documents and controls, recognizing the proper entities is of chief importance (Xie et al., 2023). Furthermore, sentiment analysis, topic modeling, and text summarization provide organizations with quick and efficient access to text readings of e-mails or other documents and information that focuses on solutions, problems, threats, etc. Finally, RPA technology enables the full spectrum of document processing technologies to be completed.

RPA bots can initiate an intelligent document processing workflow by retrieving documents from a variety of general sources, like email inboxes, or from more specific ones, like specific cases or contracts (Zhang & Wen, 2021). The next step involves routing it through CV and NLP pipelines, which perform data extraction, text recognition, and text understanding in natural language, respectively. The RPA system, capable of multiple types of action, processes and validates the insights gained from this process and the extracted data. The RPA system can populate databases, initiate approval processes, trigger alerts, create reports, and perform any other downstream process or action in the connected systems and applications, among other options. One of the key benefits of this process is its ability to create an intelligent feedback loop, which refers to the system's ability to learn (Furnham & Chamorro-Premuzic,

2004). Artificial intelligence algorithms specifically designed for machine learning underpin the learning process. The more documents the system processes, the more it learns, thereby improving its performance based on the knowledge—it becomes more accurate and efficient in extracting the necessary data or taking the needed actions. Over time, the system can adjust to the changing patterns or structures of the documents, language used, or business processes.

5.2 Applications and Use Cases

There are numerous applications of intelligent document processing because of the multiple formats of documents and the variety of document types. A prime example in the finance industry is the automated processing of invoices and contracts, where intelligent systems can extract the necessary information and perform validation based on the rules (Lillis et al., 2016). Depending on the specific type of document, the decision-making process also involves following the approval or payment process. The healthcare industry can use this process to streamline the processing of medical records, including various types of doctors' notes and forms. This method can also process handwritten prescriptions and forms (Fernández-Loría & Loría, 2022). The use cases for law firms and regulatory institutions include the processing of contracts and agreements, legal documents for due diligence, and compliance checks.

Second, applications such as human resources, where it automates the processing of resumes, job applications, and employee records, and supply chain management, where it facilitates data extraction from shipping documents, customs forms, and purchase orders, widely utilize intelligent document processing (Kant, Dixit, Memoria, et al., 2023). As businesses generate and process ever-growing amounts of unstructured data, the demand for intelligent document processing solutions is bound to increase. By leveraging the power of RPA, CV, and NLP, organizations can reach new heights of efficiency, visibility, and accuracy, helping them stay ahead of digital transformation and creating sustainable competitive advantages.

6. CONVERSATIONAL PROCESS AUTOMATION

6.1 Chatbots and Virtual Assistants Enabling Human-Robot Collaboration

There is an increasing emphasis on conversational AI technologies, including chatbots and virtual assistants, to facilitate collaboration between human workers and software robots in process automation (Khan et al., 2022). By incorporating a natural language interface in RPA, organizations can facilitate interaction between human workers and automation processes, thereby promoting human-robot interaction in the workplace. Chatbots and virtual assistants serve as intermediaries through which users may start processes, give input, and receive updates through virtual human conversations (Jędrzejka, 2019; Rai et al., 2019). Conversational interfaces use natural language processing and natural language understanding algorithms to comprehend user requests, interpret the context, and produce an appropriate response (Jędrzejka, 2019). Conversational AI makes it easier for users to communicate with automation processes using natural language because it is more human-friendly and user-friendly (Syed et al., 2020). It eliminates the need for extensive application training or immediate user interaction with complicated application interfaces.

6.2 Exceptions and Ad-Hoc Statements Handled by Conversation

Additionally, conversational AI can help the user handle exceptions and ad-hoc statements that occur while processes are running (Kedziora, 2022). When the RPA bot encounters an action that requires human intervention or judgment, it may communicate all problems using chatbots or virtual assistants. The conversational mode of addressing conditions enables the user to provide input through a user dialog without using the software interface (Ray et al., 2023). The conversational interface mediates and converts the user response into a machine-understandable condition (Czarnecki & Fettke, 2021). The conversational interaction will be context-aware, utilizing AI to allow for intellectual handling of ad-hoc statements. It improves the user experience and boosts automation utilization.

6.3 End-to-End Automation with Conversational Interface

Conversational AI not only enhances human-robot engagement and machine triggers for ad-hoc statements, but also serves as an interface to fuel end-to-end process performance (Fettke & Strohmeier, 2022). When starting the process, the user guides each step through user dialogs. A conversational interface informs RPA about the action it can take in its current state. The user triggers conversational mode instead of invoking RPA directly (Alberth & Mattern, 2017). Through conversational AI, the robot executes the corresponding steps (Murugamani et al., 2022). Human-robot collaboration through conversational AI, which handles intelligent differentiation based on user interaction, can ease user concerns while resolving any problem.

7. CONTINUOUS PROCESS OPTIMIZATION WITH AI

7.1 RPA Data

Capturing operational data from RPA robots One of the advantages of RPA is the large amount of operational data produced by the software robots performing various processes and tasks (Bellman & Göransson, 2019). This includes timestamps, system interactions, process variables, and execution logs. The data provides a rich source to analyze processes and identify potential areas for improvement (Viehhauser, 2020). When aggregated from multiple robots used in different processes and systems, it enables organizations to gain insights on the performance, bottlenecks, and points of inefficiency of the automated processes (Madakam et al., 2019). Such data forms the foundation of advanced analytics and machine learning, which are used for continuous process improvement.

7.2 Applying Analytics and Machine Learning

1. Mining and optimizing processes: Process mining is one of the data use cases, meaning the visualization and analysis of the already existing processes in the organization (Agostinelli et al., 2019). Process mining combines machine learning to identify patterns, deviations, and inefficiencies in process execution (Venigandla et al., 2023). Combining machine learning with other techniques enables organizations to gain a comprehensive understanding of their process execution. Organizations can optimize their processes based on cycle time optimization and the elimination of inefficient activities or decision points, according to the insights (Gotthardt et al., 2020). Process mining tools can also recommend a big-picture process redesign, leading to more efficient and effective processes.
2. Predictive Maintenance automation is another area where machine learning finds its application. We can use machine learning to predict potential system failures, performance degradations, or resource capacity limitations (Ribeiro et al., 2021).

Similarly, predictive maintenance also allows for proactive monitoring and intervention to reduce process disruptions and downtime (Micle et al., 2021). Machine learning models can review process system logs, resource utilization, or error messages to detect patterns that indicate an upcoming alarming signal to an issue and therefore generate alerts or smart recommendations to take action to protect that potential issue from happening.

7.3 AI-Generated Recommendations For Process Improvements

Using AI-generated recommender systems, the organization can not only extract value from historical process data analysis but also actively generate recommendations for process improvements (Jha et al., 2021). AI systems combine machine learning models and human knowledge in the domain with business rules to generate intelligent recommendations for process improvements, decision-making automation, and overall business efficiency (Reddy et al., 2019). For instance, the AI-powered recommendation system can analyze process execution data and pinpoint bottlenecks in the process by suggesting a new design or automating some activities (Hu et al., 2023). A hybrid process simply represents a process with redundant activities removed. The resequenced process is the one where activities change places to reduce cycle times, and the automated decision point is an unstaffed human decision-making workflow (Bhattacharyya et al., 2023). Moreover, AI systems can recommend which activities to assign to an RPA robot and which to a human employee, taking into account their skill level and availability, thereby minimizing bottlenecks and redeployment issues. This will also reduce resource idle time, resulting in increased process throughput (Gupta et al., 2022). Moreover, this approach also monitors processes and maintains process quality over time. This empowers the RPA process owner to modify the process to adapt to the market and technology and scale the process according to the present needs.

8. RESPONSIBLE AI AND GOVERNANCE

As organizations increasingly combine AI and RPA to enable intelligent process automation, ethical and governance considerations, along with the deployment of AI-powered systems, become more critical (Agostinelli et al., 2020). Therefore, responsible and trustworthy AI adoption should be implemented to mitigate associated risks, ensure transparency and confidentiality, and build trust among stakeholders.

8.1 Ethical Principles to Adopt AI in the Process

1. **Transparency and explainability:** One of the AI ethical principles is to ensure transparency and explainability of how people make decisions and their decision-making process. Stakeholders must be able to explain models and decision-making processes, comprehend the decision-making process, and identify the factors that influenced the final outcome (William et al., 2023). As a result, transparency equals accountability, and this allows firms to audit and

validate step-by-step reasoning in cases where things go wrong, especially in high-stakes or regulated segments.

2. **Fairness and bias mitigation:** AI models tend to inherit and affirm biases in the data or modeling process. Therefore, to ensure fairness using AI models, ethical leadership will implement bias-aware algorithms (Devarajan, 2018). Bias detection and mitigation involve reviewing data sources and feature selection to prevent bad patterns from emerging or continuing. Fair judgment in AI solves ethical, trust, and legal problems.

8.2 Governance Frameworks for AI-Powered Automation:

Effective governance ecosystems play a crucial role in responsible AI adoption and use with RPA. Therefore, organizations will form a governance ecosystem to manage risks and ensure ethical deployment.

1. **Monitoring and modeling risk management:** AI-powered automated processes handle complex tasks and decisions, thus calling for intense monitoring (Met et al., 2020). Organizations must implement monitoring for models that continuously evaluate models and organization performance accuracy and stability. Moreover, organizations should monitor models to detect outputs and pinpoint errors that adversely impact model performance (Romao et al., 2019). Therefore, we should develop processes for model validation, performance assessment, and risk evaluation to be proactive when uncertain problems arise.
2. Human-AI collaboration model: Although AI primarily automates many systems, there are many limitations to implementing a balance between human oversight and machine decision-making (Vajgel et al., 2021). Organizations should require guidelines for models and human agents. Hence, human agents should validate large decisions in businesses with uncertain environments (Chakraborti et al., 2022). Therefore, governance should be set to know the appropriate levels of human engagement in varying processes due to the risk, complexity, or regulation in the models (Villar & Khan, 2021). Additionally, organizations should invest in training their workforce to excel at human-AI formulations.

9. FUTURE OF INTELLIGENT AUTOMATION

With the ongoing convergence of AI and RPA, the field of intelligent automation has the potential to transform business processes and operations in ways never conceived before. The emergence of hyper-automation and new AI capabilities will spark a new innovation revolution and allow organizations to embed an unprecedented degree of efficiency, adaptability, and intelligence in their automation efforts.

9.1 New AI Capabilities Empowering Automation

1. Automated Reasoning and Planning: Current AI approaches predominantly excel in identifying patterns, making predictions, and executing decisions based on the analysis of historical data. The domain of automated reasoning and planning seeks to develop systems that can reason about complex issues, formulate plans, and propose novel solutions. Integrating this capability into intelligent process automation would enable the RPA robots to automatically create and execute a process flow in real time based on contextual reasoning instead of relying solely on predefined or learned models.
2. Adaptive and Learning Systems: Artificial intelligence will continue to advance to make systems adaptive and self-learning, enabling them to evolve, mature, and improve their performance autonomously. Intelligent automation tools would harness this capacity to develop and fine-tune their models, learn from a new environment, and optimize the execution dynamic rather than requiring manual interventions and retraining. A self-learning system can find and rectify performance issues, learn from a new dataset and scenario, and adjust its decision-making model on the fly, enabling truly autonomous process optimization.

9.2 The Transition to Hyper-Automation

The merger of these emergent and current AI capabilities on top of RPA creates a new routine: hyper-automation.

1. **Autonomous Process Discovery:** In an era of hyper-automation, organizations will not have to manually identify and map out their processes for automation. Using AI-driven process mining and discovery techniques, organizations can autonomously analyze operational data and reveal patterns and opportunities for automation. This autonomous process discovery will significantly reduce the time and effort required for process analysis, allowing organizations to quickly discover and implement automation opportunities.
2. **Self-Healing and Self-Optimizing Processes:** Intelligent automation in the future will feature systems that are capable of self-healing and self-optimizing, as they will resolve issues without any human intervention. To put it another way, automated processes will detect and address exceptions, performance bottlenecks, and system failures by utilizing AI-powered and self-learning monitoring, analytics, and adaptive learning. Moreover, automated processes

will continuously optimize themselves by means of real-time data and feedback, altering their execution pathways, assigned resources, and decision-making strategies to improve efficiency and productivity.

9.3 Vision for AI-Augmented Intelligent Process Automation:

Technologies such as RPA and AI are being united to introduce a revolutionary approach, known as truly intelligent process automation. According to this concept, automation solutions should be able to understand complex data, draw conclusions on a given situation, make context-based decisions, and continuously learn and adapt to changing conditions. Intelligent automation is expected to combine the best abilities of people and robots, including the ability to make joint decisions and maintain end-to-end orchestration of processes. Furthermore, conversational interfaces and human-in-the-loop models will facilitate natural interaction between employees and intelligent process automation, allowing for exception management, ad-hoc requests, and transparency into decision-making processes. Ultimately, AI-augmented intelligent process automation presents a potential for establishing operational excellence and competition for organizations. Through ethical and governance considerations, firms can leverage AI technology's power to establish extreme automation, intelligence, and operations in their processes to ensure their future in the modern digital age.

10. ROADMAP FOR ADOPTING INTELLIGENT AUTOMATION

Adopting AI-augmented intelligent process automation to its full potential requires a carefully planned and structured adoption strategy. We should develop this as a comprehensive roadmap that encompasses all technical, organizational, and cultural preparations for integrating AI capabilities with the current RPA efforts (Hofmann et al., 2020). The key elements include the following main components:

10.1 Phased Approach to AI Adoption Alongside RPA

Adopting AI into an organization's automation landscape to achieve AI-augmented intelligent process automation is of evolutionary character; it is better to do it in stages (Yatskiv et al., 2020). This allows the organization to learn from experience, mitigate risk, and ease the transition to the intelligent automation model. To achieve this goal, consider the following phases:

Phase 1: Figure out which use cases could benefit from adding AI by looking at a number of important factors, such as the complexity of the process, the amount of unstructured data that is involved, the level of autonomy, and the possible effect on business.

Phase 2: create a proof-of-concept or a pilot focused on exploring the feasibility and benefits of the AI-augmented process for the identified use case. This should include stakeholder involvement, data preparation, model building, and result evaluation.

Phase 3: Based on the lessons learned from the pilot, develop an enterprise-wide adoption strategy, including the technology roadmap as well as the governance and change management components.

Phase 4: Scale AI-powered automation across different processes and domains, monitoring and improving it using feedback loops and continuous refinement practices.

Working in this phased manner will help to mitigate the risks associated with adopting new technology, develop the in-house experience, and build trust in AI-augmented automation before scaling it across the organization.

10.2 Establishing an AI Center of Excellence

Organizational readiness includes creating dedicated staff responsible for the management of AI elements of process automation.

AI CoE General Scope of Responsibilities The key responsibilities of the AI CoE are as follows:

1. Build AI capabilities. The company achieves this by acquiring talent, providing training, and upgrading the skills of its current employees.
2. Develop and maintain AI-reusable assets. They include models, algorithms, and data pipelines, thus supporting accelerated deployment for various use cases.
3. Establish AI governance. Establish governance frameworks, ensuring ethical guidelines and best practices for AI development and deployment. This is necessary for responsible and trustworthy AI adoption.
4. Partner with business units and process owners. On the one hand, they are responsible for identifying and prioritizing AI use cases; on the other hand, they provide expertise for various implementation patterns.
5. Centre of Excellence. The CoE should build a source of expertise and act as a consultant by creating AI operations and fostering AI literacy.

Where applicable, the AI CoE should partner with the RPA CoE. It is required to drive the convergence of AI and RPA capabilities and align the two CoE initiatives.

10.2.1 Effective Change Management

Change management and workforce transformation: intelligent automation adoption is more than technology implementation; it is about changing the culture and mindset of the entire organization (Schatsky et al., 2017). Effective change management and workforce transformation initiatives are critical to ensuring successful adoption and maximizing the benefits of AI-powered automation.

1. **Communication and awareness:** Awareness is about informing others what and why is happening. Educate employees on the significance of intelligent automation, its introduction to the company, and its necessity.
2. **Training and upskilling:** It will be about making sure employees are sufficiently trained to work alongside AI-powered automation. Training curricula focus on three areas: AI literacy, data interpretation, and human-AI collaboration.
3. **Role and responsibility realignment:** Examine the impact of intelligent automation on the company's job roles and responsibilities. Let employees explore new, high-impact Executive Summary: Solving the Process Automation Puzzle [69]. McKinsey & Company McKinsey & Company has identified nine domains that enable them to increase productivity, particularly crucial given the potential involvement of humans in the process.
4. **Incentive and performance management:** Reprioritise incentives and performance management mechanisms to motivate leveraging intelligent automation solutions, as well as create a culture of improvement and integration with AI-based systems.

Such an approach to organizational change and workforce investment can foster a positive attitude toward AI-augmented automation, which, in turn, can minimize resistance and spark employee ownership and integration.

10.3 Developing an AI-RPA Strategy and Vision:

To enable a more effective alignment of AI and RPA, organizations should work on an overall AI-RPA strategy and vision that incorporates the following aspects::

1. Strategic objectives and anticipated outcomes: Companies should clearly articulate their objectives and possible outcomes of integrating an AI into intelligent automation, including enhanced efficiency, cost reduction, increased customer loyalty, or organizational competitive edge.
2. Use cases, priorities, and an implementation plan: Organizations should identify the most promising use cases based on their process maturity and AI readiness and develop a proper timeline for their implementation.
3. Technology and existing infrastructure: The robustness and complexity of an organization's AI-RPA strategy also depend on its capacity to support such an advanced AI-based regulatory environment. Organizations should understand their existing technologies and platforms and decide whether they need to automate or update their current systems.
4. Governance and ethical standards: Before implementing AI-based robotic process automation, organizations should consider such ethical and control aspects as ethical consequences.
5. Continuous improvement and innovation: Define processes for constantly monitoring, evaluating, and optimizing intelligent automation initiatives, encouraging an innovative spirit and continuous improvement that strategically enable you to accommodate ever-changing business requirements and new technology trends.

To create enduring value, a well-thought-out AI-RPA strategy and vision will help the company integrate its intelligent automation programs with strategic plans for corporate growth, prolongation, and a shift toward AI-RPA convergence.

11. CONCLUSION AND FUTURE SCOPE

The advent of AI thrusts RPA into a new era defined by a transformational shift that allows enterprises to exceed the limits of traditional automation and unleash the potential of processes intertwined with intelligence. RPA catalyzes businesses to solve intricate challenges, analyze unstructured data, automate decision-making, execute continuous optimization, and ultimately unlock unparalleled operational potential. As a myriad of technologies undergo rapid transformation and market competition intensifies, AI-powered automation becomes a prerequisite and key enabler to better enable digital transformation, optimize costs, improve customer experience, and secure competitiveness. Practically, from integrated AI to AI CoE, from a phased adoption roadmap design to adopting intelligent integration strategies, change management, and a cohesive AI-RPA uniscope within the organizational ethos, businesses need to proactively chart their course using intelligent automation. By harnessing its transformative power, they can bridge human-machine collaboration, catapult continuous improvement ahead, and secure a compelling future, while those who shy away miss out. The future scope of Intelligent RPA is increasing cognitive capabilities using higher order of machine learning, integrating IoT and blockchain, creating a scalable, cloud-based automation solutions, developing self-healing, autonomic decision-making systems, ensuring security and regulatory compliance as well as personalizing automation for a user-centric automation, and much more. Industry specific developments and developing human-robot collaboration will also drive innovation and efficiency across the sectors.

REFERENCES

Agostinelli, S., Marrella, A., & Mecella, M. (2019). Research challenges for intelligent robotic process automation. *Business Process Management Workshops: BPM 2019 International Workshops, Vienna, Austria, September 1–6, 2019, Revised Selected Papers 17*, 12–18.

Agostinelli, S., Marrella, A., & Mecella, M. (2020). Towards intelligent robotic process automation for BPMers. *ArXiv Preprint ArXiv:2001.00804.*

Aguirre, S., & Rodriguez, A. (2017). Automation of a business process using robotic process automation (RPA): A case study. *Applied Computer Sciences in Engineering: 4th Workshop on Engineering Applications, WEA 2017, Cartagena, Colombia, September 27-29, 2017. Proceedings*, *4*, 65–71.

Alberth, M., & Mattern, M. (2017). Understanding robotic process automation (RPA). *Journal of Financial Transformation*, *46*, 54–61.

Albus, J. S. (1994). *A reference model architecture for intelligent systems design.* Citeseer.

Anagnoste, S. (2018). The road to intelligent automation in the energy sector. *Management Dynamics in the Knowledge Economy*, *6*(3), 489–502. DOI: 10.25019/MDKE/6.3.08

Bellman, M., & Göransson, G. (2019). *Intelligent process automation: building the bridge between Robotic Process Automation and artificial intelligence.*

Bhattacharyya, S., Banerjee, J. S., & De, D. (2023). *Confluence of Artificial Intelligence and Robotic Process Automation* (Vol. 335). Springer. DOI: 10.1007/978-981-19-8296-5

Brynjolfsson, E., & Mcafee, A. (2017). Artificial intelligence, for real. *Harvard Business Review*, *1*, 1–31.

Chakraborti, T., Isahagian, V., Khalaf, R., Khazaeni, Y., Muthusamy, V., Rizk, Y., & Unuvar, M. (2020). From Robotic Process Automation to Intelligent Process Automation: –Emerging Trends–. *Business Process Management: Blockchain and Robotic Process Automation Forum: BPM 2020 Blockchain and RPA Forum, Seville, Spain, September 13–18, 2020. Proceedings*, *18*, 215–228.

Chakraborti, T., Rizk, Y., Isahagian, V., Aksar, B., & Fuggitti, F. (2022). From natural language to workflows: Towards emergent intelligence in robotic process automation. *International Conference on Business Process Management*, 123–137. DOI: 10.1007/978-3-031-16168-1_8

Cheong, T.-S., & Yoon, C.-H. (1999). A memory based classifier using the recursive partition averaging. *Proceedings of IEEE. IEEE Region 10 Conference. TENCON 99.'Multimedia Technology for Asia-Pacific Information Infrastructure'(Cat. No. 99CH37030), 2*, 1038–1041. DOI: 10.1109/TENCON.1999.818599

Czarnecki, C., & Fettke, P. (2021). *Robotic process automation: Management, technology, applications*. De Gruyter. DOI: 10.1515/9783110676693

Devarajan, Y. (2018). A study of robotic process automation use cases today for tomorrow's business. *International Journal of Computers and Technology*, *5*(6), 12–18.

Fernández-Loría, C., & Loría, J. (2022). Learning to Rank by Causal Effects Without Data to Accurately Estimate Causal Effects. *ArXiv Preprint ArXiv:2206.12532*.

Fettke, P., & Strohmeier, S. (2022). HR robotic process automation. In *Handbook of Research on Artificial Intelligence in Human Resource Management* (pp. 187–206). Edward Elgar Publishing.

Furnham, A., & Chamorro-Premuzic, T. (2004). Estimating one's own personality and intelligence scores. *British Journal of Psychology*, *95*(2), 149–160. DOI: 10.1348/000712604773952395 PMID: 15142299

Gotthardt, M., Koivulaakso, D., Paksoy, O., Saramo, C., Martikainen, M., & Lehner, O. (2020). *Current state and challenges in the implementation of smart robotic process automation in accounting and auditing*. ACRN Journal of Finance and Risk Perspectives. DOI: 10.35944/jofrp.2020.9.1.007

Gupta, A., Gupta, S., Memoria, M., Kumar, R., Kumar, S., Singh, D., Tyagi, S., & Ansari, N. (2022). Artificial Intelligence and Smart Cities: A Bibliometric Analysis. *2022 International Conference on Machine Learning, Big Data, Cloud and Parallel Computing, COM-IT-CON 2022*. DOI: 10.1109/COM-IT-CON54601.2022.9850656

Habibi, Y., Srifariyati, S., Hasan, H., & Subhi, M. R. (2017). Strategi pembelajaran anak usia dini berbasis multiple intelligence. *Madaniyah*, *7*(2), 237–260.

Hashemi, A. (2008). *The relationship between multiple intelligences and reading comprehension*.

Hathaway, A. P. (1982). Intelligence and non-intelligence factors contributing to scores on the Rorschach Prognostic Rating Scale. *Journal of Personality Assessment*, *46*(1), 8–11. DOI: 10.1207/s15327752jpa4601_3 PMID: 7069572

Hofmann, P., Samp, C., & Urbach, N. (2020). Robotic process automation. *Electronic Markets*, *30*(1), 99–106. DOI: 10.1007/s12525-019-00365-8

Hu, Y.-H., Fu, J. S., & Yeh, H.-C. (2023). Developing an early-warning system through robotic process automation: Are intelligent tutoring robots as effective as human teachers? *Interactive Learning Environments*, •••, 1–14. DOI: 10.1080/10494820.2022.2160467

Hutton, R. J. B., Warwick, W., Stanard, T., McDermott, P. L., & McIlwaine, S. (2001). Computational model of recognition-primed decisions (RPD): Improving realism in computer-generated forces (CGF). *Proceedings of the Human Factors and Ergonomics Society Annual Meeting*, *45*(26), 1833–1837. DOI: 10.1177/154193120104502607

Ivanov, D. (2021). *Introduction to supply chain resilience: Management, modelling, technology*. Springer Nature.

Jędrzejka, D. (2019). Robotic process automation and its impact on accounting. *Zeszyty Teoretyczne Rachunkowości*, *105*(105 (161)), 137–166. DOI: 10.5604/01.3001.0013.6061

Jha, N., Prashar, D., & Nagpal, A. (2021). Combining artificial intelligence with robotic process automation—an intelligent automation approach. *Deep Learning and Big Data for Intelligent Transportation: Enabling Technologies and Future Trends*, 245–264.

Kant, R., Dixit, A., & Memoria, M. (2023). Robots, Revolutionizing the Practice of Law: A Bright Future Ahead. *2023 International Conference on IoT, Communication and Automation Technology (ICICAT)*, 1–6. DOI: 10.1109/ICICAT57735.2023.10263664

Kant, R., Dixit, A. K., Memoria, M., & Chaudhary, T. (2023). Artificial Intelligence Based Transformation in the Field of Legal Profession. *2023 3rd International Conference on Innovative Sustainable Computational Technologies (CISCT)*, 1–6.

Katritsis, D. G. (2021). Artificial intelligence, superintelligence and intelligence. *Arrhythmia & Electrophysiology Review*, *10*(4), 223–224. DOI: 10.15420/aer.2021.61 PMID: 35106171

Kedziora, D. (2022). Botsourcing, Roboshoring or Virtual Backoffice? Perspectives on Implementing Robotic Process Automation (RPA) and Artificial Intelligence (AI). *Human Technology*, *18*(2), 92–97. DOI: 10.14254/1795-6889.2022.18-2.1

Khan, S., Tailor, R. K., Uygun, H., & Gujrati, R. (2022). Application of robotic process automation (RPA) for supply chain management, smart transportation and logistics. *International Journal of Health Sciences*, *6*(S3), 11051–11063. DOI: 10.53730/ijhs.v6nS3.8554

Kirkegaard, E. O. W., & Nyborg, H. (2021). Intelligence and general psychopathology in the vietnam experience study: A closer look. *The Mankind Quarterly*, *61*(4), 792–819. DOI: 10.46469/mq.2021.61.4.2

Lillis, L., Siverson, J., Lee, A., Cantera, J., Parker, M., Piepenburg, O., Lehman, D. A., & Boyle, D. S. (2016). Factors influencing recombinase polymerase amplification (RPA) assay outcomes at point of care. *Molecular and Cellular Probes*, *30*(2), 74–78. DOI: 10.1016/j.mcp.2016.01.009 PMID: 26854117

Ling, X., Gao, M., & Wang, D. (2020). Intelligent document processing based on RPA and machine learning. *2020 Chinese Automation Congress (CAC)*, 1349–1353. DOI: 10.1109/CAC51589.2020.9326579

Luo, H., & Luo, H. (2021). RPA and Artificial Intelligence in Budget Management Based on Multiperspective Recognition Based on Network Communication Integration. *Wireless Communications and Mobile Computing*, *2021*(1), 9723379. DOI: 10.1155/2021/9723379

Madakam, S., Holmukhe, R. M., & Jaiswal, D. K. (2019). The future digital work force: Robotic process automation (RPA). *Journal of Information Systems and Technology Management*, *16*, e201916001. DOI: 10.4301/S1807-1775201916001

Martínez-Rojas, A., Barba, I., & Enríquez, J. G. (2020). Towards a taxonomy of cognitive RPA components. *Business Process Management: Blockchain and Robotic Process Automation Forum: BPM 2020 Blockchain and RPA Forum, Seville, Spain, September 13–18, 2020. Proceedings*, *18*, 161–175.

Met, İ., Kabukçu, D., Uzunoğulları, G., Soyalp, Ü., & Dakdevir, T. (2020). Transformation of business model in finance sector with artificial intelligence and robotic process automation. *Digital Business Strategies in Blockchain Ecosystems: Transformational Design and Future of Global Business*, 3–29.

Micle, D. E., Deiac, F., Olar, A., Dren a, R. F., Florean, C., Coman, I. G., & Arion, F. H. (2021). Research on innovative business plan. Smart cattle farming using artificial intelligent robotic process automation. *Agriculture*, *11*(5), 430. DOI: 10.3390/agriculture11050430

Mohamed, S. A., Mahmoud, M. A., Mahdi, M. N., & Mostafa, S. A. (2022). Improving efficiency and effectiveness of robotic process automation in human resource management. *Sustainability (Basel), 14*(7), 3920. DOI: 10.3390/su14073920

Murugamani, C., Sahoo, S. K., Kshirsagar, P. R., Prathap, B. R., Islam, S., Noorulhasan Naveed, Q., Hussain, M. R., Hung, B. T., & Teressa, D. M. (2022). Wireless communication for robotic process automation using machine learning technique. *Wireless Communications and Mobile Computing, 2022*, 2022. DOI: 10.1155/2022/4723138

Naglieri, J. A., & Otero, T. (2011). Cognitive Assessment System: Redefining intelligence from a neuropsychological perspective. *Handbook of Pediatric Neuropsychology*, 320–333.

Paik, J., Pirolli, P., Dong, W., Lebiere, C., & Thomson, R. (2013). An ACT-R Model of Sensemaking in Geospatial Intelligence Tasks. *Proceedings of the 22nd Annual Conference on Behavior Representation in Modeling and Simulation.*

Paletz, S. B. F., Hughes, M. M., Sumer, A., Brugman, C. M., & Bunting, M. (2018). A preliminary validation of the Regional Proficiency Assessment Tool (RPAT). *Military Psychology : the Official Journal of the Division of Military Psychology, American Psychological Association, 30*(6), 528–546. DOI: 10.1080/08995605.2018.1506196

Pedretti, A., Santini, M., Scolimoski, J., Queiroz, M. H. B., de, , Toshioka, F., & Rocha, E.Junior. (2021). Robotic Process Automation Extended with Artificial Intelligence Techniques in Power Distribution Utilities. *Brazilian Archives of Biology and Technology, 64*(spe), e21210217. DOI: 10.1590/1678-4324-75years-2021210217

Rai, D., Siddiqui, S., Pawar, M., & Goyal, S. (2019). Robotic process automation: The virtual workforce. *International Journal on Future Revolution in Computer Science & Communication Engineering, 5*(2), 28–32.

Ray, J. K., Sultana, R., Bera, R., Sil, S., & Alfred, Q. M. (2023). A Comprehensive Review on Artificial Intelligence (AI) and Robotic Process Automation (RPA) for the Development of Smart Cities. *Confluence of Artificial Intelligence and Robotic Process Automation*, 289–311.

Reddy, K. N., Harichandana, U., Alekhya, T., & Rajesh, S. M. (2019). A study of robotic process automation among artificial intelligence. *International Journal of Scientific and Research Publications, 9*(2), 392–397.

Ribeiro, J., Lima, R., Eckhardt, T., & Paiva, S. (2021). Robotic process automation and artificial intelligence in industry 4.0–a literature review. *Procedia Computer Science, 181*, 51–58. DOI: 10.1016/j.procs.2021.01.104

Romao, M., Costa, J., & Costa, C. J. (2019). Robotic process automation: A case study in the banking industry. *2019 14th Iberian Conference on Information Systems and Technologies (CISTI)*, 1–6.

Schatsky, D., Muraskin, C., & Iyengar, K. (2017). *Robotic process automation. A Path to the Cognitive Enterprise*. Deloitte Consulting.

Sharma, A. (2021). Artificial intelligence in health care. *International Journal of Humanities, Arts. Medicine Science*, *5*(1), 106–109.

Shidaganti, G., Salil, S., Anand, P., & Jadhav, V. (2021). Robotic process automation with AI and OCR to improve business process. *2021 Second International Conference on Electronics and Sustainable Communication Systems (ICESC)*, 1612–1618. DOI: 10.1109/ICESC51422.2021.9532902

Siegert, S., & Völker, M. (2021). Towards Decision Management for Robotic Process Automation. *ZEUS*, 9–13.

Smith, G., & Stanley, G. (1980). Relationships between measures of intelligence and choice reaction time. *Bulletin of the Psychonomic Society*, *16*(1), 8–10. DOI: 10.3758/BF03337079

Sternberg, R. J. (1999). Intelligence as developing expertise. *Contemporary Educational Psychology*, *24*(4), 359–375. DOI: 10.1006/ceps.1998.0998 PMID: 10508532

Syed, R., Suriadi, S., Adams, M., Bandara, W., Leemans, S. J. J., Ouyang, C., ter Hofstede, A. H. M., van de Weerd, I., Wynn, M. T., & Reijers, H. A. (2020). Robotic process automation: Contemporary themes and challenges. *Computers in Industry*, *115*, 103162. DOI: 10.1016/j.compind.2019.103162

Szuba, T. (1998). A molecular quasi-random model of computations applied to evaluate collective intelligence. *Future Generation Computer Systems*, *14*(5–6), 321–339. DOI: 10.1016/S0167-739X(98)00037-5

Szuba, T. (2001). Universal formal model of collective intelligence and its iq measure. *International Workshop of Central and Eastern Europe on Multi-Agent Systems*, 303–312.

Vajgel, B., Corrêa, P. L. P., De Sousa, T. T., Quille, R. V. E., Bedoya, J. A. R., De Almeida, G. M., Filgueiras, L. V. L., Demuner, V. R. S., & Mollica, D. (2021). Development of intelligent robotic process automation: A utility case study in Brazil. *IEEE Access : Practical Innovations, Open Solutions*, *9*, 71222–71235. DOI: 10.1109/ACCESS.2021.3075693

Venigandla, K., Vemuri, N., Thaneeru, N., & Tatikonda, V. M. (2023). Leveraging AI-Enhanced Robotic Process Automation for Retail Pricing Optimization: A Comprehensive Analysis. *Journal of Knowledge Learning and Science Technology ISSN: 2959-6386 (Online), 2*(2), 361–370.

Viehhauser, J. (2020). Is robotic process automation becoming intelligent? Early evidence of influences of artificial intelligence on robotic process automation. *Business Process Management: Blockchain and Robotic Process Automation Forum: BPM 2020 Blockchain and RPA Forum, Seville, Spain, September 13–18, 2020. Proceedings, 18*, 101–115.

Villar, A. S., & Khan, N. (2021). Robotic process automation in banking industry: A case study on Deutsche Bank. *Journal of Banking and Financial Technology, 5*(1), 71–86. DOI: 10.1007/s42786-021-00030-9

Wang, Z., Zhang, L., Zhang, F., Li, W., & Yang, Z. (2021). Research and Exploration on The Application of RPA and Artificial Intelligence Technology in The Field of Electric Power Marketing. *Proceedings of the 4th International Conference on Information Technologies and Electrical Engineering*, 1–5. DOI: 10.1145/3513142.3513236

Whitesmith, M. (2020). Experimental research in reducing the risk of cognitive bias in intelligence analysis. *International Journal of Intelligence and CounterIntelligence, 33*(2), 380–405. DOI: 10.1080/08850607.2019.1690329

William, P., Choubey, S., Choubey, A., & Chhabra, G. S. (2023). Evolutionary Survey on Robotic Process Automation and Artificial Intelligence: Industry 4.0. *Robotic Process Automation*, 315–327.

Witty, P. A., & Lehman, H. C. (1931). A study of vocational attitude and intelligence. *The Elementary School Journal, 31*(10), 735–746. DOI: 10.1086/456653

Xie, L., Zhao, G., Lu, J., & Jiang, B. (2023). Research on marketing intelligence inspection based on RPA technology. *Third International Seminar on Artificial Intelligence, Networking, and Information Technology (AINIT 2022), 12587*, 170–177.

Yatskiv, N., Yatskiv, S., & Vasylyk, A. (2020). Method of robotic process automation in software testing using artificial intelligence. *2020 10th International Conference on Advanced Computer Information Technologies (ACIT)*, 501–504.

Zhang, X., & Wen, Z. (2021). Thoughts on the development of artificial intelligence combined with RPA. *Journal of Physics: Conference Series, 1883*(1), 012151. DOI: 10.1088/1742-6596/1883/1/012151

Chapter 9
Analysis of RPA Development Tools

Rohit Yadav
https://orcid.org/0000-0002-7573-8005
IILM University, Greater Noida, India

Gunjan Mohan Sharma
O.P. Jindal Global University, India

Thi T. Dang
Vietnam National University, Hanoi, Vietnam

Narayanage Jayantha Dewasiri
https://orcid.org/0000-0002-5908-8890
Sabaragamuwa University of Sri Lanka, Sri Lanka & Qasim Ibrahim School of Business, Villa College, Maldives

ABSTRACT

This chapter examines the landscape of the development tools used in RPA. Analysis includes key functionalities, usability, integration capabilities, security, and cost structure among the development tools in question. With increased attention of organizations to embracing RPA as part of efforts to cut costs and seek digital transformation, their need-to-know what tools are available becomes critical. This chapter presents a review of popular RPA platforms-UiPath, Automation Anywhere, and Blue Prism-to compare their functionalities, licensing models, and scalability. However, it also throws up the future pattern of how automation tools will have artificial intelligence, low-code/no-code environments, and cloud-based solutions as part of them. Organizations looking forward to implementing RPA solutions are given recommendations that require introspection regarding need assessment, the prioritization of security, and information being sought in industry advancements.

DOI: 10.4018/979-8-3693-4365-4.ch009

INTRODUCTION TO RPA DEVELOPMENT

Robotic process automation (RPA) has emerged as a very transformative technology in the field of business, changing the way organizations approach the automation of processes. It enables firms to automate repetitive rules-based tasks that have traditionally been labor-intensive (Agostinelli et al., 2020). From data entry and report generation to process invoices or respond to customer support queries, all such tasks can be executed with speed and accuracy by software robots, or "bots.". As a result, RPA has become the cornerstone of digital transformation activities in which more firms are improving operational efficiency and bringing down costs with increased accuracy in most business processes.

Typically, developing RPA solutions requires sophisticated tools that are designed to be light for designing, testing, and implementing bots. These RPA development tools have well-structured frameworks through which developers as well as non-technical users can design automated workflows without programming. A visual or drag-and-drop interface lets you map processes and tasks to bots for assignment, so very intuitive to develop automation solutions. This simplicity, with powerful backend capabilities, makes RPA development tools popular both for small-scale and enterprise-level automation initiatives (Agostinelli et al., 2021).

One of the major reasons RPA development is gaining momentum is because it can easily interface with existing business systems without any need for underlying infrastructure changes. Unlike traditional automation methods that require serious coding and deep system-level modifications, RPA tools work at the user interface level to imitate mouse clicks, keystrokes, and copying data. With this flexibility, businesses can automate processes across different applications even on legacy systems without massive costs and long timelines that are often associated with custom software development. In other words, RPA is non-invasive, and hence accessible to a variety of industries and use cases (Andrade, 2020).

The RPA development lifecycle begins with the process identification where the developers team up with the business stakeholders to identify the tasks that are ideal for automation. After the processes are identified, RPA developers work using specific tools to define the workflows as they configure bots toward doing the desired actions. It has many inherent features such as error handling, decision logic, and interface integration that make the automation more robust and adaptable to several situations (Asatiani et al., 2023). Finally, with the workflow designed, the bots are tested to ensure they perform as expected and then are deployed in the production environment.

As the organizations scale up the adoption of RPA, effective tools for RPA development are all the more critical. These tools do not only simplify the development process but also come with features that guarantee sustainability and scalability of automation efforts. To this end, features such as central management, version control, and real-time analytics help organizations monitor bot performance, identify problems, and troubleshoot them to optimize their automation initiatives over time. Added value and prosperity for growth are rounded into an ecosystem around RPA based on the integration of AI and machine learning extended towards even more intelligent automation, which is beyond the rule-based limit (Axmann & Harmoko, 2022).

In brief, RPA development is an ever-changing and dynamic sector that is revolutionizing the landscape of how organizations view automation. With access to friendly tools for developing, RPA has democratized itself so that even a non-technical user can participate in the design of solutions that automate tasks; thus, it is probably true that RPA development tools will be crucial in driving the next wave of digital innovation as business continues its search for the gains of efficiency they require in an increasingly competitive environment.

KEY FEATURES OF RPA DEVELOPMENT TOOLS

RPA has revolutionized the way firms can automate complex tasks in a manner whereby there would be minimal coding expertise. Such tools would be offering both technical and non-technical users a variety of robust features conceived for such applications, thus giving room for using automation in different ways, handling even the most complex business processes. For any organization, one must get to know and determine the key features of RPA development in the selection of the right platform to implement according to the specific needs of their organization (Bosco et al., 2019).

The most recognizable aspect of RPA development tools is the **drag-and-drop, visual interface** that simplifies building automation workflows greatly. Because of the intuitive nature of the design, one can now develop bots based on visually mapping steps in a process rather than writing code from scratch. Icons, such as "click," "copy," or "paste," can be dragged to the workflow canvas and make the development process much more accessible. This form of graphical programming merely makes it easier for nondevelopers so that business users and subject matter experts can be engaged in the making of bots. It speeds up the building of intricate workflows for developers thus reducing the time spent on coding and debugging (Choi et al., 2022).

The next feature is **workflow automation**. RPA triumphs in automating repetitive tasks that will follow specific rules that apply to other applications. It goes as far as desktop environments as well as web systems to simplify things. This enables functions such as screen scraping, OCR, and data entry. Hence, bots can definitely interact with a wide variety of applications. Workflow automation tools provide the ability for full execution by bots autonomously-through processing of an invoice (Gupta et al., 2019), migrating data between systems, and even answering queries by customers. The best strength of these tools is in the manipulation of multi-step processes, with minimal human intervention, which increases productivity and reduces the margin for error.

Development tools for RPA also support various **scripting languages and integration options**. Even though most RPA platforms provide no-code or low-code solutions, scripting languages such as Python, VBScript, and JavaScript are possible in case of advanced customizations. This flexibility has let the developers to create intricate automation scenarios that can require proprietary logic or even third-party system integration. Furthermore, the RPA tools also have pre-built connectors which can connect the bots with enterprise software such as SAP, Salesforce, or Oracle and thus let the easy integration of bots with the existing infrastructure (Herm et al., 2021).

Error handling and debugging are integral parts of RPA development and most of the RPA tools come up with sophisticated exception handling mechanisms. All these features ensure that the bots handle such errors like missing data or some unexpected behavior from the application without crashing or stopping the process. Debugging tools will give developers insights about what went wrong with a bot during its execution: you can step through the workflows, inspect variable values, and eventually discover wrong logic inside the process (Khan, 2020).

Most of these new RPA tools also come with integrated capabilities such as **version control and collaboration**. This encourages teamwork from different developers who would work together on an automation project. Collaboration still stays central to the enterprise environment where teams of developers collaborate during the design, share updates, and keep uniform focus upon their development pipeline. Version control tracks different iterations of a workflow so that changes can easily revert to earlier versions when necessary.

Summing Up RPA Development Tools In a nutshell, there is an array of features added to the RPA development tools that make these more user-friendly, efficient, and collaborative. The scope ranges from visual design interfaces, workflow automation capabilities, scripting support, error handling, and collaboration tools that form a base to enable production of scalable, reliable automation solutions. Each of these becomes essential for a business when it intends to harness all the functionalities of RPA in optimizing its operations (Kim et al., 2021).

COMPARISON OF POPULAR RPA DEVELOPMENT TOOLS

As the need for automation continues to grow, several RPA development tools entered the market with unique strengths and capabilities. Some of these are more popular in the industry than the others, such as **UiPath, Blue Prism, and Automation Anywhere**. They seem to be able to support heavy automation tasks but within different lines of usability, scalability, and target audience. Knowing the differences allows the selection of the right tool to support a certain need of an organization (Kim, 2023).

Users find it very easy to use and simple to work with the UiPath having a friendly, easy-to-use interface that is a drag-and-drop interface. It is one of the most user-friendly RPA platforms for both developers and non-technical users. Its visual workflow designer makes it easy for users to create bots through just dragging predefined actions and processes together. Along with this, UiPath provides rich tutorials and an extremely active community, which makes the life of novices easier in learning how to implement automation (Kronkvist, 2020). UiPath supports a wide spectrum of applications and systems ranging from web-based to legacy systems in the integration aspect. Its integrated suite encompasses AI/ML integration, which is thoroughly put into action in simple and complex automation scenarios. UiPath also has high scalability with cloud-based options for deployment, which would be beneficial for organizations looking to scale up their automation efforts across various departments or locations.

On the other hand, Blue Prism is well known for enterprise-grade security and scalability, which makes it often favored by big organizations, especially in finance, healthcare, and government. In contrast to UiPath, the interface is not very intuitive and mainly relies on coding and process logic design; it's mainly aimed at seasoned developers. However, Blue Prism compensates with its strength in high-powered automation capabilities, particularly for high-volume transactions and complex workflows. In addition, centralized management features are also available, allowing the organization to monitor, control, and optimize bots from a single place. This is a strong choice for large-scale deployment where governance and oversight are key (López-Carnicer et al., 2020).

Automation Anywhere is a middle grounder between UiPath ease and Blue Prism Enterprise strength. As with UiPath, Automation Anywhere is friendly to the user as far as drag-and-drop features are concerned, but on the other hand, it looks for more technical users through advanced scripting options. Its first benefit is that of an inbuilt bot store; it allows for ready-to-download bots for most common tasks, thus saving on development time of the in-house bot. Also, like UiPath, Automation Anywhere is tightly integrated with cloud platforms that allow for flexibility by either using hybrid or cloud-native solutions for organizations. It is capable of looking at unstructured data, for instance, email or voice commands due to its cognitive automation capabilities powered by artificial intelligence and NLP technologies (Mullakara & Asokan, 2020).

Comparing cost models, UiPath and Automation Anywhere offer a free trial with tiered pricing models with the number of bots and users. It is scalable, thus more accessible to small and medium-sized businesses. Blue Prism runs on a significantly higher price, usually demanding large upfronts-the barrier for smaller companies but an excellent opportunity for an enterprise that needs really high levels of security and scalability (Noppen et al., 2020).

In short, UiPath is suitable for those organizations seeking ease of use, quick deployment, and community support; while Blue Prism is suitable for enterprises that require robust security, governance, and scalability. Automation Anywhere brings in a middle-of-the-road approach with advanced cognitive automation, offering a broad base of fields and industries suitable for a vast number of enterprise needs. All these tools have different capabilities, and the platform must depend on the particular needs, scale, and technical capacity of the organization.

Usability and Learning Curve

The usability and learning curve of development tools for RPA, thus, are very important factors determining when organizations can unlock business value through automation. Some of them are meant for simpler user-friendly targeting non-technical users, while other RPA tools are more advanced and would thus require more skills to handle. Therefore, understanding the usability of RPA tools and associated learning curve is the key to gaining the most from the solutions (Rod Koch, 2019).

UiPath is, essentially, one of the most amiable RPA tools in existence and that allows for its use by people with no programming knowledge at all, such as business analysts and experienced developers alike. Its drag-and-drop interface allows for visually designing workflows without digging into extensive programming knowledge. In addition to that, the platform also contains many pre-built activities and templates, making work related to the automation of mundane tasks even more manageable and requiring less development time (Tiron-Tudor, 2024). Another big advantage is that UiPath provides numerous educational resources to ease the learning curve, such as online tutorials and webinars, and they even offer a free edition known as Community Edition, which reduces the hurdle for new users. The active user community, sharing insights, solutions, and best practices, enables a more effective learning experience. More broadly speaking, the learning curve is pretty low with an intuitive interface, which makes it a very promising solution to empower business users to achieve automation.

Automation Anywhere comes up with a similarly intuitive interface but is designed to work with non-technical users and developers who would require advanced functionality. Its Automation 360 platform comes along with a cloud-native environment that simplifies the deployment process and allows easy access of automation tools. Pre-built bots and workflows for novice users in RPA are offered by Automation Anywhere, which are customizable according to specific use cases (Tripathi, 2018). However, it supports more advanced scripting and coding capabilities for power users who need to create sophisticated automations. While it has a graphical user interface and the acquisition curve is relatively low in the beginning, the learning curve with respect to becoming proficient in bot orchestration, cognitive automation, and AI integration is higher. Bot Academy by Automation Anywhere has certification courses and deep training courses in every skill set so users can learn easily.

Blue Prism seems to have a slightly steeper learning curve than UiPath and Automation Anywhere and is mainly due to its emphasis on enterprise-grade automation and more technical workflows. Blue Prism is not that intuitive for a beginner user, and most builds on automations need a rather good understanding of how processes should be designed in flow and some coding concepts. Therefore, Blue Prism is typically applied by organizations with specific dedicated IT or development teams (Wewerka & Reichert, 2021). On the other hand, the strength of such a platform is compensated by powerful capabilities, especially in large-scale automation projects requiring a lot of governance and security. Blue Prism offers full training and certification programs across its spectrum of Blue Prism University for those willing to invest the time; with such, it unlocks the true power of the tool.

Summary Conclusion A user with a need for ease of use and faster onboarding will find a home at both UiPath and Automation Anywhere as they have friendly interfaces and an abundance of learning resources. Blue Prism is more complex but with scalability and security unmatched anywhere, it appeals to organizations that may have ample resource to handle a steeper learning curve when using this tool. Internal resources, therefore, are one of the elements to consider when an organization decides which RPA software to use. With that said, however, an organization should balance usability with the complexity of its automation needs (Willcocks et al., 2018).

Integration and Compatibility

The other significant factor that should be looked into while focusing on a suitable tool for the development of RPA is integration. Businesses are complex IT environments consisting of many applications, platforms, and data sources. Thus, compatibility has emerged as a critical success factor in any automation effort. RPA tools need to be versatile enough to work with diverse software such as the newer cloud-based applications and older systems and third-party platforms. Integration and compatibility features enable bots to work across various environments, thus enabling organizations to automate more processes (Yarlagadda, 2018).

UiPath is strong in terms of the capability to integrate because it has broader coverage in terms of compatibility with a series of enterprise applications and enterprise resource planning systems like SAP, Oracle, and Microsoft Dynamics. The connectors of UiPath and APIs enable communication between software across various platforms, therefore, allowing smooth interaction across systems. Additionally, UiPath supports REST and SOAP APIs which make it easy to integrate into web services thereby making it easy to adapt towards both cloud computing modern application that are going on and the older legacy systems (Ylä-Kujala et al., 2023). Moreover, AI and Machine Learning services like Google Cloud, AWS, as well as Microsoft Azure integrated into the system to enable even the most complex tasks, such as data extraction, document processing, and natural language understanding. Integration scope: This wide integration scope makes UiPath one of the most excellent choices for organizations requiring a tool that can automate tasks across multiple technologies.

The automation capabilities of Automation Anywhere are superior, particularly in cloud and hybrid environments. Its cloud-native Automation 360 means that most cloud-based services like Salesforce, Microsoft Azure, or Google Cloud seamlessly connect to it. This will be highly beneficial to organizations looking to move to cloud infrastructures or those who need to manage the automation of on-premise environments in addition to their cloud infrastructures (Agostinelli et al., 2020). Automation Anywhere provides pre-built APIs and connectors for leading enterprise applications, as well as integrations with custom applications through its API Designer. Also, with their cognitive automation feature, it is very possible for bots to interact with unstructured data, emails, and even documents. Thus, it definitely expands the possibilities of which processes are able to be automated. Businesses that want to infuse classical RPA capabilities with cognitive automation would find themselves in good hands with Automation Anywhere, providing complete integration capabilities.

Blue Prism, though not as easy to integrate because of its usage, stands out for being compatible and secure for enterprise-class usage. This technology is particularly suited for large organizations with a need to integrate automation, complex, and highly regulated IT systems. Blue Prism also provides a set of application adapters and connectors to integrate its bots with these enterprise solutions such as SAP, Oracle, and IBM systems so that bots can communicate in one environment to fetch data and to execute some tasks accordingly. Another benefit that Blue Prism offers is its ability to connect to cloud services-it collaborates with top cloud platforms, such as AWS, Microsoft Azure, and Google Cloud. Blue Prism is exceptional when dealing with legacy systems that do not have APIs or have modern interfaces. It can communicate with the applications that may be challenging to automate using screen scraping and surface automation techniques (Agostinelli et al., 2021).

In summary, UiPath and Automation Anywhere offer rich integration options which go a long way to cover the spectrum of enterprise applications, cloud services, and cognitive tools that make them highly flexible in diverse IT environments. But Blue Prism requires more technical setup for its integration and is strong enough for combining with complex enterprise systems and legacy environments, so it is popular with large enterprises that require strict security requirements and compliance needs. Businesses will need to choose the RPA tool that best fits their current systems and future visions for technology, with the integrations done so that they will work easily and grow with their business (Andrade, 2020).

Security and Compliance

Any business operating in finance, healthcare or government would most certainly engage measures to ensure security and compliance when rolling out RPA solutions. The first priority would be to confirm if the tools used meet high-security requirements, and also check if they align with the regulatory framework. Tools for RPA generally offer different security features, and an organization must analyze these features relative to what its own compliance requirement must be (Asatiani et al., 2023).

UiPath offers a full feature set of security. For an enterprise-level automation solution, UiPath provides role-based access control (RBAC), which means that the users can define in greater detail the roles and permissions of users within their organization so that only authorized people gain access to sensitive processes and data. UiPath is also integrated with both single sign-on (SSO) and multi-factor authentication (MFA) systems for added security. UiPath encrypts data at rest as well as in transit to ensure information is protected while transferring between systems. It also supports audit trails, ensuring organizations can track activities of their bots and retain that those activities were compliant with internal and external policies (Axmann & Harmoko, 2022). The other compliance certification that UiPath follows is ISO 27001 and SOC 2. Therefore, it aligns with the required needs of any organization that wants security and protection of data with global standards.

Automation Anywhere focuses on data security by offering features best suitable for cloud and on-premise environments. E2E encryption for bot-related data ensures protection at every point with proper execution and storage. Like UiPath, Automation Anywhere also supports RBAC and allows administrators to manage user permissions and access based on roles. Its Bot Insight platform is an analytics and audit capability, which is primarily required to maintain transparency and compliance with regulatory standards of governance. Additionally, Automation Anywhere aligns with various global regulations, including GDPR and HIPAA. This makes it an ideal fit for industries that process health and personal information. Automation Anywhere provides cloud-native deployments along with additional security certifications, including SOC 1, SOC 2, and ISO 27001. This provides a safe and compliant automation framework in cloud environments (Bosco et al., 2019).

Large organizations require very high security and compliance, hence the appropriateness of Blue Prism which has enterprise-grade security features. It was actually built with bank-grade security, bringing with it powerful capabilities to authenticate, monitor sessions, and audit log. It has very strong governance frameworks in place to ensure that traceability and industry compliance are assured for bot activities. Data masking and encryption: Blue Prism has also integrated data masking and encryption for preventing leakage of sensitive information, which could occur while undergoing automation processes (Choi et al., 2022). Another feature that is unique to Blue Prism includes segregation of duties, which allows organizations to divide the development, testing, and production into different environments as a way of avoiding unauthorized access or leakage of data. Blue Prism also boasts several security certificates, which include SOC 2, ISO 27001, and PCI-DSS (Payment Card Industry Data Security Standard), making the platform secure for companies that require stricter compliance and security levels (Gupta et al., 2019).

To put it simply, all three of the aforementioned platforms offer high levels of security and compliance, but differ based on their enterprise orientation. Therefore, UiPath balances its friendliness for users with security measures and enterprise-grade compliance, so it scales well to fit organizations of every size. Automation Anywhere is strongest in the area of cloud security and data protection and has a live concern for monitoring on compliance. Blue Prism is suitable for large, heavily regulated companies in a high level of governance with strict security protocols at the highest levels. The selection of the RPA tool should be taken on the basis of specific security and compliance as desired by the organization, in addition to the complexity of automation that lies (Herm et al., 2021).

Performance and Scalability

This is one of the critical aspects of most RPA development tools especially when companies look towards enhancing their automation requirements across departments and processes. With a scalable solution, one can work easily with large volumes of tasks without comprising on the number of bots supported or the smooth integration. It is more important for the RPA tool to be as effective at automation because it ensures that growth and the changing needs of a company can allow increased demand upon the platform without significant reengineering (Khan, 2020).

UiPath is highly performant and scalable and hence can be recommended to businesses that envision scaling of their automation efforts over time. UiPath supports both attended and unattended bots, and hence capable of automating all types of processes, starting from simple task automation to complex and high volume workflows (Kim et al., 2021). Its cloud and on-premise deployment options provide the flexibility of scaling up operations based on the infrastructure of the organization. More to this, Orchestrator from UiPath centralizes bot management making scheduling, monitoring, and optimizing performance pretty easier and more crucial for scaling operations across different departments or locations. One of the critical features of UiPath is its interoperability with the clouds, such as Microsoft Azure, Google Cloud, and AWS. Businesses, through it, can scale up with no high infrastructure costs. This means that thousands of processes can be automated in parallel without a consequent loss of performance (Kim, 2023).

Automation Anywhere is also strong on scalability, especially in cloud environments. The Automation 360 platform of Automation Anywhere is natively designed to be able to scale automation across global operations without a hitch. It supports having thousands of bots in both on-premise and cloud environments, hence enabling a business to easily scale. One of the most prominent features of Automation Anywhere is elastic scaling wherein the organization can add or reduce the number of bots dynamically according to demand (Kronkvist, 2020). The performance of the firms results in optimal performance with cost savings as only for the bots, they will be charged. Using the analytics tool Bot Insight, the organization would see the performance of the bots and the bottlenecks that they incur and continually optimizes its process. Additionally, Automation Anywhere bots can also do multiple complex tasks- cognitive automation based on Artificial Intelligence and Machine Learning. All these features of flexibility as well as scalability make Automation Anywhere a great option for companies that seek large-scale, high-performance automation.

Blue Prism is known to offer high-end enterprise scalability in particular in high-transaction environments where high-security measures are also required. While its setup and deployment might be somewhat more technical and time-consuming than other RPA tools, the scalability advantage of Blue Prism is unmatched for organizations that have to very complexly automate at scale. The Digital Workforce of Blue Prism can deploy and manage thousands of bots all at once, all with the watertight governance and security expected by firms (López-Carnicer et al., 2020). Due to extensibility both with legacy and modern IT systems, Blue Prism suits large organizations that require strict automation across several systems in particular. Moreover, the high-availability architecture of Blue Prism also makes sure that bots run perpetually, without interruption even in environments considered as mission-critical.

That makes it an extremely valuable system in industries like banking, healthcare, and government, where there are sizeable financial risks of operations if they shut down for even a small period of time (Mullakara & Asokan, 2020).

All three tools, UiPath, Automation Anywhere, and Blue Prism, are designed to support large-scale automation projects. Where both UiPath and Automation Anywhere can offer more flexibility related to cloud-native scalability, rapid deployment, and scale fast organisations or hybrid IT infrastructures, Blue Prism will be set up more at the outset but will have unparalleled performance and reliability for a high-volume complex automation need in large enterprises. In the end, it would solely depend on the size, complexity, and goals of the organization's automation strategy (Noppen et al., 2020).

Cost and Licensing Models

The cost and licensing models of development tools, representing RPA, are another parameter upon which an organization can base its decision to choose the platform within its organization. The cost structure can significantly affect the total cost of ownership and return of investment of an RPA initiative. Licensing models basically differ due to the number of bots, numbers of users, and feature types, and some RPA tools offer more flexibility in pricing to fit the different needs of an organization (Rod Koch, 2019), thus once a choice is made regarding which tool is appropriate, the cost implications can then be evaluated so that its automation goals are given in line with that specific budget constraint.

UiPath offers flexibility in licensing models and caters to businesses of all sizes, from the smallest of startups to the most massive enterprises. It is based on a subscription pricing model, and its determination depends on the kind of bot-attended or unattended-and the other features depending on UiPath Orchestrator and AI capabilities (Tiron-Tudor, 2024). Attended bots, which can aid users in completing the work assigned to them, are usually more affordable than an unattended bot, as it functions independently with no human interference. More advantageously, UiPath also offers an enterprise as well as small business and individual friendly Community Edition, to be used free of charge on a non-commercial basis, to give the users an opportunity to experience its RPA features free of charge. For enterprises, customized packages are available depending on the number of bots or the extent of deployment (Tripathi, 2018). However, for the organisation that intends to scale up their automation deployment to a significantly higher level, the costs do shoot way up, especially with complex implementations involving features such as AI Fabric and Task Mining. The flexibility of the licensing options and the seamless vast capabilities available within UiPath make it a very cost-effective solution for diverse automation needs.

Automation Anywhere, while in no way less traditional, mirrors a subscription pricing model in which costs depend on the number of bots and deployment options: cloud or on-premise. Automation Anywhere is particularly renowned for its Bot-as-a-Service offering, allowing organizations to adopt bots on a pay-per-use model. That cloud-native model provides much-needed flexibility and happens to be an ideal choice for businesses that need to grow their automation footprint fast with minimal upfront investment in infrastructure (Wewerka & Reichert, 2021). The cloud-based platform of Automation 360 has added the free trial option, so companies will be able to test the system in action and see all its capabilities before a full deployment. For organizations that need to automate vast quantities of processes, the subscription costs may climb, particularly when including more advanced features like IQ Bot with AI to automate tasks using unstructured data. However, its cloud-based pricing structure does tend to compensate for a lot of the upfront costs as it is more scalable in terms of growth, hence ideal for companies looking to scale up quickly in terms of automation (Willcocks et al., 2018).

Blue Prism tends to run on a significantly higher price point than that of UiPath and Automation Anywhere because of the enterprise focus it caters to and the resultant highly developed features. Its licensing model also follows the subscription-based model, but this time for large organizations that need high automation capabilities across departments and regions. The pricing of Blue Prism is on the number of bots with extra costs for features like Blue Prism Cloud, digital workers, integrations with AI and machine learning technologies, and more. The pricing of Blue Prism tends to be towards the higher side. However, governance, security, and scalability place the offering at a justified cost for organizations operating in regulated industries, such as banking, healthcare, and government. A free edition or version of free trial, like the ones offered by UiPath or Automation Anywhere, would not compete with smaller or medium-sized business's test the application and solutions with much less budget for RPA. However, if one has big organizations with relatively complex automation requirements, the features set and massive scalability give good reasons for investment in Blue Prism (Yarlagadda, 2018).

UiPath and Automation Anywhere are more flexible and affordable licensing models and suitable for businesses of all sizes. UiPath Community Edition and Bot-as-a-Service from Automation Anywhere are great entry points into RPA for businesses that do not have their base within RPA. Blue Prism is expensive, thus aimed at high-scale enterprises with more automation and security.

Thus, the choice of the right tool will depend upon the automation potential in the near term and far term, coupled with the cost at which the organization can sustain it. The perfect RPA tool needs to provide the best-suited balancing act between price and functionality for the desired level of automation (Ylä-Kujala et al., 2023).

FUTURE TRENDS IN RPA DEVELOPMENT TOOLS

RPA is transforming rapidly, led by technologies and a surging requirement for even more intelligent, scalable, and integrated automation solutions. As the newest wave of business adoption of digital transformation takes hold, RPA will become an advanced and flexible tool, whose scope will extend to areas far beyond process automation. Several emerging trends are shaping the future of RPA development tools by making them more accessible, intelligent, and transformative.

1. Integration of Artificial Intelligence and Machine Learning:

The biggest trend in RPA development is the deeper integration of artificial intelligence (AI) and machine learning (ML) technologies. While traditional RPA tools have focused on automating rule-based, repetitive tasks, in the future, there will be a need to integrate the concept of RPA with cognitive capabilities that allow the bot to understand unstructured data and make decisions and learn from certain patterns. Businesses might be able to automate even more complex tasks, for example, natural language processing, sentiment analysis, and predictive analytics through intelligent automation (Asatiani et al., 2023). The AI-enabled RPA bot extracts insights from emails and documents, images, and takes some kind of recommendations and actions based on that. The convergence of AI and RPA is often reflected in the term hyper-automation, which would influence the future of industries: that is to say, creating more autonomous systems requiring less human intervention.

2. Increased Adoption of Cloud-Based RPA

More and more organizations shift their operations to the cloud, and cloud-native RPA platforms start gaining traction. Cost efficiency, flexibility, and scalability tend to be better offered by the cloud-based systems than on-premise ones. From there, enterprises can dynamically respond to fluctuating workloads by rolling out, managing, and scaling bots via the cloud without significant investment in infrastructure. Platforms like Automation Anywhere's Automation 360 already drive this trend; the pay-as-you-use model of RPA-as-a-service enables businesses to simply pay for the number of bots they require to perform tasks rather than investing in expensive infrastructure. Also, seamless integration with other cloud-based applications and services simplifies the use of the cloud. This facilitates all kinds of end-to-end automation solutions across one's digital ecosystem for businesses (Axmann & Harmoko, 2022).

3. Low-Code/No-Code Development

The third trend emerging in RPA is that related to low-code and no-code development environments. With democratization of automation, more RPA tools are focused on equipping business users and non-tech employees to design and deploy automation workflows. Low-code/no-code platforms offer drag-and-drop interfaces, templates, and reusable components; it is relatively easy for users without much programming capability to create bots (Choi et al., 2022). This is a trend that accelerates the development and deployment of automations but, at the same time, reduces their dependency on IT teams, and businesses can respond more rapidly to changing process automation needs. For instance, UiPath and Automation Anywhere, some of the leading RPA vendors, have low code solutions that open up participation channels for citizen developers in the automation journey. Thus, automating applications becomes more widespread within the organization.

4. Expansion of RPA Use Cases Beyond Traditional Processes

Tools for RPA will transform and will be capable of affecting a wider number of business applications beyond just the routine back-office services covering finance and HR. Even more and more, it applies to areas like customer service, supply chain management, sales and marketing, due to its capabilities of developing AI and integration (Herm et al., 2021). For instance, bots can currently be utilized in helping to analyze customer feedback, predict sales trends, and optimize logistics in real-time within the supply chain. RPA has also been applied in the IT operations where standard processes are automated, such as updating software, security patching, and systems monitoring. As firms look for new use cases of RPA, the scope of automation is broadly expanding and will allow organizations to drive greater efficiency and innovation across departments.

5. Enhanced Focus on Security and Compliance

With increased automation, the demand for security and compliance-enabled features in RPA tools is also arising. Organizations are increasingly concerned with the data privacy, cyber threats, as well as with regard to compliance requirements and regulations like GDPR and HIPAA. In response, the RPA vendors are weaving more security features than before-including end-to-end encryption, role-based access control (RBAC), and audit trails-so that bots operate under strict security and compliance guidelines. Future of RPA RPA will see further advancement in this area as bots increasingly handle sensitive and confidential data. This would give businesses confidence in the security and compliance of their automated processes with increased scale across the enterprise (Khan, 2020).

6. RPA and Process Mining Integration

Another significant integration trend is that of process mining tools with the RPA platforms. Process mining helps businesses visualize and analyze their workflows, pointing out areas of inefficiency and bottlenecks that can be optimized using automation. The benefits of the combination of process mining with RPA can allow organisations to automate the right processes for maximum efficiency and impact with a data-driven approach. This trend is finally helping businesses refine their automation strategy and improve effectiveness of RPA deployments in general. For instance, UiPath has already incorporated process mining capabilities into its platforms to offer businesses end-to-end automation solutions (Kim et al., 2021).

The future development tools for RPA will therefore be an impact of the advancements made in AI, cloud computing, low-code development, and more secure measures. As all these technologies keep on advancing, RPA will become even more powerful, thus allowing businesses to automate the more complex tasks, scale more efficiently, and expand into new areas. The rising integration of RPA with process mining and cognitive capabilities will further add to the true value of automation, so organizations can drive innovation, optimize operations, and thrive in today's fast-paced changing digital landscape (Kim, 2023).

CONCLUSION AND RECOMMENDATIONS

This is the explosion of space in Robotic Process Automation; given the fact that more businesses are realizing its potential in maximising efficiency, reducing operational costs, and driving innovation. Modern RPA development tools are changing as is the organizational need. The new-generation tools that implement the latest technologies, such as AI and ML, will eventually produce more intelligent automation solutions in RPA software. As businesses begin to consider their RPA opportunities, understanding the core features, cost structures, and near-term trends in RPA tools will help guide solid decision-making that goes along with their organization's strategy for automation.

In nutshell, RPA has massive prospects for organizations that would like to streamline processes, enhance productivity, and encourage innovation. Still, the actual benefits from proper use of RPA can only be reaped when such careful consideration is taken in choosing the best development tool suited to each business's specific needs and contextual operation. Factors such as scalability, usability, integration capabilities, and security should serve to guide the choice. Organizations should also keep abreast of emerging trends and technologies in the RPA space, hence influencing future automation paths and deciding how RPA tools are to be exploited for competitive advantage.

Recommendations

1. Organization Needs Analysis

The initial step of identifying the right RPA tool for the organization is to carry out a good needs assessment of the automation need within its business process. It means considering repetitive and rule-based processes which can be automated by either identifying the need or understanding technical requirements and infrastructures for the existing systems. Proper engagement with various stakeholders of departments will assure the representation and support that an organization seeks in its chosen RPA tool.

2. Features and scalability of the tool

The features chosen in RPA tools depends on the usage applied to the organization, for example, AI integration, cloud deployment options, easy usage, and so on. Scalability will be important because the tool is supposed to scale up according to the growth requirements of the organization and expand across multiple departments, which will have long-term benefits.

3. Adoption of Low-Code/No-Code solutions

RPA tools which offer low-code or no-code development environments should be implemented for automation. It empowers non-technical users in contributing towards automation efforts and hence limits dependence on IT teams and delivers automation initiatives at a much faster pace. Democratization of automation fosters innovation and leads to quicker reaction towards business needs.

4. Security and Compliance

More organizations are automating sensitive processes. In such a scenario, it is of utmost importance that security and compliance features are in focus so that RPA tools you choose have good security measures like encryption, access controls, and audit trails that can ensure automated processes conform to a corresponding high level of regulatory standards and maintain sensitive data well protected.

5. Monitor and Improve

On Automation Efforts: Implementing RPA solutions requires continuous monitoring of the performance and outcome. Metrics analysis with the feedback would give areas that need improvement on the efforts towards automation, and thus businesses could orient them appropriately for return on investment.

6. Industry Trends

Keeping pace with emerging trends in RPA and related technologies will be required by organizations to make appropriate strategic decisions on automation strategies for their establishments. Organizations will realize the inroads of the development of AI, process mining, and cloud computing and position themselves to exploit new capabilities to enhance the effectiveness of these automation initiatives.

In sum, the future promises of RPA development tools seem quite tremendous, with tremendous leaps on the way forward in changing the basic ways process automation is carried out within organizations. With thoughtful approach in selecting their tools for RPA and embracing best practices about automation, RPA could be fully unleashed for organizations toward operational excellence and innovation.

REFERENCES

Agostinelli, S., Lupia, M., Marrella, A., & Mecella, M. (2020). Automated generation of executable RPA scripts from user interface logs. In *Business Process Management: Blockchain and Robotic Process Automation Forum: BPM 2020 Blockchain and RPA Forum, Seville, Spain, September 13–18, 2020* [Springer International Publishing.]. *Proceedings, 18*, 116–131.

Agostinelli, S., Marrella, A., & Mecella, M. (2021). Exploring the challenge of automated segmentation in robotic process automation. In *Proceedings of the Conference* (pp. 38-54). https://doi.org/DOI: 10.1007/978-3-030-75018-3_3

Andrade, D. (2020). Challenges of automated software testing with robotic process automation RPA—A comparative analysis of UiPath and Automation Anywhere. [IJICR]. *International Journal of Intelligent Computing and Research, 11*(1), 1066–1072. DOI: 10.20533/ijicr.2042.4655.2020.0129

Asatiani, A., Copeland, O., & Penttinen, E. (2023). Deciding on the robotic process automation operating model: A checklist for RPA managers. *Business Horizons, 66*(1), 109–121. DOI: 10.1016/j.bushor.2022.03.004

Axmann, B., & Harmoko, H. (2022). Process & software selection for robotic process automation (RPA). *Tehnički Glasnik, 16*(3), 412–419. DOI: 10.31803/tg-20220417182552

Bosco, A., Augusto, A., Dumas, M., Rosa, M., & Fortino, G. (2019). Discovering automatable routines from user interaction logs. *In Advances in Intelligent Systems and Computing* (pp. 144-162). https://doi.org/DOI: 10.1007/978-3-030-26643-1_9

Choi, D., R'bigui, H., & Cho, C. (2022). Enabling the gap between RPA and process mining: User interface interactions recorder. *IEEE Access : Practical Innovations, Open Solutions, 10*, 39604–39612. DOI: 10.1109/ACCESS.2022.3165797

Gupta, S., Rani, S., & Dixit, A. (2019, October). Recent trends in automation—a study of RPA development tools. In *2019 3rd International Conference on Recent Developments in Control, Automation & Power Engineering (RDCAPE)* (pp. 159-163). IEEE.

Herm, L., Janiesch, C., Steinbach, T., & Wüllner, D. (2021). Managing RPA implementation projects. *Robotic Process Automation*, 27-46.

Khan, S. (2020). Comparative analysis of RPA tools—UiPath, Automation Anywhere, and Blue Prism. *International Journal of Computer Science and Mobile Applications, 8*(11), 1–6. DOI: 10.47760/ijcsma.2020.v08i11.001

Kim, K. T., Jeong, S. N., & Lee, S. H. (2021). Development of RPA with information extraction module. In *Proceedings of the Korean Society of Computer Information Conference* (pp. 435-436). Korean Society of Computer Information.

Kim, S. H. (2023). Development of evaluation criteria for Robotic Process Automation (RPA) solution selection. *Electronics (Basel), 12*(4), 986. DOI: 10.3390/electronics12040986

Kronkvist, J. (2020). Development of robotic process automation enhancement tools: Platform centralization of RPA key aspects.

López-Carnicer, J. M., Del Valle, C., & Enríquez, J. G. (2020). Towards an open-source logger for the analysis of RPA projects. In *Business Process Management: Blockchain and Robotic Process Automation Forum: BPM 2020 Blockchain and RPA Forum, Seville, Spain, September 13–18, 2020* [Springer International Publishing.]. *Proceedings, 18*, 176–184.

Mullakara, N., & Asokan, A. K. (2020). *Robotic process automation projects: Build real-world RPA solutions using UiPath and Automation Anywhere*. Packt Publishing Ltd.

Noppen, P., Beerepoot, I., Van De Weerd, I., Jonker, M., & Reijers, H. A. (2020). How to keep RPA maintainable? In *Business Process Management: 18th International Conference, BPM 2020, Seville, Spain, September 13–18, 2020* [Springer International Publishing.]. *Proceedings, 18*, 453–470.

Rod Koch, C. M. A., & CSCA, P. (2019). Can RPA improve agility? *Strategic Finance, 100*(9), 68–69.

Tiron-Tudor, A., Lacurezeanu, R., Bresfelean, V. P., & Dontu, A. N. (2024). Perspectives on how robotic process automation is transforming accounting and auditing services. *Accounting Perspectives, 23*(1), 7–38. DOI: 10.1111/1911-3838.12351

Tripathi, A. M. (2018). *Learning robotic process automation: Create software robots and automate business processes with the leading RPA tool–UiPath*. Packt Publishing Ltd.

Wewerka, J., & Reichert, M. (2021). Robotic process automation—A systematic mapping study and classification framework. *Enterprise Information Systems, 17*(2), 1986862. Advance online publication. DOI: 10.1080/17517575.2021.1986862

Willcocks, L., Hindle, J., & Lacity, M. (2018). Keys to RPA success. *Executive Research Report, Knowledge Capital Partners, USA, Technical Report.*

Yarlagadda, R. T. (2018). The RPA and AI automation. [IJCRT]. *International Journal of Creative Research Thoughts*, *2320*(2882).

Ylä-Kujala, A., Kedziora, D., Metso, L., Kärri, T., Happonen, A., & Piotrowicz, W. (2023). Robotic process automation deployments: A step-by-step method to investment appraisal. *Business Process Management Journal*, *29*(8), 163–187. DOI: 10.1108/BPMJ-08-2022-0418

Chapter 10
Intelligent Robotic Process Automation for Healthcare Service Efficacy at the Grassroot Level

Animesh Bondre
https://orcid.org/0009-0000-9840-545X
Optum, India

Sanjukta Ghosh
Srishti Manipal Institute of Art, Design, and Technology, India

ABSTRACT

Healthcare services in rural areas face numerous challenges due to limited resources, infrastructure, and awareness. This chapter examines the use of Intelligent Robotic Process Automation (RPA) to address the inefficiencies in rural healthcare delivery. An intelligent RPA system aims to streamline operational processes, enhance patient intake procedures, optimise resource allocation, and improve the overall efficacy of healthcare services at the grassroots level. By integrating intelligent RPA with existing digital healthcare infrastructure and information systems of telehealth, the chapter explores the potential impact of this technology in transforming rural healthcare, ultimately reducing the chances of human errors and improving access to quality healthcare for the rural population.

DOI: 10.4018/979-8-3693-4365-4.ch010

INTRODUCTION

The healthcare landscape of the nation showcases very highly contrasting services. The major share of healthcare infrastructure and resources are concentrated in urban areas. Even if several government programs for the growth of rural healthcare have been initiated, the procedural delay in implementation leads to its ineffectiveness. Various contagious diseases like diarrhea, amoebiasis, typhoid, infectious hepatitis, worm infestations, measles, malaria, tuberculosis, whooping cough, respiratory infections, pneumonia, and reproductive tract infections badly affect rural areas of the country. Lack of basic health and hygiene awareness has made the condition even worse. The poor state of the primary health care system in rural areas of India is a consolidated outgrowth of multiple degraded systems and most often lack of coordination among various stakeholders in the system. It signifies not only lacunae in existing policy and infrastructure but blockage in the development of an efficient primary medical service at the grassroots level.

On the other hand, urban healthcare services are flooded with state-of-the-art medical facilities while rural India continues to grapple with a myriad of challenges. As per the 2022 report, India's rural population is around 908.8 million whereas urban India has a population of around 508 million. This shows the majority of the population in the country faces unique obstacles regularly that hinder access to quality healthcare. Many villages are suffering from major challenges related to limited access to healthcare facilities, shortage of human resources, inadequate medical supplies and equipment, limited awareness and education; infrastructure and logistics, etc. To address these issues the system needs to keep a continuous track of various activities and processing data in real time. The data will be generated from a variety of internal and external sources ranging from labs, apps, insurance agencies, etc. Robotic Process Automation is emerging to be a promising technology to tackle the health care system through efficient processing of information. This can integrate existing digital infrastructure or information systems to generate effective healthcare service and reduce the chances of human error in certain touch points. This can autonomously collect and combine data from multiple sources through the integration of different kinds of systems through software robotics. This will allow the workforce to invest their talents in areas that require more human touch and optimize the healthcare budget and human resources in various mundane mechanical jobs. This will enhance the speed and efficacy of healthcare systems in rural India.

This chapter focuses on the Intelligent Process Automation system's various scopes and opportunities in the healthcare domain, along with some use cases and scenarios. The applicability of these cases and scenarios will be explored, considering the larger population at the grassroots level.

RPA IN HEALTHCARE

A Business has multiple processes and requires integration for its effective functioning. The process is a core asset and is a collection of connected tasks, jobs, and responsibilities. It integrates, systems, data, and resources and after completion delivers the service/product and achieves the business's goal. To ensure that these processes are consistent and businesses can take advantage of improvement opportunities in the field of Business Process Management has emerged that offers methodologies, tools, and techniques to achieve it. (Grigori, D. et al.,2004 ; Chakraborti, T et al 2020)

According to (Business Process 101 : Definition, Steps, and Example, Guide for 2024,2024) business processes are organized into three categories: operational business processes, supporting business processes, and management business processes.

Operational Business

These processes add direct value and are in contact with the customers. In the healthcare industry, these processes are related to healthcare delivery and can be categorized into Patient/Customer Intake, Health Care Delivery, and Patient/Customer Discharge.

Supporting Business Process

These support the operational processes and are not in direct contact of customers. Important categories of these processes in healthcare are Human Resources Management and Financial Management

Management Business Processes

These control and regulate the operational & Supporting business processes.

Examples of this process in healthcare are related to governance, quality assurance, and strategic planning.

The Healthcare Business process is highly dynamic, complex, and ad-hoc, and has a multi-disciplinary nature, so to simplify the scope let us focus on an example of an Operational Business Process of Patient/Customer Intake where a patient needs to schedule medical appointment scheduling process as defined by (Dumas, M et al 2018) the ingredients of a business process are:

Events and Activities Or Tasks

A medical appointment requisition is an Event since it happens automatically and has no duration, this has led to a series of activities for eg: When the administration assigns appointments to the request queue by checking medical records or services.

Here confirming the date /time in the system can qualify as a task as it is rather simple can be seen as one single unit of work and does not require many steps.

Decision points

This affects the way a process is executed for example checking if the available time slot is within the clinical time or not if it is not the process changes to checking more specialty physicians and in case it is within the available time slot it moves to the task of confirming the date/time

Actors & Customers, Physical Or Immaterial Objects

Here there are three human actors Admin staff, patient, physician, and physical and immaterial objects can include the software used to check the availability and book the appointment and electronic record of the appointments & patient

Outcomes

Finally, this process leads to the completion of the appointment schedule by setting the slot after confirming the date and time.

In the process mentioned in (Martinho, R., Rijo, R., & Nunes, A, 2015) healthcare administrative staff has to select physicians from a specialty in case there was no available slot within the clinical time but the process mentioned in the literature is improved by updating the system to suggest the next available timeslot for all the physicians in the specif specialty without selecting each physician this reduced the complexity in the system. Every business wants to find such opportunities to increase growth and eliminate time-consuming tasks. Business Process Automation (BPA) seeks to improve the efficiency of business processes in terms of cost, resources, and investment by automating the management of relevant information and data, the time spent by team members, and the execution of logic (Chakraborti, T et al 2020; Antwiadjei, L. (2021)).

Robotic process automation is an emerging technology in BPA that creates software robots that perform tasks by automating repetitive, labor-intensive, and time-consuming tasks previously done by humans and improve the efficacy of business processes (Jain, R., & Bhatnagar, R. 2019; Barla, N. H., Almeida, S. M., & Almeida, M. S., 2023).

The healthcare industry is burdened by the increasing volume of patients, high operating costs, stringent compliances, and slow and redundant processes. Many of the industry's current processes are manual, such as staffing, data management, administrative duties, claim management, and more. Many of the tasks are repetitive and time-consuming for eg: Administrative work like creating notes, reviewing medical records, and documentation of patient records (*Understanding the Role of RPA in Healthcare | MuleSoft*, n.d.; Administrator, 2024; Williams, 2021). According to the literature (Woolhandler, S., & Himmelstein, D. U. 2014; 1Rivet. 2024, March 12), these tasks consume one-sixth of physicians' working hours. Since these repetitive tasks involve human intervention they can lead to overworked physicians leading to burnout, low productivity, operational inefficiencies, and processes can be prone to errors ultimately leading to a poor patient experience. To address these challenges, integrating healthcare processes with Robotic Process Automation (RPA) can automate these tasks and increase productivity, improve patient experience and employee satisfaction. The healthcare industry is the largest industry that has seen rapid growth in leveraging RPA, where it is ahead of other industries. In fact, 50% of US healthcare providers are predicted to invest in RPA (Fivesdigital, n.d.; Barla, N. H., Almeida, S. M., & Almeida, M. S. 2023).

CASE STUDIES OF RPA IN HEALTHCARE

Literature (Deloitte, 2017), mentions the key advantages like cost reduction, increased productivity, increased accuracy, minimization of errors, enabled compliance, and the elimination of risks of using RPA. Let's explore some use cases that have these outcomes:

Increasing System Productivity To Address The Surge In Healthcare During Covid

Turku Welfare Division reported a sudden increase from 2500 phone calls per day to 6000 phone calls similarly in the United States managing patient data became difficult as the number of patients increased beyond 2000 per day. In response to this increasing volume, there was a growth and evolution in the usage of RPA to combat healthcare needs as the virus spread rapidly.

One successful example of RPA implementation was done by the Turku Welfare Division in Finland (Kedziora, D., & Smolander, K., 2022). They automated various processes using robots it search for laboratory tests done in Finland acknowledge the results and upload it to a web service where citizens can access them. Along with the results, the robot also adds medical records and subscriptions. The results are informed to the citizens through SMS and spending on it asks the infected people to take safety precautions and the health professional reach out to them.

Similarly application of automating the data collection and collation processes was done by the National authorities in the United The ground staff and health workers after data collection updated the National Authority portal but this became difficult to collate state data from the portal because of increasing volume. Here the robot collated the data for each state portal in real time.

Turuku welfare division also automated the referral process by using digital assessment tools for COVID symptoms and guiding affected citizens to testing centers, here a robot finds the affected people and creates a laboratory appointment booking and examination referral and also creates a report after the examination for the doctors.. Additionally Turuku welfare division also had robots for Reserving vaccination times and signing up with nurses., Scan test results for border authorities and guide people to testing centers based on the results

Thus here RPA was leveraged due to easy and faster integration, and the ability to amend/change the logic due to regulatory and compliance changes which helped in reducing the the impact on the economy and battling the immediate effect of addressing the sudden increase in the volume for healthcare for the general public.

Reduced Risk of Human Errors Due To Manual Data Entry

The healthcare industry collectively spends $2.1 billion every year on provider data management where tasks are performed poorly, prone to error, and are manual. Errors can range from incorrect billing of the patient is billed incorrectly to receiving the wrong medication or incorrect treatment.

RPA is an efficient technology that can be leveraged to reduce these errors. Healthcare IT companies like Optum (United Health Group) are leveraging RPA for Medicaid assessment. To ensure people get benefits from Medicaid, assessors have to conduct interviews that involve filling out forms and answers in the database to determine eligibility. These interviews can generate data with as many as 1000 questions. Entering the data is a labor-intensive and time-consuming task that can lead to errors and poor internet connectivity can hinder real-time data entry. However, Optum has automated this process using bots, where accessors no longer have to re-enter the data into the systems as it uses the filled-out PDFs by the accessors and auto-fills the content in the database. Because of this automation accessors save a lot of time, errors are avoided and they can concentrate on interviewing more beneficiaries. Similarly administrative errors like syncing data, form of data entry for claims or new patient forms, completion checks, and extracting patient information from electronic health records (EHRs).

Another example is of Helse Vest which provides health services to west Norwegian countries where doctors have to same information register information into different systems which have a potential of errors (the journal system DIPS, the research database, and reports to the Cancer Registry). Here a system Robbie Ves was designed to automate the systems and eliminate manual entry reducing the burden on cancer doctors and increasing the accuracy of patient data and reducing errors.

Compliance Adherence For Patients And Providers

Both the patients and providers have to comply with guidelines where patients are asked to adhere to medications or post-discharge guidelines and healthcare organizations have to meet regulatory compliances set by agencies like the Health Insurance Portability and Accountability Act (HIPAA), Centers for Medicare and Medicaid Services (CMS), etc. and RPA can benefit both the use cases.

First lets tak a look at RPA's application for patient compliance as It becomes difficult for healthcare providers to track and monitor the patient after they are discharged and if patients don't comply with health requirements it can lead to re-admission HCAHPS (Hospital Consumer Assessment of Healthcare Providers and Systems) scores. Here RPA solution by UIPath ensures the accuracy of discharge instructions and sends patients reminders about prescriptions and appointments.It also notifies the health providers if patients need further assistance or any further requests.

Now for compliance for healthcare providers, OptumServe has automated the process of compliance and tracking of external materials review as currently, it is a manual and time-consuming process, lacks comprehensive tracking abilities, and also lacks visibility in the review process. It has developed a bot that takes documents via email along with a form for review. The bot assess the form and identifies if there is any missing information, in case there is none the form gets uploaded to the database

When it comes to the compliance external materials review, the current process for many clients is manual, time-consuming and inconsistent. Current practices also tend to have a lack of comprehensive tracking capability and complete visibility into reviews performed. Optum Serve expedites the compliance and tracking review process with RPA. We developed a bot that intakes documents for compliance review from an Outlook inbox and leverages an existing source of truth, including a keyword list. The process results in over 90% of the work being automated. The Bot then asses the document using a taxonomy workbook based on which it identifies the correct document reviewer and informs the submitter along with the reviewer.

Depending on the reviewer's decision the bot records it in the database and notifies the submitter. This has helped Optum save e 9,465 hours and $950,000 over five years.

Thus automating such a process can help in identifying and alerting healthcare professionals to potential compliance issues which further can be analysed to identify patterns and anomalies. This can help healthcare professionals address them before it becomes an issue.

Compliance and regulatory requirements are very important in clinical settings here RPA's ability to audit and trace enables organizations to maintain the records.RPA solutions can also help in data access control, encryption, and security measures to comply with regulations like the Health Insurance Portability and Accountability Act (HIPAA) and the General Data Protection Regulation (GDPR).

Similarly, organizations like Kofax, HelpSystems, Appian, SS&C Blue Prism, UiPath, and more are automating labor-intensive healthcare processes like self-service scheduling, medical record management, healthcare inventory management, and more.(*Understanding the Role of RPA in Healthcare | MuleSoft*, n.d.; Administrator, 2024; Williams, 2021)

Case Studies of Intelligent RPA in Healthcare

Traditionally RPA is used for automating repetitive tasks i.e 'Doing'. However, with advancements and integration with emerging technologies like Artificial intelligence Machine learning, and natural language processing the scope has broadened beyond automating standardized repetitive tasks to more complex tasks that involve decision-making, understanding information, insight analysis, and coordinating with multiple automation solutions i.e 'Thinking' as well as 'Doing' (Chakraborti, T et al 2020; Intelligent Automation in Healthcare & Cognitive Computing, n.d.; Prism, 2024). We saw some of the above case studies where it has enabled the integration of systems which has helped to ease the load and increase the efficiency of the business process, By adding thinking to the process automation the systems become smarter and this field is called Intelligent Robotic process automation, it takes RPA a step further where it automates more complex tasks that need decision making, analysis composition, coordination based on the insights, and collaboration between multiple IPA solutions. It automates the entire lifecycle and also identifies opportunities to automate based on the results of monitoring the performance of the system.

Thus AI & ML approaches have made it possible to extract information, classify, associate, optimize, group, predict, and identify patterns. Thus interaction of this with Robotic process automation that mimics repetitive tasks performed by humans using software robot's application interface (UI) now can gain cognitive abilities and perform analysis with unstructured data input, such as images, text, videos, and vocals. Chatbots are a well-known application of Intelligent process automation. But with widespread developments in the field of Artificial intelligence and machine learning like handling large datasets in real-time, context-aware computing, and powerful graphic processes RPA can benefit from it Agostinelli, S., Marrella, A., & Mecella, M,2020). Let us explore some emerging applications of IPA in healthcare:

Automating Prior Authorization Of Medication

Prior authorization is also known as pre-authorization, it is a process done before providing care to the patient. This process involves the exchange of information like medical necessity, the cost of the patient, and whether payers will reimburse the providers for the patient's treatment and procedures. Prior authorization is necessary to lower healthcare costs but it involves a lot of time-consuming paperwork which providers could spend more on taking care of the patients. The process is long and Errors in this process can further lead to delays in the care received by the patients.

Here Intelligent Robotic process automation can be leveraged to overcome the excess time required to gain authorization. It can take 8-10 days for the authorization process to Intelligent Process automation has the potential to reduce it to 24-48 hours by optimizing authorization workflows, ensuring a reduction of claim denials from health insurers, relieving common patient access challenges, speeding up patient registration, and automate critical revenue cycle tasks.For eg: Nvideous automated the prior authorization process for an eye care group by using bots to extract patient details from legacy applications, automated search queries on insurance carrier websites, downloaded eligibility details in the required format, updated internal portals, communicated exceptions, and sharing details with patients. This resulted in 1500 staff hours saved per month along with 100% data accuracy (Hester, 2024; Preauthorizations Intake - RPA Component | UIPATH Marketplace | Overview, n.d.; SS&C Blue Prism, 2023).

Revenue Cycle Management

Revenue Cycle Management (RCM) handles claims adjunction, generates revenue for healthcare providers, and negotiates claims payments for payers. It is the process that tracks patient information right from registration to final payment that settles outstanding balances and It is also an important process for a healthcare organization's business to stay viable (Operationally & Financially). However, there are several challenges that payers providers and patients face which lead to slow and inefficient processes affecting productivity, cash flow, and patient experience, billing errors, coding errors, claims denial, unpaid claims, and more.

Here intelligent process automation is leveraged to overcome these challenges and optimize the workflows by eliminating manual processes like accounts receivable, prior authorization, insurance verification, claims processing, and data entry. Intelligent RPA can impact and befit:

Claims Department: Here it can help optimize the throughput and prior authorization which improves payer communication through access to claim status, creating follow-ups and quicker reimbursements

Back-Office: Bots can be deployed for premium processing plan enrolments, determining eligibility, and filing expense reporting.

Finance & Accounting: Bots automate and streamline the counts of payable and accounts receivable transactions, payment posting, billing, and reimbursements and eliminate manual processes.

Hospital Records: Intelligent automation can be used for speeding patient enrollment, appointment requests, and updating and extracting information (Suke, S. G., Kosta, P., & Negi, H. 2015; SS&C Blue Prism, 2024).

Intelligent Pharmacovigilance Processes

Pharmacovigilance (PV or PhV) stands out as one of the most demanding processes within the pharmaceutical industry, characterized by its heavy reliance on labor and the intricate landscape of stringent regulations and deadlines, all centered around ensuring drug safety. Traditionally overseen by healthcare professionals, the core responsibility of PV entails identifying and categorizing adverse events, generating reports for regulatory bodies, and proposing necessary actions. In recent times, the surge in reported incidents, fueled by the proliferation of reporting channels including social media, has exacerbated the labor-intensive nature of PV.

Both clinical and post-market PV present distinct challenges that conventional tools and technologies struggle to address, prompting pharmaceutical companies to seek cost-effective and efficient alternatives such as RPA. Furthermore, the exploration of AI for enhancing operational efficiencies in adverse event processing, encompassing tasks like medical review, causality assessments, signal, and risk management, is underway.

Intelligent RPA holds the promise of simplifying and expediting PV processes by automating repetitive, rule-based tasks and ensuring the quality and accuracy of generated reports. Often, RPA requires supplementation with Natural Language Processing (NLP), Image Processing, Optical Character Recognition (OCR), and rule-based prioritization techniques to handle unstructured datasets effectively. Some prevalent applications of RPA in this domain include managing acknowledgment emails and their follow-ups, organizing source documents, mapping WHO products, validating date chronologies in reports, and cross-verifying duplicate cases for completeness (Suke, S. G., Kosta, P., & Negi, H.,2015)

HEALTHCARE SERVICE LIFECYCLE IN INDIA

Inequality in Healthcare At The Grassroots

The Indian healthcare industry is a complex system of public, private, and informal healthcare sectors where the private sector has more hospitals compared to the public sector. The complexity is increased with the rapid integration of information technology. The vast network consists of primary healthcare which is the first point of contact, secondary healthcare which focuses on specialty services, and tertiary care for advanced medical services across the three sectors.

Public care services are managed by the central and state governments. It is divided into three levels:

1) Primary health centers (PHC) for increasing awareness and preventative care
2) Community health centers (CHC) to link PHCs and general hospitals
3) District hospitals for final referrals from the PHC and CHC

Whereas the private sector consists of two types of sub-systems 1) Institutions (hospitals, nursing homes 2) Independent practitioner-run clinics. The individual practitioners follow both modern and traditional medications. Hospitals are of two types for-profit and not -for profit. For-profit hospitals are generally assumed to be quality and efficient where they have trained medical staff and physicians conscious the urban regions are more dependent on these health services. Not-for-profit hospitals are run by charitable and community-based organizations that are separate from the government (Naithani, C., Sood, S. P., & Agrahari, A. 2023).

Because of the scattered services, the cost of care in private hospitals is generally high. This gap is further increased as most of the healthcare facilities are available in the urban regions leading to inequitable distribution of care services. 75% of the doctor population is available in the urban regions but 68% of the population in India lives in rural regions. Thus in rural regions has informal sector plays a major role in healthcare needs i.e. community health workers. They work towards promotion, awareness and immunization, pregnancy and elderly care, and more.WHO recommends a doctor-population ratio of 1:1000 but currently we have a deficit in the number of doctors which is leading to a ratio of 0.62:1000 (Gangolli, L. V., Duggal, R., & Shukla, A.,2005).

The challenges faced by the Indian healthcare system can broadly be stated as Inadequate infrastructure, Shortage of healthcare professionals, Urban-rural disparities, Financial constraints, and health insurance, Insufficient public healthcare funding, Fragmented healthcare system inequity in access to care, Growing burden of non-communicable and communicable diseases. Various studies show that a healthcare system that has strong comprehensive primary health care (CPHC) delivery systems can result in better health outcomes for people and reduce the nation's economic liability on health care. Strengthening CPHC in terms of increased access, financial protection and improved quality of care is especially relevant for low- and middle-income countries like India. Studies by the World Health Organisation estimate that scaling up primary healthcare interventions across low- and middle-income countries can save 60 million lives and increase average life expectancy by 3.7 years by 2030. According to WHO and UNICEF, comprehensive primary health care has three major components:

(a) Meeting people's health needs through comprehensive promotive, protective, preventive, curative, rehabilitative, and palliative care throughout the life course,
(b) Systematically addressing the broader determinants of health and
(c) Empowering individuals, families, and communities to optimize their health (WHO & UNICEF, 2018).

Every corner of the country should have access to effective primary healthcare facilities in its proximity, there should be enough continuous guidance and support provided by medical practitioners. This can then create a meaningful impact on the health and well-being of the population. Our current primary health care system is going through various challenges to provide quality primary health care.

Thus to make up for it telemedicine is partly being leveraged, it becomes a very useful tool since the rural regions are separated by distance from healthcare facilities majorly situated in the urban regions. However, there is a difference in the implementation of telemedicine between the public and private sectors. Telehealth broadly helps in the areas of consumer health information, health and medical education, and clinical medical services. Here the public sector lacks clinical medical services compared to the private sector because telemedicine follows a top-down structure in the public sector where it is designed based on the technology and not based on the user's needs. Also, the Implementation is quicker in the private sector due to available infrastructure, skilled workforce, and research push and resources. (Sood, S. P., Negash, S., Mbarika, V. W., Kifle, M., & Prakash, N, 2007)

Let's consider a sample use case of a telemedicine service e-Sanjeevani that provides remote patient care and bridges geographical distances. Understanding the care journey of a patient in this ecosystem will help to identify different processes and touchpoints that have the potential to get automated and further increase the efficiency using Intelligent RPA.

The simple journey of a patient in India (Venkateswaran, 2020; Kumar, A, 2023; Talib, F., & Rahman, Z, 2013), and its touchpoints is:

Awareness of the Condition

The patient self-assesses their conditions and symptoms and the understanding of the disease is shaped by the patient's own explanation, trust established with medical providers, and past experiences of family members. Depending on the availability of digital sources patients also carry out their research and reach out to online communities by posting questions on social media etc.

Identify the Right Provider Or Specialist

If the patients are not able to self-medicate they make their first contact with the health system where they reach out to the nearest medical healthcare facility. The majority of the patients contact private care providers where private secondary care hospitals are accessed by the middle class and tertiary care by the highest income class. but the. This can be done via call center, email, mobile, etc But patients often describe their first contact with the health system as not helpful because it doesn't guide them clearly toward the next step

Screening & Diagnosis

The patient's ability to get testing and monitoring done depends on the cost and distance of the lab test centers. The Government provides free tests but they are generally far away from the patients or they have to pay an informal fee, hence most of the time patients are self-medicated based on diagnosis and part prescriptions. Depending on their health conditions patients seek additional or specialty help in this phase of their journey. When the diagnosis report is ready the testing facilities inform the patients where they have to visit the facility to collect the report. Depending on the relationship of the doctor with the facility they can be informed about the report as well or the patient visits the doctor with the generated report to get treatment.

Treatment

The ability of patients to get treatment for the diagnosed disease often depends on the cost of treatment, the government is addressing this issue by making medication available at affordable rates. Based on the diagnosis report the patients are given treatment but there is a lack of time for providers to explain the treatment which can influence their decisions.

Adherence through Behavioral/Lifestyle Change -Patents have to follow the prescribed changes and medication to avoid readmission and have better health. Patients also fail to adhere to medication treatment due to complex plans and poor communication. They also have to deal with various social relations eg: The elderly have to depend on their family members to travel to health facilities

Ongoing Care/Post-Operative Care/Palliative Care

The patient manages their care between clinical visits. The health system fosters engagement between the patient & physicians to enable the patient to better manage his/her own care.

E-Sanjeevans offer interventions/processes across these seven stages by leveraging them for education and telemedicine where medical peripherals like medical film scanners, video microscopes, electronic stethoscopes, and PC-based ECG machines are connected and integrated into the application to support screening and diagnosis. It is further used for Doctor to Doctor (D2D) telemedicine for patient treatment i.e. a doctor seeking advice from another specialist. This works on a hub and spoke model where the health and wellness center can remotely access specialist care available in the nearest city. The e-Sanjeevani application connects three tertiary care hospitals that have good telecommunication services and facilitates consultation by integration with integrated services digital network (ISDN), Plain Old Telephone System (POTS), and Very Small Aperture Terminal (VSAT) for medical data related to radiology, pathology, and cardiology.

E-Sanjeevani also supports ongoing & post-operative care care through Electronic health record management. The e-sanjeevani application includes demographic, physiological, and clinical details rate, and temperature;—medical images, patient history, and clinical examination reports. The data is recorded through medical instruments are entered manually in the form of images, digitized readings, etc. Clinicians & Doctors can create, view, edit, and annotate necessary information of the patient record, radiographs and pathological reports, treatment charts, generate prescriptions, plan doctor visits and appointment booking, etc (Naithani, C., Sood, S. P., & Agrahari, A, 2023).

LIFECYCLE AUTOMATION OF HEALTHCARE SERVICES

The lifecycle of telehealth as we saw in the above section starts from awareness of the condition to ongoing Care/Post-Operative Care/Palliative Care. Below are a few applications across the life cycle where Intelligent RPA for pharmacovigilance comes & data storage and management comes under Care/Post-Operative Care/ Palliative Care and Automating Teleophthalmology is an example of Diagnosis and screening.

Intelligent RPA for Pharmacovigilance

As seen in the above sections pharmacovigilance detects, assesses, understands, and prevents medicines to prevent any adverse effects/problems adverse effects. Thus here e-sanjeevanee's EMR data set can support the process. In India, a major challenge faced by pharmacovigilance is underreporting of adverse drug effects due to a lack of medical expertise in drug administration, and inadequate nationwide awareness. India follows a spontaneous or voluntary system for Adverse drug reporting and at times the effect and the drug cannot be related or linked or doctors might feel guilt, and fear of litigation which leads to underreporting. Thus there is a need for intensive monitoring, identification, detection, and prediction of the Adverse drug reactions and improving the knowledge and attitude for reporting it. Here integrating e-sanjeevani's process can help potentially use intelligent RPA for Analyzing and predicting adverse drug events.

RPA can be leveraged to automate repetitive tasks involved in PV like checking the generated reports and processing unstructured data. E-Sanjeevani has unstructured data generated from medical peripherals like medical film scanners, video microscopes, electronic stethoscopes, and PC-based ECG machines. Thus here intelligent RPA can support processing this data when paired with NLP, Image Processing, and Optical Character Recognition (OCR). This can help in identifying, monitoring, and surveilling any potential Adverse drug reactions that can be under-reported due to the failure of the medical teams to identify it or correlate it with biochemical, pathological, or radiological abnormalities. This can further be linked to automated periodic E-mail, and SMS alerts to increase awareness about the identified Adverse drug reactions among doctors and practitioners Thus RPA can help in analyzing existing data but it can also help in creating new knowledge by predicting Adverse drug reactions (Parikh, D. P., Sattigeri, B. M., & Kumar, A, 2014; Hester, 2024)

Automating Teleophthalmology

Glycoma causes irreversible blindness and hence there is a need to detect it early. There is an increasing acceptance of tele-ophthalmology that helps in remote screening and eye consulting. E-sanjeevami provides specialist consultation in ophthalmology and here intelligent RPA can help in automating repetitive human tasks of the ocular screening process. When a patient visits a local hospital or fundus images can be captured and stored in the database which can later be retrieved if needed. Here RPA can act as an intermediary between the patient and the medical staff. The fundus images and previous medical history can be automatically gathered and submitted for glaucoma analysis. The application can detect glaucoma and if it shows that it is a severe condition we can have a D2D consultation with an eye specialist for timely treatment. The ML models within the application and adjusting parameters like OD size help the doctors to screen and make the glaucoma diagnosis remotely. After the diagnosis is confirmed the application can be further leveraged for updating relevant information in the electronic health records of E-sanjeevani and further support follow-up appointment scheduling and payments.(Thainimit et al., 2022)

Data Storage, Management, And Retrieval

Currently, E-sanjeevani stores patent records, radiographs and pathological reports, and treatment charts, and generates prescriptions. However, doctors have to manually create, view, edit, and annotate necessary information. This process has the potential to be automated and made intelligent through Intelligent bots leveraging Natural Language Processing (NLP) that can recognize the context and extract content from prescriptions, health documents, and stored electronic health records gathered from doctor-patient interactions and maintain the data infrastructure. This can help in improving the diagnostic process by enabling predictive analysis which supports early detection and enables healthcare providers to intervene in the early stages. For eg: based on the data that is captured in the electronic health records of E-sanjeevani like age, lifestyle, and genetics intelligent RPA processes can support forecasting the risk of diabetes or cardiovascular diseases and thus enable healthcare providers to provide preventative and personalized measures.

Further intelligent RPA can also be leveraged for automating administrative work. After a patient is done with teleconsultation with a doctor, the doctor has to document their visit for the electronic health record. This leads to a lot of administrative work and increases their workload. Here Intelligent RPA can support doctors by automating the tasks of entering the patient information where a bot can leverage the medical notes or recordings to understand the context fill required documents, and forms, and create needed prescriptions which later can be shared with the patient by the bot. Prescriptions and decisions generated from doctor-patient interactions can be integrated into insurance, pharmacy management, and diagnostic services. Bots can automate this process, eliminating the need for patients and doctors to re-enter the data generated by their respective service providers

FURTHER SCOPE & CHALLENGES OF INTELLIGENT RPA IN HEALTHCARE

Even though there are a lot of benefits of intelligent RPA and it has benefits to revolutionaries healthcare it has challenges that can broadly be categorized as organizational and technical Healthcare technologies often fail to serve their purpose when they do not consider the complex social structures they are intended to serve as this can lead to challenges (Tak, A; Hall, J. L, & McGraw, D, 2014; Venigandla, K. 2022). For example:

Technical Challenges (Data Security, Privacy, Autonomy)

With growing digitization considerations for healthcare data become important especially in the context of healthcare as it consists of highly sensitive data that needs to be regulated, encrypted, stored securely, controlled access, and data masking, to protect confidential information from unauthorized access, disclosure, or misuse. In the use case of telehealth, we saw in the above section where E-sanjeevami involves the bidirectional collection and communication of sensitive health information among healthcare providers and patients. This data when combined with RPA has applications for analysis and predictions which creates a risk for privacy and security. Privacy risks include a lack of controls or limits on the collection, use, and disclosure of sensitive personal information. Application employed for PV needs medical data to be transferred from the device it is collected from which can be accessible by people beyond the healthcare provider and can be used by third-party advertisers which patients did not consent to and hence creates a need for privacy protocols to be in place. Hence the systems needs to be compliant with healthcare regulations and should have regular audits but it become difficult to maintain the train due to automation in the healthcare processes.

Similarly, there are also security concerns for the communication that is facilitated by the RPA model where there can be unauthorized access to the functionality of supporting devices as well as to data stored on them; and untrusted distribution of software and hardware to the patient. Hence employing intelligent RPA there should be regulatory compliance including data privacy laws, and consumer protection regulations. Further, RPA can automate the processes but it can come at the cost of reducing the autonomy of patients and healthcare providers. Hence there is also a need for informed consent from the patients about the role of automation in their care experience and to enable participants' participation in health decisions. Also, healthcare providers should have the reliance to make care judgments based on their experience as providers have raised concerns of over reliant behaviors on the automated system. Thus for mitigating automation dependency, there is a need for oversight of humans i.e. Healthcare providers and patients, Compliance audits, Transparency & documentation for how automated systems are operating, and stakeholder engagement.

Furthermore, healthcare is a complex system that has already existing technical systems for eg: E-sanjeevani electronic healthcare data management across three hospitals but there might be different legacy systems in respective hospitals when scaled. Thus there is a need to ensure compatibility with existing systems as the software on which the automation/intelligent RPA systems need to be built is up to date but upgrading systems can be costly and time-consuming as well and there could be challenges for interoperability because of a lack of standardization of data formats across respective health facilities.

Organizational Compatibility (Displacement of Workforce, Resource Management, Skill Training)

Intelligent RPA can be seen automating a lot of administrative tasks like data entry, scheduling, and billing as seen in the above sections. This creates resistance among the healthcare workers to adopt it due to fear of their jobs being redundant and leading to displacement of the workforce. Jobs like Radiology & Pathology where automation is predicted to be high but it is argued that it will not displace the workforce because currently, RPA is automating identifying and interpreting images but there are a lot of other tasks that their Job roles require them to do like treating illness, image-guided procedures, providing technical requirements for image processing connect test results and medical records and more. .Thus when implementing RPA it needs to be such that it augments the work of the healthcare providers rather than replacing it. There needs to be a change in the management and enable adapting of Intelligent RPA within the workforce. The stakeholders should convey the benefits and objectives of automation as well as enhance the skills of the workforce so that they can adapt to the automation.

Thus for the successful integration of intelligent RPA in healthcare, there needs to be staff training and skill development. This will enable humans to oversee the automation process reduce any potential biases and ensure fairness. Also for scalability across healthcare systems, there is a need for resource allocation of time & budget and also to make sure there is system stability as there can be multiple bots that can emerge as the implementation scales and hence there needs to be efficient management and ensure the bots function harmoniously. This can be avoided by ensuring data standardization and reducing data siloes by storing information in isolated systems.

In conclusion, the use of Intelligent Robotic Process Automation (RPA) in the healthcare sector has the potential to address the inefficiencies in rural healthcare delivery. By integrating intelligent RPA with existing digital healthcare infrastructure, we can streamline operational processes, enhance patient intake procedures, optimize resource allocation, and ultimately improve the overall efficacy of healthcare services at the grassroots level.

Rural areas face numerous challenges due to limited resources, infrastructure, and awareness. Implementing intelligent RPA can contribute to the transformation of rural healthcare, reducing the chances of human errors and improving access to quality healthcare for the rural population. This technology has the potential to revolutionise the way healthcare is delivered in underserved areas, ultimately contributing to better health outcomes and improved quality of life for rural communities. Therefore, it is important to responsibly design intelligent RPA solutions for healthcare to deliver an efficient health care service at grassroot level . The technology can be potentially used as a powerful tool to minimise the gap in current healthcare industry and move towards an equitable health care system in the country .

REFERENCES

Administrator. (2024, April 23). AI & RPA in Healthcare: Benefits, use cases, Trends 2024. AutomationEdge. https://automationedge.com/blogs/ai-and-rpa-in-healthcare/

Agostinelli, S., Marrella, A., & Mecella, M. (2020). Towards intelligent robotic process automation for BPMers. arXiv preprint arXiv:2001.00804.

Antwiadjei, L. (2021). Evolution of Business Organizations: An Analysis of Robotic Process Automation. Eduzone: International Peer Reviewed/Refereed Multidisciplinary Journal, 10(2), 101-105.

Anywhere, A. (2022, April 11). Revenue Cycle Management-RCM Automation. Automation Anywhere. https://www.automationanywhere.com/solutions/healthcare/rcm-automation

Barla, N. H., Almeida, S. M., & Almeida, M. S. (2023). Rpa revolution in the healthcare industry during covid-19. In *Confluence of Artificial Intelligence and Robotic Process Automation* (pp. 199–229). Springer Nature Singapore. DOI: 10.1007/978-981-19-8296-5_9

Business Process 101 : Definition, Steps and example [Guide for 2024]. (2024, May 15). https://kissflow.com/workflow/bpm/business-process/

Chakraborti, T., Isahagian, V., Khalaf, R., Khazaeni, Y., Muthusamy, V., Rizk, Y., & Unuvar, M. (2020). From Robotic Process Automation to Intelligent Process Automation: –Emerging Trends–. In Business Process Management: Blockchain and Robotic Process Automation Forum: BPM 2020 Blockchain and RPA Forum, Seville, Spain, September 13–18, 2020, Proceedings 18 (pp. 215-228). Springer International Publishing.0

Deloitte. (2017,). Reimagining business process reengineering in the age of automation.https://www2.deloitte.com/content/dam/Deloitte/us/Documents/public-sector/us-fed-reimagining-business-process-reengineering-in-the-age-of-automation.pdf

Dumas, M., Rosa, L. M., Mendling, J., & Reijers, A. H. (2018). Fundamentals of business process management. Springer-Verlag.L

Fivesdigital. (n.d.). How the healthcare industry is benefitting from Robotic Process Automation. Copyright 2023 @ fivesdigital.com. https://www.fivesdigital.com/blog/how-the-healthcare-industry-is-benefitting-from-robotic-process-automation

Gangolli, L. V., Duggal, R., & Shukla, A. (2005). Review of healthcare in India. Mumbai: Centre for enquiry into health and allied themes.

Grigori, D., Casati, F., Castellanos, M., Dayal, U., Sayal, M., & Shan, M. C. (2004). Business process intelligence. *Computers in Industry*, *53*(3), 321–343. DOI: 10.1016/j.compind.2003.10.007

Hall, J. L., & McGraw, D. (2014). For telehealth to succeed, privacy and security risks must be identified and addressed. *Health Affairs*, *33*(2), 216–221. DOI: 10.1377/hlthaff.2013.0997 PMID: 24493763

Hester, A. (2024, May 3). Automated prior authorization with intelligent automation. Nividous Intelligent Automation Company.https://nividous.com/blogs/automated-prior-authorization

Intelligent automation in healthcare & cognitive computing. (n.d.).https://www.optum.in/thought-leadership/library/intelligent-automation.html

Jain, R., & Bhatnagar, R. (2019). Robotic process automation in healthcare-a review. *International Robotics & Automation Journal*, *5*(1), 12–14. DOI: 10.15406/iratj.2019.05.00164

Kedziora, D., & Smolander, K. (2022). *Responding to healthcare emergency outbreak of COVID-19 pandemic with Robotic Process Automation*. RPA. DOI: 10.24251/HICSS.2022.752

Kumar, A. (2023). The transformation of the Indian healthcare system. *Cureus*, *15*(5). Advance online publication. DOI: 10.7759/cureus.39079 PMID: 37378105

Martinho, R., Rijo, R., & Nunes, A. (2015). Complexity analysis of a business process automation: Case study on a healthcare organization. *Procedia Computer Science*, *64*, 1226–1231. DOI: 10.1016/j.procs.2015.08.510

Naithani, C., Sood, S. P., & Agrahari, A. (2023). The Indian healthcare system turns to digital health: eSanjeevaniOPD as a national telemedicine service. *Journal of Information Technology Teaching Cases*, *13*(1), 67–76. DOI: 10.1177/20438869211061575

Parikh, D. P., Sattigeri, B. M., & Kumar, A. (2014). An update on growth and development of telemedicine with pharmacological implications. *International Journal of Medical Science and Public Health*, *3*(5), 527–531. DOI: 10.5455/ijmsph.2014.020320141

Preauthorizations Intake - RPA Component | UIPATH Marketplace | Overview. (n.d.). UiPath Marketplace. https://marketplace.uipath.com/listings/solution-accelerator-preauthorizations-intake4369

Prism, S. B. (2024, April 24). What is Intelligent Automation? SS&C Blue Prism. https://www.blueprism.com/guides/intelligent-automation/

1. Rivet. (2024, March 12). RPA in Healthcare: What it is, Benefits, and Real-World Use Cases | 1Rivet. https://www.1rivet.com/insights/rpa-healthcare

Sood, S. P., Negash, S., Mbarika, V. W., Kifle, M., & Prakash, N. (2007). Differences in public and private sector adoption of telemedicine: Indian case study for sectoral adoption. *Studies in Health Technology and Informatics*, *130*, 257. PMID: 17917199

SS&C Blue Prism. (2023, September 21). Intelligent automation for provider prior authorization. https://www.blueprism.com/resources/blog/healthcare-prior-authorization-automation/

SS&C Blue Prism. (2024, April 3). Optimizing Revenue Cycle Performance with Automation | RCM Automation.https://www.blueprism.com/resources/blog/healthcare-revenue-cycle-automation-rcm/

Suke, S. G., Kosta, P., & Negi, H. (2015). Role of pharmacovigilance in India: An overview. Online journal of public health informatics, 7(2). Tak, A. ISAR Journal of Multidisciplinary Research and Studies.

Talib, F., & Rahman, Z. (2013). Current health of Indian healthcare and hospitality industries: a demographic study. Talib, F., and Rahman, (2013), 1-17.

Thainimit, S., Chaipayom, P., Sa-arnwong, N., Gansawat, D., Petchyim, S., & Pongrujikorn, S. (2022). Robotic process automation support in telemedicine: Glaucoma screening usage case. *Informatics in Medicine Unlocked*, *31*, 101001. DOI: 10.1016/j.imu.2022.101001

Understanding the role of RPA in Healthcare | MuleSoft. (n.d.). MuleSoft. https://www.mulesoft.com/resources/api/rpa-in-healthcare

Venigandla, K. (2022). Integrating RPA with AI and ML for Enhanced Diagnostic Accuracy in Healthcare. Power System Technology, 46(4).

Venkateswaran, V. (2020,). Patient Experience in India.https://healthcare-in-india.net/wp-content/uploads/2020/08/Patient-Experience-in-India.pdf

Williams, P. (2021, September 21). Amazing Ways That RPA Can be Used in Healthcare - IBM Blog. IBM Blog. https://www.ibm.com/blog/amazing-ways-that-rpa-can-be-used-in-healthcare/

Woolhandler, S., & Himmelstein, D. U. (2014). Administrative work consumes one-sixth of US physicians' working hours and lowers their career satisfaction. *International Journal of Health Services*, *44*(4), 635–642. DOI: 10.2190/HS.44.4.a PMID: 25626223

Chapter 11
Natural Language Processing for Capability Enhancement of RPA

Mohit Yadav
https://orcid.org/0000-0002-9341-2527
O.P. Jindal Global University, India

Tanupriya Choudhury
University of Petroleum and Energy Studies, India

Ajay Chandel
https://orcid.org/0000-0002-4585-6406
Lovely Professional University, India

Majdi Anwar Quttainah
https://orcid.org/0000-0002-6280-1060
Kuwait University, Kuwait

ABSTRACT

This chapter elaborates on how NLP pushes RPA to tasks beyond rule-based activities, to areas involving complex processing such as document analysis, real-time decision-making, automation of customer services, etc. Key areas of enhancement pertain to the integration of sentiment analysis, chatbots, and intelligent extraction of data. Technical challenges and general implications of such integration are also covered in the chapter, like infrastructure requirements, avoiding biases, and responsible deployment. Changes in the workforce and equitable access to technology are two examples of managerial and social impacts. Understanding the potential and limitations of NLP-enhanced RPA will position an organization strategically to employ such technologies and optimize their operations toward innovation.

DOI: 10.4018/979-8-3693-4365-4.ch011

INTRODUCTION

In the recent years, Robotic Process Automation (RPA) functionalities have really led to its establishment as one of the transformative technologies that help businesses streamline business operations through automating repetitive and rule-based tasks. RPA has traditionally been effective in executing such structured and manual tasks that take a well-defined procedure, like transaction processing, manipulating data, and answering basic questions. However, as businesses continue to generate vast amounts of unstructured data in the form of emails, documents, and customer feedback, traditional RPA really suffers by its inability to adapt to unstructured data (Chakraborti et al., 2022). This is where NLP comes into play. NLP stands for Natural Language Processing; it is a subfield of AI focused on the interaction between computers and human language. NLP refers to the capabilities of machines pertaining to processing, identifying, and recognizing human language. Being able to interact with computers easily, this benefit of RPA is improved by the integration of NLP-a giant step in the world of automation. Businesses can eventually automate previously intricate tasks for traditional automation tools as enhanced RPA systems with prowess to process unstructured data are gradually developed. NLP will enable the RPA bot to read documents, understand emails, and even interact with customers using chatbots and virtual assistants. The development will make it possible for businesses to automate more processes, to make operations more efficient, and to respond faster and more accurately to customer needs (Cheligeer et al., 2022).

Increased demand for automation solutions that can manage both structured and unstructured data is one of the major driving factors behind this trend. Structured data processes easily on traditional RPA systems – numbers and dates, for example. However, unstructured data, including text, images, and videos, make up a large portion of business data and inherently requires more processing time (Jain et al., 2024). NLP enables upgrading the capability of RPA bots to process unstructured data, and therefore an organization may automate document classification, sentiment analysis, as well as information extraction on the basis that some data is written in texts. This also means there is reduced as well as increased human input, with increased acuteness and speed in the processing.

Integration of RPA and NLP also creates opportunities for the automation of customer services. Using NLP-powered chatbots and virtual assistants, a company can pass on the real-time, personal replies to customer inquiries from a human agent, enabling the latter to indulge in more complex work. Advanced NLP models allow conversational agents to interpret customer queries, analyze relevant information, and provide human-like responses to further engage the customers. This enables the systems to understand and improve customer satisfaction while lessening the work burden on service teams (Kotagiri & Yada, 2024).

In conclusion, it was found that the possible convergence of NLP and RPA marked an important advancement in the field of automation. As a result, it opens business entities to the possibility of moving beyond simple task automation into the realm of cognitive automation that would incorporate NLP into RPA, hence enabling organizations to process and analyze unstructured data, more complicated workflows and improve customer experience. This chapter will further elucidate how NLP complements and aids RPA, technical complications associated with their integration, and what the prospects are for this powerful combination in changing the way business processes take place.

FOUNDATIONS OF RPA AND NLP

RPA and NLP are the powerful technologies, which means the integration of both may lead to business automation. RPA works mainly to automate structured and repetitive tasks, while NLP does well in understanding and processing human language for structured and unstructured data interaction by machines. Therefore, for this purpose, the background of both is necessary to understand how their interaction may add value to automation capabilities (Jain et al., 2024).

Robotic Process Automation (RPA)

RPA is one technology enabling automation of any repetitive, rule-based activities that are often undertaken by humans, including processing transactions, inputting data, generating reports, and answering simple questions. The RPA simply emulates the behavior of a human operator in computer systems using software "bots," thus it makes it possible to execute tasks more quickly, error-free, and reliably. It is a very versatile business process automation tool that works across different applications and platforms.

What characterizes RPA fundamentally is the possibility of automating structured tasks—tasks with given inputs, processing activities, and outputs. For example, in finance, RPA can automate the invoice processing activity. RPA bots will suck information from structured data fields into an accounting system and generate reports. In customer service, RPA bots can take up basic inquiries, such as checking an order or account status (Kotagiri & Yada, 2024).

However, traditional RPA cannot handle unstructured data, including free-form text, audio, and images. Such data does not have rules or formats under stiff compulsion, so automation systems fail to handle that. Therefore, RPA must be combined with the emergent technology like NLP that deals with the unstructured data and also more complex business tasks.

Natural Language Processing (NLP)

Natural Language Processing is the branch of artificial intelligence (AI) that gives machines the ability to understand the human language. NLP enables some machines to be able to process natural language data-from text, speech, or writing-between humans and computers. NLP can use linguistic algorithms, machine learning models, and statistical methods to complete the most varied sets of tasks such as language translation, sentiment analysis, text summarization, and speech recognition (Leopold et al., 2018).

NLP generally functions through four stages of lexical analysis, syntax analysis, semantic analysis, and in some cases, advances through more sophisticated processes such as sentiment analysis, named entity recognition, and language generation. NLP systems can handle both structured and unstructured data, making them valuable in those scenarios where voluminous quantities of text-based information need to be processed, such as emails, contracts, social media posts, and customer reviews.

At the core, it is advanced machine learning models, but deep learning algorithms, for instance, transformers, are those that power modern-day NLP tools, for example, OpenAI's GPT and Google's BERT. These models can learn patterns from vast data sets of human language and produce highly accurate interpretations or predictions. NLP has progressed rapidly in the understanding of contextual, sentiment, and nuances of language to become an indispensable tool for enhancing the capabilities of automation systems, such as RPA (Ling et al., 2020).

The Interplay of RPA and NLP

RPA and NLP are merged to become a powerful force that can automate structured as well as complex workflows involving unstructured data. Though RPA succeeds best in the well-structured and rule-based tasks, NLP offers the feasibility of automating processes which may require natural language interpretation. This integration enables the cognitive automaton, so that the system can read, understand, even respond to human language (Molla, 2024).

For example, with NLP, an RPA bot can read and extract information from contracts, classify emails according to their content, or maybe evaluate sentiment based on customer feedback forms. With the addition of NLP within the process of automating, new avenues within business operations streamline, enhance the basis for informed decisions, and increase the quality of customer experience, all of which are strictly imperative in today's digital economy since gigantic amounts of unstructured data are produced every day, and business today engages to find ways to extract value from this massive data.

To sum it up, RPA excels at automating structured tasks while NLP allows the machines to process and comprehend unstructured data expressed in natural language. It is a synergistic solution that much expands what is achievable within the boundaries of automation. The subsequent sections will discuss how this amalgamation works, associated technical challenges, and practical applications of NLP-enhanced RPA across different industries (Nalini et al., 2024).

KEY AREAS WHERE NLP ENHANCES RPA

With the integration of RPA with Natural Language Processing (NLP), it greatly expands the automation field, making businesses capable of handling complex tasks involving unstructured data. Traditional RPA is limited to structured and rule-based processes, while NLP enables the systems to interact with human language and unlock new dimensions of automation efficiency and accuracy. The major areas that NLP enhances RPA capabilities in include the following:

1. Textual Data Processing and Understanding

One of the major areas where NLP further adds value to RPA is for the purpose of processing and understanding data in the form of text. Traditional RPA cannot process information that cannot be put in a structured format within emails, documents, or reports. NLP enables the meaning and processing of data so that it can now read what it wanted to extract, classify, and report in the current information. For example, an NLP-enriched RPA bot may automatically extract key details of contracts or legal documents. More importantly, invoices, receipts, and feedback in various languages can be processed, thus improving automation capabilities further in global operations (Oyewole, 2024).

2. Document and Email Automation

Indeed, much of the communication does take place through documents and emails in many business environments. NLP enables RPA bots to not only read and classify the document but also understand the intent and content. For example, in a customer service situation, an NLP-enabled RPA bot can scan incoming emails determine the problem and send it to the correct department or even draft an auto-response, for example. For instance, it may allow NLP to automatically classify and archive documents or retrieve relevant information on anything and ensure adherence to the regulations due to its ability to process and store large amounts of textual data (Qi, 2024).

3. Sentiment Analysis and Customer Service Automation

Another significant boost to RPA systems, especially customer service, is the capacity that NLP has in performing sentiment analysis. Sentiment analysis is the capability of the company to know what the customer's feelings are as well as opinions based on unstructured data such as social media postings, online reviews, or emails. RPA bots with NLP can look into customer feedback and classify it as negative, positive, or neutral, so for attention fields, it can place them according to urgency. For example, if a customer complains about a product in the review, the bot automatically flags it and brings it to urgent notice. This ensures the discovery of problems in advance as well as enhances customer satisfaction at a minimal level of human intervention (Rainey et al., 2017).

4. Chatbots and Conversation Interfaces

NLP is an essential feature that ensures the creation of conversation AI: chatbots, virtual assistants, which are steadily integrated into RPA. These language-powered bots can naturally interact with users, understand the question by making sense of it, and respond to that in real-time. Business can automate interactions with customers at scale on the simplified level of answering repeated questions up to the intricate process of troubleshooting a product by adding NLP on top of RPA. For example, an NLP-based RPA chatbot can identify customer requests, fetch more information such as order number and provide status of their orders without human interference. This cuts down human customer support and smoothen the service deliverables (Rane et al., 2024).

5. Extraction from Unstructured Sources

Such business processes widely involve extraction of information through unstructured sources, including scanned documents and PDFs, to web pages. In the scenario described above, it will allow RPA bots to interpret and extract specific information. For instance, an RPA bot in financial services can use NLP to interpret information from a loan application, perform categorization using predefined rules, and then automatically fill the relevant fields in the system. Similarly, with NLP, RPA bots can retrieve information such as patient information, diagnosis, and treatment; these capabilities not only are error-free but also save a lot of time in data entry (Schmitz et al., 2019).

6. Workflow Automation Using Language Commands

NLP enables workflow automation with the support of language commands because instead of having to focus on fixed criteria, natural-language-based triggers are being used. This feature is very helpful in environments where non-technical users must interact with RPA systems. A manager could command a bot to generate a report or schedule a meeting by using simple language commands, and the RPA bot would grasp and execute such a request. This pushes automation closer to users who are not necessarily involved with RPA and boosts productivity by doing away with the need for complex programming or manual input.

But in regard to both the scope and depth of the possible automation, NLP really integrates with RPA because it can handle better more complicated activities involving unstructured data. As such, NLP has streamlined the efficiency and effectiveness of automation by supporting RPA systems, among others, on the processing of textual data, automatic document workflows, sentiment analysis, conversation, and extracting information from unstructured sources. Further, as these technologies grow, the avenues for further sophisticated and intelligent automation solutions will open, paving the way for businesses to streamline their operations and create ever better customer experiences (Siderska et al., 2024).

TECHNICAL INTEGRATION OF NLP AND RPA

In technical integration of NLP and RPA, we can see an important step for altering business automation in complex workflows. RPA is developed as the technology that would be able to automate such repetitive, rule-based tasks. In the case of NLP, processing and interpretation of human language can be carried out by a bot for handling unstructured data. The integration of these two technologies involves several layers. On one hand, we are talking about building the machine learning models for NLP through the integration of the same with platforms that implement RPA in a manner that goes ahead to create solutions that are powerful as well as intelligent automation. This section dwells into the technical process of integrating NLP with RPA.

1. Development and Training of NLP Model

At the core of NLP lies the model development and training-high-level abstractions that could understand and interpret human language. In practice, such models are built using large datasets of natural-language inputs, including text documents, emails, and spoken language (Sowjanya & Vijaya Chamundeeswari, 2024). The training algorithm uses algorithms that might learn patterns in language: namely syntax, semantics, and context. Some key types of models are:

- Tokenization and Parsing Models: The model is decomposing text into smaller units (tokens) and analyzing sentence structures to understand the relationships between words.
- Named Entity Recognition (NER): The model identifies and categorizes entities in the text, including names, dates, locations, and product details.
- Sentiment Analysis Models: These classify text based on emotion or opinion; for instance, positive, negative, or neutral.
- Advanced NLP Models: Models like BERT and GPT based on transformers' architecture that is applied understanding the context and generating very precise language predictions.

Once such models are trained, they can be fine-tuned to fit specific business needs, for example, automation of customer service or document processing.

2. Integration of NLP APIs into RPA Workflows

NLP capabilities are usually integrated into the RPA platforms via the NLP APIs, which provides pre-trained NLP models or allow for custom models to be designed and used by RPA bots for their workflows. Quite a few cloud-based NLP services-from Google Cloud's Natural Language API to Microsoft Azure's Text Analytics API and OpenAI's language models-provide powerful NLP functionalities that may be incorporated into RPA systems (Szabó, 2020).

One of the functionalities in the RPA workflow is how a bot interacts with an NLP API by providing the data, such as in this case, text from an email or documents, for processing in an NLP engine. The output is then returned to the RPA bot, and it takes a subsequent action based on the information extracted. For instance, if an RPA bot needs to process e-mails with customer complaints, it can use the NLP API to dig deeper into details such as names of the customers, type of complaint, and if it is urgent, which would help the bot classify and respond to each e-mail appropriately.

3. Integration of NLP with RPA via Middleware

To ensure seamless communication between NLP and RPA systems, middleware is often used as an intermediary in data exchange and workflow orchestration. It acts as a connector through which RPA bots could send requests to a corresponding NLP model and receive their responses based on real-time computation. This component is very helpful especially when large volumes of data need processing or when multi-NLP models need to be included within one RPA workflow (Valluri, 2024).

Middleware is also utilized for other important operations, such as managing errors, structuring data, and ensuring the scalability of the solution for supporting multiple languages or even big data sets. Middleware solutions can also provide additional optimization for fully integrated RPA and NLP solutions by having models used in NLP deployments optimized and whether they reside on-premises or via the cloud dependent on an organization's infrastructure requirements.

4. Processing Unstructured Data

Traditional RPA systems are adversely challenged by any unstructured datatype like Emails, reports, social media posts, and scanned documents, for example. NLP enables RPA bots to make sense of such "unstructured" data by extracting valuable information and converting the data into structured formats that can be used further in workflows. Some technical steps are involved:

- Data Preprocessing: It needs to preprocess the data by any NLP engine. All the irrelevant information present in the data must be removed which includes punctuation and stop words. Then normalization needs to be done either by lemmatization or stemming, and finally, the text must be tokenized to make meaningful units (van der Aa & Leopold, 2021a).
- Entity Extraction: Techniques such as Named Entity Recognition (NER) are used to identify important entities in the text? be it names, addresses, product details, or monetary values. The extracted information will be structured in formats, either table or JSON, which will later on be consumable by the RPA bot.
- Sentiment Analysis and Categorization: NLP enables customer service to categorize texts by sentiment, through which the bots of RPA can prioritize tasks or send information to the appropriate department. Many cases of this nature depend upon activities like sentiment analysis and text classification in such triaging of customer inquiries or processing social media feedback (van der Aa & Leopold, 2021b).

5. NLP-Fuelled Models of Decision-Making

The inclusion of machine learning models in the RPA system is another feature of the technical integration of NLP with RPA. More importantly, NLP models especially the deep learning-based models further promote better decision-making capabilities because they allow bots to understand contextual and qualitative languages. Therefore, if one is automating claims processing in insurance, for example, RPA bots can use NLP to read claim documents, extract relevant information from them, and then use predictive models to draw out information about the validity or risk levels of the claims. This enables a more-informed decision by the bot and reduces cases of manual intervention (Zhu et al., 2024).

6. Feedback Loops and Continuous Learning

Another concept that will enhance the work of integrated NLP and RPA is a system-based feedback loop for continuous learning. Of utmost importance, as the bot processes more tasks and more data, the system must be able to learn from its errors or from a change in user feedback or patterns of data. Continuous learning may be achieved through retraining models NLP from new datasets, updating the flow of RPA workflows to accommodate new business rules, or analytics monitoring bot performance.

This integration aspect ensures that the RPA bots remain agile and responsive to changing language usage, customer expectations, and business situations. For instance, as the customer service chatbot engages in conversations with users, it learns to comprehend even more frequent queries, slang words recently entering usage, or perhaps product-related jargon over time (Chakraborti et al., 2022).

7. Security and Compliance Considerations

Since NLP and RPA systems mostly handle sensitive data like personal customer details or financial records, ensuring security and compliance is key in technical integration. This means both NLP models and RPA bots need to be compliant with data protection regulations such as GDPR, HIPAA, or CCPA. In that regard, this involves encryption of data, controls to access data, and regular audits to ensure that data is handled securely throughout the automation pipeline (Cheligeer et al., 2022).

Further, when availing the cloud-based NLP services, ensure that data privacy is maintained and all the information fed to the third-party NLP service providers are anonymized or secured.

The integration of NLP with RPA involves several levels such as model development, API, integration, data processing, decision making with the help of machine learning, and continuous learning. Unlocking new developments, this integration would unlock the business automation capabilities ranging from unstructured data processing to improving the decision-making process. NLP models and RPA technologies will continue to progress. This means that businesses will be able to reap more intelligent, adaptable, and secure forms of automation (Kotagiri & Yada, 2024).

LIMITATIONS AND CHALLENGES

While the integration of NLP with RPA has myriad advantages, it poses some challenges and limitations that are derived from both technological and operational complexities. Understanding these challenges is important for business as it would help in optimizing the deployment of NLP-enhanced RPA systems and maximizing their benefits. The following are key challenges and limitations to this integration (Jain et al., 2024).

1. Complexity of Unstructured Data

The greatest challenge in the attempt to combine NLP and RPA is the management of unstructured data. Unstructured data comes in a tremendous form and content variation from structured data, in which all records have all a specific predefined format-like emails, social media posts, and scanned documents-to purely unstructured NLP engines can't even come close with the subtlety of the human language like slang, idioms, sarcasm, and ambiguity. Therefore, there are possibilities of errors and inaccuracies in the interpretation and understanding of NLP models from unstructured data. In this case, an automation may be hindered, and it could delay a workflow or be reviewed manually (Leopold et al., 2018).

2. Training and Updating NLP Models

NLP model training is time-consuming and resource intensive. The development of top-notch models that interpret the context, semantics, and domain language requires vast datasets and strong computational power. Moreover, NLP models require always updating and retraining because language dynamics are time-evolving. That is, terms and trends may be introduced, or industry-specific jargon may evolve. It then has to be adapted in the model. This ensures that such models come with a challenge for organizations, especially for ones that do not have in-house data science expertise (Ling et al., 2020).

NLP models developed for a particular language do not become any stronger to use in another language and, as such, present significant constraints to organizations in carrying out their businesses in multilingual environments. It makes the implementation more complex in multi-lingual NLP models, but still maintains accuracy across languages.

3. Data Privacy and Security Issues

NLP models involve processing sensitive or confidential data such as personal information, financial records, or legal documents. This further raises serious issues of data privacy and security. Most organizations rely on third-party NLP services, often using cloud hosting, which presents the risks of a data breach or unauthorized access. For instance, ensure that all data gets processed safely either on-premises or in the cloud and complies with laws like GDPR or HIPAA.

Anonymization, encryption, and safe storage of data become a pre-requisite, and these may add further complexity to technical integration of NLP and RPA. Strict privacy control mechanisms also may limit further the scope of what tasks can be safely automated by balancing efficiency in automation (Molla, 2024).

4. Weaknesses in Contextual Comprehension

NLP systems are still relatively weak in contextual comprehension and inference even at this stage of advancement. Unlike humans, who can interpret the meaning based on a tone, body language, or some prior knowledge, NLP models rely purely on text and patterns found within the data that they have been trained on. This may sometimes end in misinterpretation of context, especially in situations regarding complexity and ambiguity. For example, in the case of customer service automation using an NLP-augmented RPA bot, it may miss out the subtle nuances in any complaint or decipher the wrong intent from the customer leading to inappropriate or incomplete responses.

This resonates the role of human interventions or monitoring in such scenarios where the understanding of the language must be quite precise. It also brings out the fact that an RPA model needs continuous improvement through the inclusion of feedback loops and retraining mechanisms (Nalini et al., 2024).

5. Huge Initial Investment and Resource Requirements

NLP and RPA Integration imply heavy infrastructure, software, and talent expense that demands an upfront investment. The organization would need to commit dedicated resources for the development, training, and maintenance of NLP models while ensuring integrated models run unobtrusively with workflows existing in RPA. High performance computing, cloud services, and specialized expertise in data science and machine learning will attract significant cost pressures, especially for smaller to mid-sized businesses (Oyewole, 2024).

Moreover, the rollout can take several months to totally integrate and optimize. Such delay can disrupt the operational activities since the business should operate in parallel by using manual methods and automated procedures during the transition period

6. Scalability Issues

More importantly, NLP-boosted RPA systems suffer from scalability problems where the growth of organization size and complexity of processes is involved. Scaling down many RPA bots to handle higher volumes of unstructured data or more language variants or handle increasingly complex workflows demands the underlying NLP models are also scalable to work efficiently. Most the NLP models break down when scaled up to large amounts of real-time data processing.

Moreover, NLP models will degrade in performance when they encounter data outside their training scope, creating limits to the scalability of the system across different domains or use cases. Organizations may need to invest in customized NLP models for each domain or continuously fine-tune existing models, all requiring vast amounts of time and resources (Qi, 2024).

7. Lack of Standardization

One of the major difficulties when integrating NLP and RPA is the lack of standardization within models across NLP and platforms across RPA. NLP models may be optimized for different purposes-from text classification to sentiment analysis or entity recognition, among others, while integration in not feasible on all the RPA platforms. Similarly, different RPA tools may allow for different levels of flexibility and support to integrate NLP. Lack of standardization has a significant problem in integration, causing customers to either create proprietary solutions or rely on third-party middleware for their connectivity (Rainey et al., 2017).

With the rapidly growing market for NLP and RPA technologies, the need for standardized protocols and frameworks that simplify the integration process and ensure compatibility between different tools and models is urgent

8. The Traction of High-Quality Data

Quality training data, in many ways, boils down to the success of NLP models. Labelled or biased data may result in inaccurate predictions and flawed automation workflows. For instance, biased customer service will lead to failure in an NLP model's ability to give unbiased responses, such that customers are not satisfied with service, or the quality and fairness are questionable (Rane et al., 2024). It is a very challenging task to ensure that datasets are diverse, representative, and high quality because data collection, labeling, and preprocessing require much effort.

In the absence of good quality data, the precision needed to improve the RPA systems would not be provided by NLP models, and businesses would end up getting unreliable or inconsistent outcomes from automation, which would affect the overall performance of their systems and business operations in general.

While the integration of NLP with RPA opens up vast possibilities for automating intricate, language-based tasks, it has its challenges as well. Issues such as the handling of unstructured data, ensuring data privacy, understanding context, and high investment in the beginning are factors that limit the scope of functioning of this integration. Another factor is the potential for scalability, being not eligible for standardization, and having a huge dependency on data of quality (Schmitz et al., 2019).

FUTURE TRENDS AND OPPORTUNITIES

NLP-RPA integration will continue to be developed further with advancements in AI, machine learning, and cloud technology. These innovations open broad avenues for business applications that could not be possible with unaided RPA systems. Some of the key future trends and opportunities through which the development of NLP-RPA integration may be formed for the years ahead are discussed as follows:

1. Conversational AI and the Emergence of Chatbots

Improving the NLP technology will bring conversational AI and chatbot abilities much closer to being truly sophisticated. RPA bots can, at present, perform the predefined tasks as per structured commands; however, in the future, they will be programmed to deal with much more complex interactions. Advanced NLP models will make it possible for chatbots to clearly understand and process human language in real-time and thus better interpret nuances, such as tone and context, with greater accuracy (Siderska et al., 2024).

It would also enable the automation of higher-level customer service functions, including dispute resolution and product recommendation and more personalized flows of conversation. Moreover, conversational AI-powered RPA bots will prove to be good for the customer experience and get human workers freer for strategic tasks, thereby bringing efficiency to the operations.

2. NLP for Better Document Processing

Document processing remains one of the main areas where NLP combined with RPA is driving further automation, and future innovation will continue to enhance these abilities. The capability of NLP in extracting and interpreting data from unstructured text will become essential in automating tasks like contract management, invoice processing, or even reviewing legal documents (Sowjanya & Vijaya Chamundeeswari, 2024).

Some recent trends feature NLP-powered RPA with more complex forms of documentation, in addition to legal texts, medical records, and technical manuals. Improvement of NLP models to perceive domain-specific terminologies and industry jargon will enhance document automation processes. Accuracy and reliability will also be increased as processes accelerate and RPA bot speeds up to perform processing without human oversight due to their ability to understand context and intent.

3. Multilingual Skills

Globalization has increased business operations cross borders. Consequently, the requirement of automation tools for various languages in businesses is rising. NLP is getting better day by day in processing multi-lingual data, which is expanding fields by which the businesses can automate multi-lingual-based activities such as translation, communication, and support services in different languages (Szabó, 2020).

Later, NLP-empowered RPA bots will be able to freely carry-on multilingual communications with customers, business associates, and suppliers without human translators being involved. This will be a crucial asset to industries like e-commerce, customer service, and legal where multilingual communications are required.

4. NLP-Powered Predictive Analytics in RPA

Such a conjunction of NLP, RPA, and predictive analytics will revolutionize the decision-making paradigms integrated into the business circles of all industries. NLP can be applied to unravel deep insights from the tremendous unstructured data - customer feedback and market trends, social media interactions, and many more such similar information that would otherwise remain inaccessible. The insights obtained from this may be integrated with models of predictive analytics to let RPA systems become proactive with more forecasting capabilities, such as predicting market trends, identifying customer sentiments, and predicting bottlenecks in operation (Valluri, 2024).

For instance, an NLP-enriched RPA robot can scan history to be able to make forecasts about when there might be potential supply chain disruptions and, therefore, prevent problems from happening before they arise. Similarly, it can provide real-time suggestions for business operations improvement, allowing for more informed strategic planning.

5. Cognitive RPA Driven by AI

It is cognitive RPA, the meaning of which is an infusion of AI-driven technologies such as NLP, ML, and computer vision that will incredibly expand the horizon of automation possibilities. This is because the systems based on cognitive RPA will be programmed to simulate human reasoning and, therefore will be built for tasks considerably more complex than those currently being offered and which by definition must be perceived in context, judged, and then decided upon (van der Aa & Leopold, 2021a).

For example, future versions of the cognitive RPA bots may independently handle end-to-end processes in the domain of insurance claims processing as well as fraud cases and legal compliance. Such bots will be able to scan voluminous data, make inferences, take decisions, and develop based on the predefined rules and the emerging trends. In this respect, cognitive RPA will be a transformative technology that changes industries through the automation of valuable knowledge-based tasks that have until now called for human intervention to execute.

6. NLP and RPA in Compliance and Regulatory Automation

The other headache for organizations in nearly all the sectors, but financially, health-related, and pharmaceutical sectors the most, is compliance towards legal and regulatory mandates. As periodic updates of regulations imposed by the governments are introduced through legislation, manually monitoring and managing diverse business processes' compliance is an arduous task (van der Aa & Leopold, 2021b).

NLP-enhanced RPA will pave the way for streamlined compliance and automated extraction, analysis, and reporting of regulatory data. Future development will be geared towards NLP models that can interpret tax codes, privacy regulation, and financial reporting standards in complex regulatory texts. Business entities will have the automation of checks for compliance, make processes compliant with current changes in regulatory changes, and minimize the risk of noncompliance.

7. Hyper-automation with NLP and RPA

Hyper-automation is an accelerator term used to describe the use of advanced technologies, including AI and RPA, to automate everything possible. NLP plays a key role here because it makes it possible to automate complex, unstructured tasks for which the older versions of RPA were out of reach. As more organizations deploy hyper-automation strategies, they are likely to become increasingly dependent on NLP to draw meaning from unstructured data sources like emails, chat logs, and documents (Zhu et al., 2024).

With hyper-automation, the possibility of automating entire workflows involving more than one department-what was once impossible with customer service and HR to finance and legal-can now be done. Organisations will never again miss an opportunity to reach for unprecedented levels of operational efficiency and scalability by integrating NLP and RPA seamlessly on hyper-automation frameworks.

8. Edge Computing and NLP-RPA Integration

The convergence of edge computing with NLP and RPA brings quite a fascinating future prospect. Edge computing is about processing data much closer to where it originates, thus evading latency and enabling decisions to be made in real-time. This trend will enable NLP enhanced RPA bots to become much more effective in environments where raw data is processed almost in real time-such as in manufacturing, logistics, and IoT-enabled industries (Chakraborti et al., 2022).

For instance, edge-enabled RPA bots can assess natural language inputs and respond in real time for industrial locations, thus speed up the decision-making process and make it more accurate. This would be priceless, for instance, when the situation requires an immediate response, such as predictive maintenance in factories or a timely customer service response.

The integration of NLP and RPA holds so many great promises for the future, considering how AI and machine learning are continually blurring the boundaries in automation. Business opportunities for NLP-enhanced RPA range from more sophisticated conversational AI and enhanced document processing to cognitive RPA and predictive analytics. Such trends will just extend the frontiers of automation but also open new areas into business improving operational productivity, cost reductions, and enhancements for customer experience. Organizations need to be agile and flexible by updating on the developments in NLP and RPA technologies, with the ever-increasing dynamic nature of intelligent automation (Cheligeer et al., 2022).

IMPLICATIONS OF NLP AND RPA INTEGRATION

The implications of NLP-RPAs can be widespread, ranging from practical to managerial, and ethical to social. It helps the organizations to understand and navigate the business environment where such technology has been deployed.

1. Practical Consequences

From a practical point of view, integration of NLP and RPA can revolutionize the way business processes are executed in relation to handling unstructured data as well as human language. Some of the practical implications have been:

- Improvement in Efficiency: Automated processes like document processing, customer service interaction, and emails result in greater speed in terms of execution while having fewer rates of mistakes. This involves better efficiency in operational functionality and minimizes the involvement of humans in repetitive operations (Jain et al., 2024).
- Reducing Costs: The replacement of manual labor with automation by organizations can greatly reduce operating costs. The complexity of language-based processes such as responding to customer inquiries or reporting compliance-- processes which were earlier reliant on a large workforce--requires little in terms of expense once automated.
- Scalability: NLP-enabling RPA lets businesses scale their capacities more smoothly, especially in handling language-intensive processes such as support operations. While generally carrying minimal overhead added on top, companies can take care of vastly increased volumes with no sacrifice to quality (Kotagiri & Yada, 2024).

However, businesses will have to invest in infrastructure development and specialized skills to gain all the real-world benefits.

2. Managerial Implications

For managers, the integration of NLP and RPA has massive implications to decision making, workforce management, and operational strategies. Some managerial implications include the following:

- Strategic Decision Making: The insights found by NLP about information in unstructured data such as customer feedback and social media interactions provide valuable information to the managers about making strategic decisions. Predictive analytics enabled by NLP may give actionable insights into emerging trends both in markets, customer behavior, and even areas of operational bottlenecks (Leopold et al., 2018).

- Workforce Transformation: Since RPA will take care of routine tasks, what managers need to do is rethink workforce allocation. They should reskill the jobless for more complex tasks involving human judgment, creativity, and interpersonal skills. The change will require investing in upskilling and workforce development initiatives (Ling et al., 2020).
- Data-Driven Management: With NLP and RPA, the managers are increasingly reliant on data-driven management practices. NLP recovers meaning from humongous amounts of unstructured data, which leads to better, more informed decisions in customer service, marketing, and risk management.

Managers will need to factor into the technical implementation of NLP-RPA systems with change management processes to ensure that teams can adapt to it easily.

3. Ethical Implications

It would be unethical if the integration of NLP with RPA were done without considering the possibility that business processes using AI technologies may be less transparently fair and accountable. Key considerations include:

- Bias in NLP Models: NLP models might unwittingly perpetuate biases that are housed in learning materials. The NLP system creates biased choices or biased recommendations if it is trained with customer interaction data that are biased. Basing organizations' decisions on such outputs would mean unfair outcomes. It is necessary for organizations to audit their NLP models with due ethical care and eliminate biases and harmful stereotypes from NLP models (Molla, 2024).
- Job Displacement: Processing of tasks based on language may displace some jobs that workers are doing in terms of customer service operations, data entry, and content processing. However, positively, with the setting up of automation, many more positions will be opened up the ladder up, and management will have to ensure that employment loss for the workforce is lost to a minimum level.
- Data Privacy and Security: NLP-enhanced RPA systems process sensitive information, like personal details, financial records, and legal documents. This calls for data privacy issues and ethical use of personal information. Companies have to ensure that NLP-RPA solutions comply with the regula-

tions regarding privacy, like GDPR and ensure the secure handling of data (Nalini et al., 2024).

- Transparency of Decision-Making: As the systems take more independence of NLP, there should be transparency in the process of decision-making. Users should explain how decisions are being made, and provision should be provided for challenging or correcting decisions taken by automated systems.

Businesses have to build strong ethical rules and framework to regulate the application of NLP-supported RPA for fairness, accountability, and adherence to the higher level of legal standards (Oyewole, 2024).

4. Social Implications

The convergence between NLP and RPA carries broader social implications than those bounded within the workplace. Changes here translate to impact human interaction patterns. A few of the social implications include:

- Workforce Shifts: The more that NLP-RPA is embedded in processes, the more there will be a societal shift in the nature of jobs. In the longer term, automation can set free people from mundane, repetitive work; however, it does create a requirement for new jobs in AI management, data science and process optimization. The adverse effect on moving towards reducing inequality in this respect can arise when the less-skilled worker is not really supported to acquire the new skills (Qi, 2024).
- Human communication; such automated systems used in customer service expose the business unit to eventual loss of the human touch. Though the NLP enriched RPA can attend to the regular issues, the complicated and emotional ones fall straight into the expert court where tactful advice is paramount in allowing human-like problems to be solved suitably. Organizations must strike a balance between automation and human touch in the relationships that prevail between customers and their organization.
- Availability of Automation Technology: Through NLP and RPA, social inequality can be magnified because not all big, resource-rich organizations have been able to harness these technologies. Small organizations or even developing nations cannot keep abreast with such technologies and widen the gap between business entities that are automated and those that are not (Rainey et al., 2017).
- Language and Cultural Boundaries: NLP-RPA integration can cut across the language barriers, hence becoming part of an increasingly interconnected global world. Social benefits of this technology include its role in helping

businesses and governments translate services provided to a multilingual level, hence making it possible for people who do not speak a particular language to utilize various offered services (Rane et al., 2024).

Social responsibility will make businesses consider the wider impacts of automation on society, thus making sure that the benefits of technology are shared fairly and equitably.

CONCLUSION

Therefore, NLP with RPA stands for an enormous step in intelligent automation because it allows combining the best strengths of both technologies to allow organizations to automate more complex and language-dependent tasks than what had been beyond the scope of traditional RPA systems. The outcome is an efficient, scalable, and responsive automation platform that can work with unstructured data and understand human language to help make timely decisions. This results not only in faster workflows but allows businesses to explore new areas of innovation while taking customer experiences to the next level, with extra cost savings.

However, NLP-based RPA does have its own challenges and considerations. Technical integration requires adequate planning, strong infrastructures, and expertise in both AI and RPA technologies. Beyond the technical, these organizations would also face ethical challenges associated with automation: displacement of employment; biased decision-making, and data privacy. The responsible use of these technologies will thus ensure automation transition is fair, transparent, and not against societal values.

The integration of NLP into RPA also requires a change in approaches toward leadership and staffing from a management perspective. Human workers will have a different role with routine tasks being automatically completed, and thus a reskilling and upskilling focus is required to better train workers who can move into more strategic, higher-value roles. Leaders should leverage data-driven insights that NLP can bring to decision making while defining business operations that optimize the delivery of value.

The greater social implications of NLP and RPA adoption in this particular social context induce several problems associated with equity, access, and the future of work. While automation promises efficiency and growth from it, this must equally be distributed across all business firms and different regions.

In conclusion, NLP and RPA integration provides a huge opportunity for various companies to enhance their capacities and streamline their operation. However, it needs to be approached with technical, ethical, and social impact consideration.

Organizations that would successfully navigate these challenges would be in a prime position to capitalize on the transformative potential of NLP-enhanced RPA, drive innovation, and lead the competitive spectrum of an increasingly automated world.

REFERENCES

Chakraborti, T., Rizk, Y., Isahagian, V., Aksar, B., & Fuggitti, F. (2022, September). From natural language to workflows: Towards emergent intelligence in robotic process automation. In *International Conference on Business Process Management* (pp. 123-137). Cham: Springer International Publishing. DOI: 10.1007/978-3-031-16168-1_8

Cheligeer, C., Huang, J., Wu, G., Bhuiyan, N., Xu, Y., & Zeng, Y. (2022). Machine learning in requirements elicitation: A literature review. *Artificial Intelligence for Engineering Design, Analysis and Manufacturing*, *36*, e32. Advance online publication. DOI: 10.1017/S0890060422000166

Jain, A., Paliwal, S., Sharma, M., Vig, L., & Shroff, G. (2024). SmartFlow: Robotic process automation using LLMs. *arXiv preprint arXiv:2405.12842.*

Kotagiri, A., & Yada, A. (2024). Crafting a strong anti-fraud defense: RPA, ML, and NLP collaboration for resilience in US finance's. *International Journal of Management Education for Sustainable Development*, *7*(7), 1–15.

Leopold, H., van Der Aa, H., & Reijers, H. A. (2018). Identifying candidate tasks for robotic process automation in textual process descriptions. In *Enterprise, Business-Process and Information Systems Modeling: 19th International Conference, BPMDS 2018, 23rd International Conference, EMMSAD 2018, Held at CAiSE 2018, Tallinn, Estonia, June 11-12, 2018* [Springer International Publishing.]. *Proceedings*, *19*, 67–81.

Ling, X., Gao, M., & Wang, D. (2020, November). Intelligent document processing based on RPA and machine learning. In *2020 Chinese Automation Congress (CAC)* (pp. 1349-1353). IEEE. DOI: 10.1109/CAC51589.2020.9326579

Molla, M., Ahmad, J., & Wan Kadir, W. M. N. (2024). A comparison of transforming the user stories and functional requirements into UML use case diagram. *International Journal of Innovative Computing*, *14*(1), 29–36. DOI: 10.11113/ijic.v14n1.463

Nalini, M., Dhanraj, R. K., Balusamy, B., Abirami, V., & Kavya, K. (2024). Intelligent assistants using natural language processing for hyperautomation. In *Hyperautomation for Next-Generation Industries* (pp. 91-126). DOI: 10.1002/9781394186518.ch4

Oyewole, A. T., Adeoye, O. B., Addy, W. A., Okoye, C. C., Ofodile, O. C., & Ugochukwu, C. E. (2024). Automating financial reporting with natural language processing: A review and case analysis. *World Journal of Advanced Research and Reviews*, *21*(3), 575–589.

Qi, Y. (2024, August). Application research of RPA technology in the construction of financial robot business automation processing platform. In *2024 3rd International Conference on Artificial Intelligence, Internet and Digital Economy (ICAID 2024)* (pp. 340-345). Atlantis Press. DOI: 10.2991/978-94-6463-490-7_37

Rainey, S. K., Brown, B., & Kirk, D. B. (2017). Bots, natural language processing, and machine learning. *Tax Executive*, *69*, 39.

Rane, N. L., Choudhary, S. P., & Rane, J. (2024). Artificial intelligence-driven corporate finance: Enhancing efficiency and decision-making through machine learning, natural language processing, and robotic process automation in corporate governance and sustainability. *Studies in Economics and Business Relations*, *5*(2), 1–22. DOI: 10.48185/sebr.v5i2.1050

Schmitz, M., Stummer, C., & Gerke, M. (2019). Smart automation as enabler of digitalization? A review of RPA/AI potential and barriers to its realization. In *Future Telco* (pp. 349–358). Successful Positioning of Network Operators in the Digital Age. DOI: 10.1007/978-3-319-77724-5_31

Siderska, J., Aini, S. N. B. M., & Kedziora, D. (2024, March). Complementing robotic process automation with generative artificial intelligence (ChatGPT), case of Robocorp. In *Future of Information and Communication Conference* (pp. 37-53). Cham: Springer Nature Switzerland.

Sowjanya, S., & Vijaya Chamundeeswari, V. (2024, February). Information extraction using RPA and generative AI from unstructured documents: A case of invoices. In *International Conference on Computational Intelligence in Data Science* (pp. 250-264). Cham: Springer Nature Switzerland. DOI: 10.1007/978-3-031-69986-3_19

Szabó, K. (2020). Bridging natural language processing AI techniques and corporate communications: Towards an integrative model. *Információs Társadalom*, *19*(4), 100. DOI: 10.22503/inftars.XIX.2019.4.7

Valluri, D. (2024). Integrating augmented reality, gesture recognition, and NLP for enhancing underwater human-robot interaction. *International Journal of Science and Research Archive*, *11*(2), 956–968. DOI: 10.30574/ijsra.2024.11.2.0509

van der Aa, H., & Leopold, H. (2021). Supporting RPA through natural language processing. In *Robotic Process Automation: Management, Technology* (p. 187). Applications.

van der Aa, H., & Leopold, H. (2021). Automatically identifying process automation candidates using natural language processing. In *Blockchain and Robotic Process Automation* (pp. 77-86). DOI: 10.1007/978-3-030-81409-0_7

Zhu, R., Liu, H., Xu, X., Lin, L., Chen, Y., & Li, W. (2024). A-PGRD: Attention-based automatic business process model generation from RPA process description. *Concurrency and Computation*, *36*(8), e7940. DOI: 10.1002/cpe.7940

Chapter 12
Design and Development of Automatic QR Code Printing on Turbine Housing Component Using PLC Laser Technology and Machine Learning

Adik Takale
Sharad Institute of Technology College of Engineering, India

Rakesh Mote
Indian Institute of Technology, Bombay, India

Sumit Mangave
Sharad Institute of Technology College of Engineering, India

Milind Kognole
Sharad Institute of Technology College of Engineering, India

Sudesh Mane
Sharad Institute of Technology College of Engineering, India

ABSTRACT

A solution for an industry-specific problem by implementing the poka-yoke tech-

DOI: 10.4018/979-8-3693-4365-4.ch012

nique in a laser printer system. This automation is designed to detect the keyway slot from the component and then marking the QR code of that component with specific number of the component. The primary goals of this solution in industry are to minimize human intervention and reduce product rejection rates. The same component is fitted into three different machines, each component has the keyway slot at a different angle. To address this, used photoelectric sensors to detect the keyway slot and then relay this information to printer to print the final result on the component. The sensor is connected to a Programmable Logic Controller (PLC), which is then connected to a processor. This information sends back to PLC, which finally communicates with the laser printer to execute the printing of the QR code and component number. By integrating this system, the aim to streamline the process, enhancing efficiency and accuracy in the component marking phase of production without rejection of the component.

1. INTRODUCTION

It also explain the Intelligent Robotic Process Automation (RPA) is further expanded in this chapter and introduces a new PLC laser method with machine learning to automate QR code printing TURBINE HOUSING COMPONENTS. The solution is developed using Poka-yoke principles and photoelectric sensors designed to read printed data on boxes, guided by a PLC-driven automation framework that ensures zero errors in the entire process, achieving high precision matching IRPA's goals of enhancing productivity and reducing human error. Through the implementation of Machine Learning algorithms, it allows more accuracy as they make the system adjusts to skew in keyway slot position over time. The automatic order is a solution regarding deficit points of classical manual indication systems, such as human error and efficiency, but also example further advanced robotic procedures construction. The fourth one driven by sensor networks, machine learning (ML), and the internet of things (IoT) just started this decade. These digital tools are enabling a new generation of end-to-end automation across many sectors such as supply chain, quality control, preproduction and postproduction. The trend of technological advancement is increasing, which results in a rise burden of obsolete machinery and equipment, causing the need for more building supplies. With 4.0 emerging as the new buzz in academia and practice, it is leveraging upon AI and ML to enhance production processes aiming at building resilience and sustainable growth. This involves a transition to ML-powered robots for physical labor and sensor networks for real-time data acquisition (De Simone et al., 2023). Current with the aim to decrease human effort and increase efficacy and decision-making. What can be surmised from that the outcome, is indicative of how ML and IoT are paramount to industry 4.0 as it showcases some ways these technologies can convert traditional industrial process into intelligent interlinked systems. The objective of this system is to provide better quality leads for conversions as it uses machine learning to predict the probability of a lead converting into a customer primarily aimed at small & medium enterprises (SMEs) in the education sector. The actors of the last industrial revolution are the intersection between machine learning (ML), sensor networks and Internet of Things. That is, the supply chain, quality assurance, post-production and pre-production processes that are carried out in the industrial setting are already automatizing.

Artificial intelligence (AI), which is mainly driven by machine learning and deep learning, one of the core concepts proposed in Industry 4.0. Artificial intelligence has been playing a great role in changing the industrial-based systems, applications, and processes. As they help in increased efficiency, maintenance prediction, and customization of the process for industry. They can revolutionize multiple industries from the ground up. A real problem, however, is the complexity of deep learning models. These are also sometimes referred to as "black boxes" and they lack transparency into the decision making of the models and this further degrades the confidence in them by both business and customer; Furthermore, because these models are both complex and opaque to begin with, it becomes more challenging to combine them in a manner that promotes synergistic behavior. By using off-the-shelf office-grade equipment and common consumables3, this method supports the wide spectrum of devices without requiring expensive infrastructure or custom-built instrumentation. The research describes a circuit that preprocesses signals using an ARM network-on-chip and a deep learning algorithm, which sends these to a laser. To compare the character set present in the laser transmission speech to that of the prerecord speech cipher, a similarity comparison method was established. In the trials, the results showed a word error rate (WER) of 14.06% for English speech (Guo et al., 2023). This method is laser-proof, making it more resistant to breakdown and unlocking security.

Today industrial world in Industry 4.0 era: automating everything with reducing human interaction and improves the accuracy and precision of the industrial activities. For real-time data collection and analysis, the incorporation of technology such as IoT (equipment and sensors) is a must. Compiling a structured literature review in the realm of Industry 4.0-related technology and automatic production within automotive industry Break down applicable Industry 4.0 technologies like robots and 3D printing (Jandyal et al., 2022). This has become a critical frontier in today's production processes automation and robotics. These technologies not only speed up industrial activity and bring intelligence into the device, but also improve efficiency. Therefore, the use of a multitude of control systems means an extensive amount of sensor fusion necessary to realize fully automated processes. The printer will mean a smaller can be designed, still using a decent laser diode module for the same high power. Similarly, this method increases the utility of the workspace and also installs the effector easily for required tasks.

Rapid Prototyping (3D Printing) has also become increasingly popular in the field of additive manufacturing. This revolutionary nature makes it easier for manufacturers in Industry 4.0 (Sweeney et al., 2023) to produce complex shapes and geometries. Full characterization of such components and bottom-up investigation for thermally enhanced 3D-architecture designs were conducted, despite the large challenges presented from the strongly reflecting optical properties due to laser printing(marking step).

To illustrate the effectiveness of using surface compensation functions to enhance measurement accuracy in a typical 2D machine layout, this work presents an error correction solution for field distortion errors based on a laser marking system (Chen et al., 2009). It is the rejects and mistakes non-trained personnel made when using computers on last finish stages of machined components that have plagued the industry. The heart of the problem is a certain part which has an abnormality it is almost the same size as its counterparts, except for a really crooked keyway slot. For many fields, physical sensors are an essential component and with the increased need for flexible yet low-cost, energy-efficient alternatives laser-induced graphene (LIG) physical sensors have gained significant momentum (Kaidarova & Kosel, 2020).

It uses experimental results to test the performance of the system in reading generated code and performing print processes. They are so essential from everyday life applications to industrial application because they provide "simple processes" technology and give us the information in real-time. Sensor R&D boom over the past 25 years (Maddipatla et al., 2020): a revolution akin to silicon technology in the 1980s. This has given rise to the demand for reliable, low-cost and small-footprint sensors capable of monitoring a variety of parameters; hence there is an obvious requirement for novel sensing devices possessing advanced functionalities. The components are identifiable by different codes and a number which is depend on the keyway slot present on the component, employees are unsure of which one to print. Collaborative efforts between these diverse operators could lead to risks and flaws not present in the current industrial processes, whether they are automated or human.However, discrete material matrix selection is the sole capability offered by existing multi-material printing technologies.

In order to get over this problem, the laser printing process was automated using a photoelectric sensor and a PLC (Programmable Logic Controller).The primary innovation is the use of a photoelectric sensor to sense of the keyway slot. Whichever sensor is engaged determines which signal is sent to the PLC. The marking software is set up to respond to input from the PLC by printing the correct QR code and specific number on the component. By ensuring that every component is correctly identified and drastically reducing the chance of human error, this automated method increases overall manufacturing efficiency and accuracy. Also, with the help of machine learning the counts of components like rejected, ok, CR, MR can be monitored.The advancements in science, equipment manufacture, and application area growth are reviewed in this article along with the prospects, problems, and needs for future development (Diao et al., 2024).

Machine Learning (ML): An Overview

Machine Learning: What Is It?

Within the field of artificial intelligence (AI), machine learning (ML) is the study and use of statistical models and algorithms that allow computers to learn from and make decisions based on data. ML algorithms identify patterns and insights from data, which they utilize to generate predictions or judgments without needing to be explicitly written for every circumstance. This is in contrast to traditional programming, where precise instructions are coded to accomplish a task.

Machine Learning Types:

a) Supervised Education:

This collection of input data (features) as known labels marking such data is called a labelled dataset on which we train the system, because in supervised learning task —input must be associated with an adequate output. It is important for the algorithm to learn the mapping between inputs and outputs then predict new data correctly.

Some examples include regression (e.g., estimating property prices) and classification (e.g., spam detection in emails).

b) Unmonitored Learning:

The data Supervised learning you will provide the algorithm labelled data and in unsupervised learning, you are going to give the algorithm unlabelled data. Aim: Finding underlying structures or patterns in the data i.e. to group related data points from same structure.

This encompasses items such as being able to perform dimension reduction (i.e., reduce the number of features without letting variance slip away), and clustering (e.g. consumer segmentation. Constraint Enforcement: Allows for various ways in which a datum according to some pre-defined rules).

c) Reinforcement Learning:

Reinforcement Learning is the training of an agent with making decisions, by performing actions in underlying environments based on rewards. This feedback signal is used by the agent to learn since it uses rewards or penalties.

For example, these tasks can be in robotics (e.g. autonomous navigation) or gaming (e.g. AlphaGo).

Common Machine Learning Algorithms:

1. Linear Regression: Linear regression is a linear procedure for modelling the relationship between one dependent variable with one or more independent variables.
2. Decision Tree: A decision tree model for classification or regression. It splits the data across branches based on feature values to form predictions.

Support Vector Machines (SVM): this is an algorithm for classification that searches for the optimal hyperplane to fit in between data points of differentiating classes.

3. K-Nearest Neighbours (KNN): A simple classification algorithm, K-Nearest Neighbours (KNN) labels a data point with the majority label of its k-nearest neighbours in feature space.
4. Neural Networks: A set of algorithms, modelled loosely after the human brain, that are designed to recognize patterns. They interpret sensory data through a kind of machine perception. So, a Neural Network consists of an input layer, output layer and one or more hidden layer/s. applied in deep learning: and at other places like natural language processing and picture identification.

Machine Learning Applications:

1. Predictive Maintenance: ML can also be utilized within industrial environments to predict when machinery will fail, facilitating a more proactive form of maintenance that minimizes downtime.
2. Quality Control: Identify defects of items on an assembly line by analysing pictures, or sensor data using ML models.
3. Demand Forecasting, Inventory Optimization, and Logistical Streamlining: By calculating promising demand analytics the required inventory can be made available in advance to streamline your logistics.

4. This includes — Automation as in Machine Learning (ML) helping the cause of robotics and process automation, making systems more fluidic and intelligent.

Steps to Implement Machine Learning:

1. Collect data – Get all relevant data that may include pictures, sensor data, historical records or any other important input.
2. Data pre-processing consists of organizing and cleaning the data, handling missing values, scaling features, selecting useful predictor attributes.
3. Choose the model: Choose the ML model suitable for problem type (classification, regression…)
4. Training: Given the training data available to you, for teaching the model new structures split the information into a training and testing set.
5. Conclusion: It is used to check the memory usage, accuracy, precision of the model by passing it a testing data.
6. Deployment: Once the model shows satisfactory performance, it should be deployed in its desired environment (e.g. a IIoT system for predictive maintenance).
7. Training Monitoring—Make sure that your model continues to perform, be ready to refresh it or the training data to ensure you have an accurate model over time.

The PMMA-coated sensors provide excellent sensitivity with low detection limit and cycle stability. Touch sensors, heart rate monitoring, gait analysis and underwater operations. The simple fabrication process permits easy mass manufacturing and customisation (Kaidarova et al., 2020).

2. LITERATURE REVIEW

1. Patel et al. In their research paper, (2021) investigated the innovative impact of Programmable Logic Controllers (PLCs) on industrial automation quoting important real-world case studies. According to the abstract for this paper, it investigates the development of PLCs from humble relay-based control systems all the way through exactly what are essentially industrial-strength digital computers. This evolution, fuelled by the greater efficiency, accuracy and freedom demanded in manufacturing or production processes areas, means an incredible technological revolution (Patel et al., 2021).
2. In a more recent study of Mellado and Núñez (2022) advances on industrial automation using the integration between PLC and IoT. This makes the containerized IoT-PLC very well-connected, highly scalable and flexible one — suitable for

Industry 4.0 use-cases. IoT-PLC that provides the control and flexibility needed to make automatic QR code printing systems a reality, with the quality of results being top-notch. This paper provides a very important view for understanding the combination of PLC and IoT in industrial IEDs (our study objects as discussed earlier) (Mellado & Núñez, 2022).

3. Sehr et al. In this paper, authors from Tillmann's group (2020) revisit the role of PLCs applied in the environment of Industry 4.0 demonstrating that they are increasingly performing and becoming compatible with newer industrial approaches. The paper addresses the way by which PLCs (initially employed for discrete control in industries) are adapting themselves to meet up with data-driven processes, smart manufacturing, and improved connectivity related needs of Industry 4.0 era of automation. The authors then review changes in the evolution of PLCs from simple control devices to complex networked systems that can communicate with other components via the industrial Internet of things (IIoT). They point out the importance of existing PLC flexibility, real-time data handling and interoperability for new industrial processes and higher automation levels. Case studies and real-world applications featured in the paper illustrate the interconnectedness of PLCs with other digital technologies such as cloud computing, big data analytics, correct predictive maintenance, and AI. These linkages allow for predictive maintenance, improved production processes, and the measurement of real-time environmental conditions. Writers also touch upon the need for standardized protocols in order to enable proper communication between devices and systems, and cyber security threats as well as the complexity of a system.

Results from this paper were useful for further research in the development of advanced PLCs that can process complicated control functions, combined with laser marking technology and machine learning algorithms for great levels of accuracy, and to maximize efficiency in the case of an automatic QR code printing system on components. Thus, it could help in the development of automated QR code printing systems, which is important when producing high volumes of QR codes as required by many industrial settings (Sehr et al., 2020).

4. Sethy and Dadhich (2023) discussed the adoption and use in QR code technology at management information services and libraries extensively their paper. This study has essentially focused on the dynamics of the QR codes, how have they evolved over time in increased efficiency and simplicity of usage, access to information and services impulses. Authors look at potential library uses for QR codes — marketing, ease of use with library catalogues and access to digital resources. They highlight that QR codes allow instant access to details,

minimize physical contact with objects as well as reinforce the provision of contactless amenities something we all noticed was needed even more in light of the Covid-19 outbreak. The article also includes the technical sides of how to plan and physically print QR codes, as well as the hardware and software requirements for implementing QR code systems in libraries. The report mentions an assessment of various methods that are used for printing the QR codes concentrating on accuracy and robustness- In order to assure QR codes are going work for a long time, two things you have to do. This paper too throws the light on How QR code technology can be useful? It underscores the need for reliable methods of printing QR codes and for integrating them to state-of-the-art automation system like PLCs and laser technology (Sethy & Dadhich, 2023).

5. Kumar et al. In their paper (2021), they provide a systematic review of the recent advancements as well as applications of laser-based surface texturing devices. This study underlines the profound scope that exists in modifying surface characteristics with laser technology to augment material performance across a wide range of industrial applications (Kumar et al., 2021). The first investigate a few exact laser texturing strategies like femtosecond laser texturing, laser-instigated occasional surface structures (LIPSS), and immediate laser impedance patterning (DLIP). Because they give researchers precise control over surface morphology, these methods lead to micro and nanoscale roughening of the surface which can improve characteristics such as wettability, wear resistance, and friction.

Case studies are also provided for a variety of industries including automotive, aerospace, biomedicine and electronics to illustrate the diverse application possibilities of laser texturing. Laser texturing, in fact, is utilized to create antimicrobial surfaces and bring heat exchanger performance enhancement as well as better coating adhesion. While the writers stress on advantages of laser texturing, including its non-contact/material-independent high precision solution. It also examines the challenges of laser texturing, including the requirement for advanced control systems, potential impact on material heat and tooling. This, however, is a key resource to develop modern industrial applications as phone-readable QR codes require high levels of repeatability and accuracy in printing them on-site by the use of PLCs chained with laser texturing systems.

6. Monteiro et al. (2023) present a comprehensive framework to enhance industrial sticker data understanding using state-of-the-art object detection and optical character recognition (OCR) techniques. This work tackles the pressing need for maintaining an accurate and precise information extraction in industrial settings, where data reliability is essential for smooth operation and traceabil-

ity. To accurately detect and interpret information from industrial stickers; the authors present a system that positions advanced Optical Character Recognition (OCR) algorithms with trustworthy object detection proliferation. Many times, the stickers include essential information required for inventory control, quality assurance, regulatory conformity and other purposes This information includes product identifier codes, batch numbers and manufacture dates.

It is used as inventory tracking for automatic inspection systems and product traceability improvement in the supply chain. Based on the results of this investigation, it becomes necessary to add new age identification methods along with robust control mechanisms in place as far as printed QR codes are concerned in the specific scenario. By doing this way, the performance and utility of automated QR code printing systems can be notably enhanced because they can incorporate object detection and optical character recognition (OCR) features. This ensures a reliable system in industry (Monteiro et al., 2023).

7. Moshayedi et al. In their work, Apud et al. The evaluation is based on the history of how the attendance systems have evolved over time, from their manual methods to sophisticated automated solutions utilizing modern technologies like machine learning, artificial intelligence (AI), and internet of things (IoT). Telescope on advantages of auto attendance, which includes better data management, lesser cumbersome administration work and increased precision as mentioned in the study. They also point out the shortcomings that these systems can hide, such as concerns about privacy, the need for robust cybersecurity and technical problems (Moshayedi et al., 2021).
8. Ayvaz and Alpay (2021) most recently sought to deploy machine learning in a predictive maintenance system for production lines in manufacturing. This is the topic of their paper: "Predictive maintenance system for production lines in manufacturing: a machine learning approach using IoT data in real-time" By leveraging real-time Internet of Things (IoT) data, their primary goal is to improve efficiency and minimize manufacturing equipment downtime by predicting potential failures or maintenance needs. The research examines how predictive maintenance can reduce operational costs and improve productivity in a manufacturing environment. So here the authors propose a possible way to do things in an intelligent manner by appropriate analysis and interpretation of IoT Data Streams with some of the latest machine learning algorithms. This means any equipment problems or performance deterioration can be caught in the bud, allowing plants to step in and prevent an outage. In addition, the course provides a basic understanding of applying machine learning methods to manufacturing for improving quality control, optimizing processes and processing

data. In this work (Ayvaz & Alpay, 2021) we hope to provide a foundation, as well as support implementing the integration of machine learning methods in system using this theory.

9. The research paper is "The experimental application of popular machine learning algorithms on Predictive Maintenance and design of IIoT based condition monitoring system" which is conducted by Cakir, Guvenc and Mistikoglu from the Computers & Industrial Engineering Journal (2021). In this paper, a study is conducted by the authors to explore applicability of some popular classifiers that are part of standard machine learning libraries for prediction problems in industrial predictive maintenance. In particular, they focus on the design and implementation of an IIoT based condition monitoring system. The study seeks to reduce the practical challenges with regards to predictive maintenance by evaluating several machine learning algorithms on their functioning in realistic environments. Through experimentation the authors evaluate different algorithms to predict equipment breakdowns and performance degradation. We empirically show that machine learning algorithms are indeed useful and effective for predictive maintenance in industrial settings, which is then used to inform on how to develop and deploy your proposed system (Cakir et al., 2021).

3. MATERIALS AND APPARATUS USED FOR MANUFACTURING THE SYSTEM

Figure 1. Printing machine setup

The materials and equipments used to manufacturing the system is as shown in fig.1, Here's a breakdown of the equipment and their potential role in such a system:

1. **Photoelectric laser sensor:** This sensor could be used to detect the presence of a key way slot on the component which is placed on the fixture for the QR code printing at a designated work station. Image of photoelectric sensor is shown in fig.2.

Figure 2. Photoelectric sensor

2. **Programmable Logic Controller (PLC):** A PLC is a computer used to automate industrial processes. In this system, the PLC would likely control the overall process, including triggering the laser printer, controlling the movement of the component, and ensuring the QR code is printed correctly.
3. **Fixture:** A fixture is a device that holds or positions a part during manufacturing. In this system, the fixture would likely hold the component in place while the QR code is being printed.

Figure 3. 3D printed fixture design.

4. **Control panel:** The control panel would allow an operator to interact with the system, such as starting and stopping the process or monitoring its status.
5. **Computer / Processor:** A computer might be part of a more complex system that integrates machine vision or other quality control measures. It is less likely that a separate computer in addition to the PLC would be needed for the system. Also the data of the component is stored in the computer for the industry record.
6. **Laser printer Machine:** This is unlikely to be the machine used for printing QR codes on components. Industrial QR code marking typically uses lasers for permanent and tamper-proof marking.

4. WORKING PRINCIPLE OF THE SYSTEM

The programmable logic controllers (PLC) and laser printing mechanisms are integrated with modern sensor and technologies like poka-yoke in laser printer system. Machine learning applied to laser printing can support flexible and effective procedures with help of components count data. By automating the process of post-machining component of tagging using QR codes and identifying numbers, this system aims to minimize mistake rates and human interaction. The working of the system is shown in fig.4. Also, the detailed information of working principle of the system is given below in phase wise.

Figure 4. Process flow chart

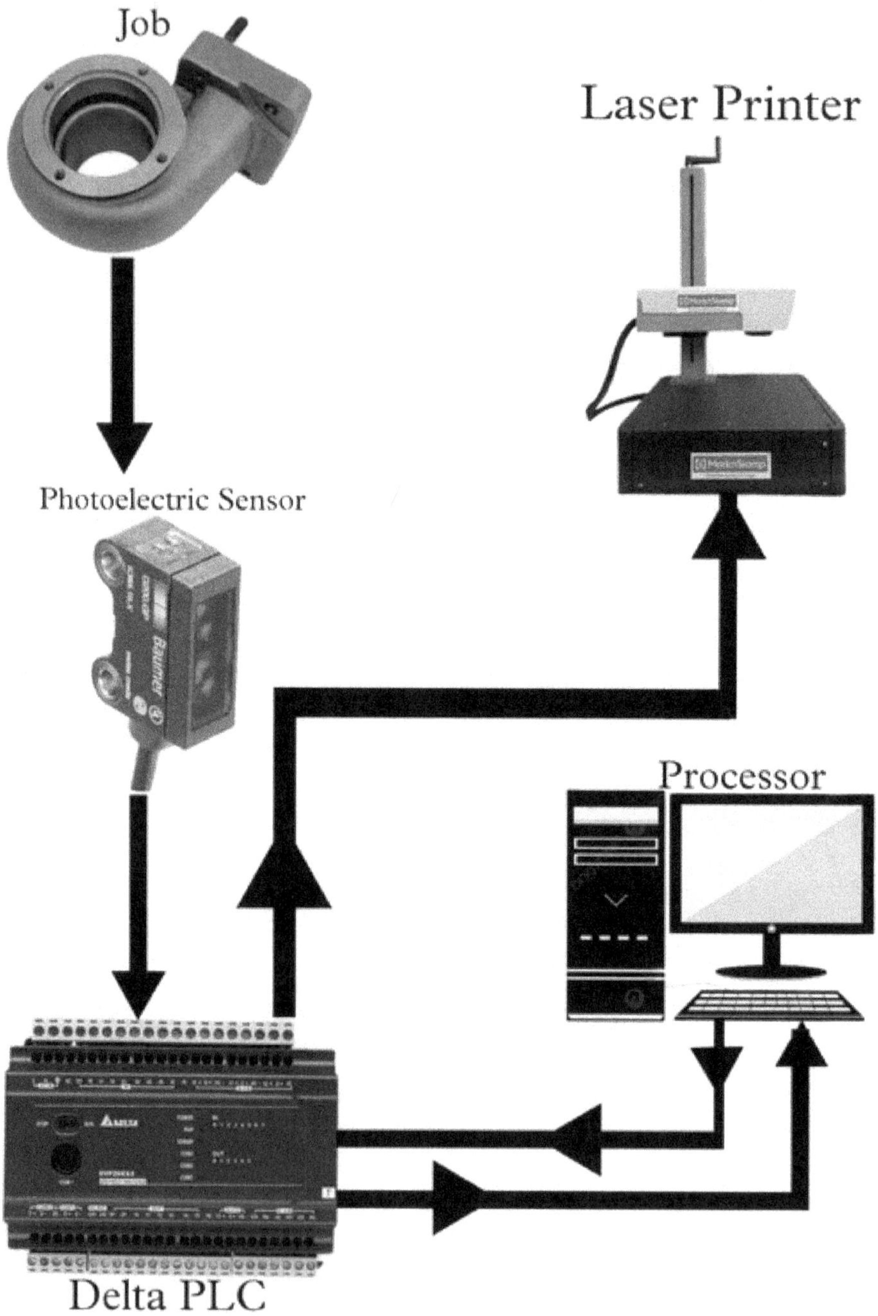

Phase 1-Key Way Slot Detection:

After machining, every component is placed on the fixture. Then photoelectric sensors pick up the keyway slot's angle and position, which is a special characteristic of each part that establishes its machine compatibility. When orange light is on in the photoelectric sensor then keyway slot was not detected and when orange light is off and green light is on in the photoelectric sensor then keyway slot was detected on the component as shown in fig.5 & 6.

Phase 2- Communication and Data Processing:

The PLC receives the sensor data pertaining to the keyway slot. After processing this data, the PLC determines each component's precise identifying parameters and orientation. The processer system receives this processed data after which it verifies it using the unique QR code and the identification number.

Phase 3- QR Code Printing:

The PLC instructs the laser printer once the data has been confirmed and finalized. The component number and QR code are correctly printed by the laser printer on the component with appropriate region when it follows these directions.

Phase 4- Assurance of Quality and Feedback Loop:

To make sure the printed information is accurate, a brief verification is performed after printing. The procedure is immediately stopped in the event of any discrepancies or system failures, indicating the need for manual inspection or system recalibration.

Phase 5- Tracking and Integrating Data:

Each printed component's details can be entered into the processor system by machine learning technique. For the purpose of quality assurance, inventory management, and following the components through the many phases of the manufacturing process, this data is essential.

Figure 5. Keyway slot not detected

Figure 6. Keyway slot detected

5. SYSTEM APPLICATION OVER EXISTING OF MODEL

By integrating this poka-yoke technique, the system aims to streamline the component marking process, minimize human intervention and product rejections, and ultimately enhance efficiency and accuracy in production. For that purpose, there are some changes over existing model as described below:

Integration with Current Manufacturing Processes: Make sure the system is seamlessly integrated into the current manufacturing line. Precisely align photoelectric sensors to determine each component's keyway slot angle.

Sensor Configuration and Calibration: Carefully calibrate photoelectric sensors to distinguish between keyway slot orientations, which are essential for parts that are used in various machines.

PLC Programming: Set up the PLC to precisely interpret sensor data and make decisions in real time. This includes figuring out the keyway slot's proper angle and indicating the QR code and number that should be printed.

Laser Printer Integration: To provide durable and excellent prints that are essential for traceability and quality control, set up the laser printer to precisely print the component number and QR code on the component.

Data Traceability and Logging: Keep track of every action, including discrepancies or mistakes, to facilitate data traceability and support ongoing manufacturing process improvement.

User Interface and Monitoring: Provide an easy-to-use interface for maintenance, checks, manual overrides, and system monitoring

Future Looking Concept Compared To Current Models

Reduction in Human Error: By automating the process of detecting and labelling, there is a significant reduction in human error. This instantly impacts the consistency as well truthfulness of the product.

Enhanced Efficiency: It improves throughput by being faster than anything done manually or even outsourced to human hands. This efficiency can lead to more output and less labour cost.

Better Quality Control: To labelling of every component which ease the quality control to handle. This technology reduced the reject rate by ensuring that only machined parts are fed for further processing.

Data Management and Accuracy: They also provide more reliable and consistent data collection than manual approaches. All of this data is useful for predictive maintenance, process optimization or quality audits.

Scalability and customization: The system also has a level of versatility that some human procedures might not be able to beat, as it can be configured to deal with different kinds of items or changes in the manufacturing process.

Decreased Rework and Waste: It eliminates the drawbacks of rework or waste as the system ensures that each part is labelled appropriately, making it beneficial for industry and more economical.

6. SYSTEM DESIGN ASPECT

Automation is key to improving production efficiency and quality in advanced manufacturing. This chapter detail on designing of PLC (Programmable Logic Controller) based a laser system very less maintenance to the use which is being provided automated QR code printer for turbine housing components with machine learning part. At the heart of this architecture is the successful implementation Poka-Yoke principles, Photoelectric sensors and real-time data feedback loops which have significantly reduced both marking times & defects while making them error-free.

Key Design Features

1. Incorporation of Photoelectric Sensors

Photoelectric Sensors – Photoelectric sensors, the role of it is used to locate wavier slot in turbine housing components. Each of these sensors emits a beam of light, which will be blocked off if there is a slot present providing accurate feedback of the component orientation.

Sensor Placement and Configuration:

- Top Protocols - ILT placement Best position for sensors: location of sensor influence on detection and elimination less targeted objects and environmental factors.
- Calibration: to keep a track of variations in component sizes and shapes, regular calibration checks are required for sensors so that the detection will not lose its accuracy.

2. Processor Interface and Communication

An effective processor interface is needed to handle data obtained from photo-electric sensors. This is like the interface between our sensors and the PLC, and it helps to get data from input port and then process on it.

Data Interpretation:

- Real-Time Processing: The sensor data is interpreted on the processor interface which allows it to know the real-time condition of position and orientation of every part.
- Error Handling: In the system, it has error detection mechanisms to track and fix any discrepancies in data which makes sure for reliability.

3. Programmable Logic Controller (PLC)

PLC Role: The PLC acts as a master controller to coordinate the functions of the sensors, processors and laser printer. It reads the data from sensors and manages the print operation.

Communication Protocols:

- Industrial Protocols: Communication protocols such as Modbus, Profibus, or Ethernet/IP that are widely used in the industry is implemented to ensure reliable data transmission between the PLC and other system component.
- Real time feedback: PLC gives real time feedback for sensor and processor to do instant changes or corrections.

4. Laser Printing System

Laser High-Precision: The high-precision laser technology makes sure that you need to get the kind of marking are very important when it comes time to produce a QR-code print job. This positions and marks each component accurately by using the PLC data.

Control Systems:

- Laser Control: High-end control systems ensure precise and consistent marking by managing laser power, speed, and focus.
- Feedback Mechanisms: Live feedback mechanisms track the printing procedure, adjusts to real-time changes and maintain quality.

5. Poka-Yoke Implementation

Error Prevention: Poka-Yoke or error prevention is built in the system to kill human errors and ensure no further occurs. The mechanisms such as sensors that check if the components have been correctly placed before printing and self-error detection are significant features.

System Safeguards:

- Verification Steps: Step through a number of verifications to make sure each component is correctly placed and aligned before printing.
- Automatic Halting: if an error occurs, the system automatically stops the process without letting defective products continue.

6. Machine Learning Integration

Machine Learning: Machine learning algorithms make the system better adaptable and efficient. Those algorithms that learn how to detect lesions more accurately as they analyse historical data and those are able adjust the printing parameters for optimization.

Continuous Improvement:

- Data Analysis: Here, machine learning models consume data from printing process and analyse to make sense of patterns and anomalies.
- Optimization: This is a system that automatically tunes parameters based on lessons learned from machine learning insights, which allows itself to keep getting better at being accurate and efficient.

Designing and implementing a PLC controlled QR code automated printing system combined with laser technology and machine learning will become a milestone in manufacturing processes. The system provides high accuracy, reliability and efficiency by integrating photoelectric sensors with real-time data feedback using Poka-Yoke principles. This chapter offers rigorous examination of the system's design elements and potential to deliver superior efficiency in manufacturing, as well as product quality across a spectrum of industrial uses.

7. VALUE ADDITIONS IN SYSTEM

Applying the poka-yoke principle to industry-specific laser printer technology for reducing human intervention and product rejection rates. Photoelectric sensors, connected to a PLC and processor interface allow for precise detection of keyway slots in components before the application of QR codes or unique identifiers for component tagging. That way each part is recognized and labels exactly even if installed on machines with different rotational orientation of keyway slot. This results in efficiency in production, which is of added value since it saves time by doing things right from the beginning so as not to make any mistakes. Ensure constant and error-less component marking with seamless communication between the sensors, PLC, processor and laser printer. This creative approach strengthens quality control procedures and increases operational efficiency, which optimizes the whole manufacturing process.

8. PARAMETERS ON WHICH SYSTEM PERFORMANCE IS TESTED

1. Accuracy of Keyway Slot Detection

Significance: Detection of keyway slot forms the basis in aligning and placing during QR code printing. This allows supervising the accuracy of print also so that QR codes are printed in exact positions they should be printing off, avoiding the misalignment and scanning difficulty.

Methodology:

a) Machine Vision Systems: Use high-resolution cameras along with Image processing algorithms to find keyway slots.
b) Machine Learning Algorithms: These are used to improve detections by learning from a collection of images of components, such as (convolution neural networks) CNNs.
c) Calibration: this includes frequent calibration on the vision system to allow for changes in lighting and positioning.

Evaluation:

a) Accuracy Metrics: Evaluate the performance of keyway slot detection using metrics like precision, recall, or f1 score.
b) Test: Thoroughly test with many different components to ensure the solution is robust and that it returns accurate results.

2. Reliability of PLC Communication

Key points to remember: The careful coherency between sensors, actuators and the laser printing module installs reliable PLC communication and ensures that all elements in the process can interact seamlessly.

Methodology:

a) Network Protocols – Use robust industry communication protocols such as Modbus, Profibus, or Ethernet IP for higher availability
b) Error Handling: Include error detection and correction mechanisms to preserve the integrity of data.
c) Redundancy: Build communication systems with redundancy so they can tolerate failures without interrupting service.

Evaluation:

a) Uptime: Time to track how often the system is up and also note them if they are down.
b) Latency:- Analyse the Communication latency to satisfy the Real-Time processing.
c) Stress Testing: Running stressful loads to appropriately evaluate how the system will behave when too many requests are made and/or network is congested.

3. Printing Precision

Report Summary: The crux of QR code printing is to make the codes scannable and deliver the right information with no errors at all.

Methodology:

a) Laser Tech: Print QR codes using fine-detail capable high-performance lasers.
b) Control Systems: Incorporate sophisticated control systems that oversee and regulate laser power, speed, and focus effort.
c) Feedback Mechanisms: Implement in-cloud feedback mechanisms to monitor and adjust printing parameters on-the-fly.

Evaluation:

a) Dimensional Accuracy:: TEST OBJECTIVE: dimensional accuracy of printed QR code SHALL be measured by high precision measurement tools.

b) Quality Inspection: An optical character recognition (OCR) and other such vision-based inspection system to confirm the quality of QR codes and that they are readable or not.
c) Consistency Tests: do a comparability trial in which the same photograph is printed multiple times.

4. Reduction in Rejection Rate

Significance: Lowering the rejection rate is important to raise the production efficiency and reduce trash, being profitable overall.

Methodology:

a) Process Optimization: Optimize printing process parameters continuously w.r.t feedbacks and insights from machine learning.
b) Detect Defects: Utilize machine learning techniques to detect potential defects and predict in advance before printing the 3D object.
c) Quality Control: Implement strict quality controls measures in the production line to detect problems and correct them soon as possible.

Evaluation:

a) Defect Rate: Monitor and assess the defect rate before and after implementation of the new system.
b) Cost Analysis: quantify the financial consequences of reduced rejection rates using cost-benefit analysis.
c) Continuous Improvement: Apply statistical process control, SPC tools to observe and improve the process.

9. COMMERCIAL VIABILITY OF THE SYSTEM

The automation system for keyway slot detection and laser printing includes Poka-Yoke technology (Al Ayyubi et al., 2020) as manufacturing process up gradation by PLC & Machine learning. This not only makes your drill press more accurate and less prone to error, but also greatly increases the efficiency of production and the quality of each component. Market analysis, cost-benefit analysis and possibly technological challenges are all important factors to consider when assessing the potential for commercialisation of this system.

a) Industrial Machineries

The use of these things is very important otherwise you will not be then able to make any advanced industrial machinery and it does not matter from where you get the companies that are manufacturing it, offshoring or nearshoring. Automated systems streamline keyway slot detection, delivering a highly accurate and reliable production line. High-resolution cameras and sophisticated image processing algorithms from the manufacturer help achieve near-perfect alignment and component positioning. Such precision is essential in order to meet the strict standards as set by traditional QR codes and, with that much at stake, make it highly important not to waste resources.

This easy communication and integration between different parts of the production line is possible due to PLCs being implemented. Because of this, PLCs have a reputation for being durable and reliable in industrial settings, especially when they are used for real-time control and monitoring purposes. This will make any alteration or deviation in the process to be immediately known and checked, thus resulting to an equally effective and reliable system.

b) Quality Control and Assurance

Manufacturing depends more than ever on quality control, and this is especially true for sectors such as aerospace or automotive: both areas demand extremely high standards of precision and reliability during production. When the laser printing is combined with Poka-Yoke technology, there is additional protection. By the knowledge of Poka-Yoke this is all about mistake-proofed; means that designing a system to make it impossible for individuals to commit errors, and if errors cannot be eliminated entirely; then design them in such a way as they are instantly detected and link before delivery.

In the case of laser printing, Poka-Yoke methods may include tools to confirm the correct placement of parts prior to printing, sensors for identifying and recording alignment or defects, or self-regulating control mechanisms that stop the process, in case an error would be detected. This would greatly minimize the chances of there being any bugs and will mean that only the best products actually get sent to the manufacturing stage.

c) Reducing Product Rejection Rates

The other most important advantage of this automated system is the decrease in product rejection rates. Manufacturers can reduce the number of defective products by improving keyway slot detection accuracy and incorporating fail-safe alarms. It leads to production efficiency which perennially helps in the cost savings.

Lower rejection rates, in addition to having reduced costs of materials and labour for elimination of rework and scrap, do give the overall image a manufacturing process that functions cleanly. Product quality with fewer defects performs better leading to more satisfied customers who result in greater sales and market share.

d) Economic Feasibility

A detailed market analysis is a must for testing the economic feasibility of this system. This would include identification of high-precision manufacturing systems demand, competitive environment analysis and market segments ripe for this technology. Sectors such as aerospace, automotive, medical devices, and electronics are expected to benefit greatly from implementing this system owing to their high-quality standards.

The cost-benefit analysis is another thing that will require doing to see if it is worth putting something like this in place. The costs of the initial capital investment, ongoing maintenance, and any savings from lower rejection rates or higher schedule efficiency and faster plate output due to improved product quality should be determined in a cost analysis. In most scenarios, the long-term advantages as well as costs savings surpassing the initial investment of funds and making it commercially viable practice with a proven success rate.

e) Technological Challenges

The advantage is obvious, but there are associated risks as technological issues have to be solved together. Specific challenges are also present in the capacity for integration of machine learning algorithms with PLCs, reliability of automated systems under tough industrial environmental conditions and degradation of keyway slot detection accuracy over time. Solving these problems are normally collaborative efforts involving teams of engineers, data scientists, and industry experts to build production quality solutions that scale.

Finally, the whole system supported by PLCs and machine learning as well ensures a high viability from a commercial point of view. This system would comply with the requirements of advanced manufacturing by improving accuracy, minimising errors and enhancing production output. A complete evaluation of market opportunities, cost analysis, and potential technological challenges is all you need to be convinced this would make a worth investment for industries looking to follow the trend and improve their manufacturing processes.

10. RESULT

Photoelectric Sensors Keyway Slot Detection in Automation Systems, this white paper covers some of the latest developments on the use of photoelectric sensors to keyway slot detection in automation systems. In addition, the photoelectric sensor can emit light to detect the presence or absence of objects, so it is an indispensable output for automated control system. The Machine Learning is used to collect minimum information or data of components that are laser printed. This feature plays a significant role in error reduction and overall system performance optimization, in addition to improving the efficacy of automated operations which benefits to the industries. A revolutionary change in the efficiency, accuracy, and security metrics of QR code printing for components is highlighted by the introduction of the PLC-controlled laser marking system with machine learning integration. This system's automation results in significant improvements in production throughput and efficiency, lowering labour costs and production time while maintaining flexibility and responsiveness to meet rising demand. Under the direction of the PLC, the laser marking system prints QR codes with extreme precision, reducing mistakes and rejected units. This maximizes resource efficiency and improves overall product quality, which in turn boosts customer satisfaction and brand reputation.

Figure 7. Component rejection data

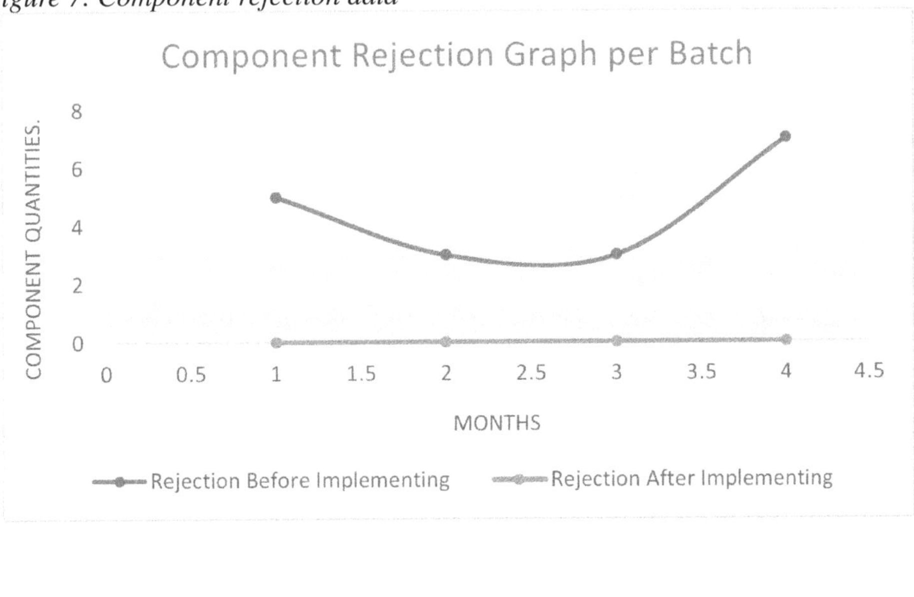

Above fig. 7 graph shows the number of components rejected per batch which is of 200 quantities over a period, presumably before and after the implementation of a system. There are two data series plotted on the graph, represented by square markers. The data series labelled "Before Implementing" is plotted in red and the data series labelled "After Implementing" is plotted green. For this graph data is collected of eight-month, four-month data of before implementing & four-month data after implementing. It appears that the system resulted in a decrease in the number of rejected components per batch.

Furthermore, the fundamentally data-rich characteristics of QR codes, when combined with strong security protocols, strengthen data integrity and improve supply chain traceability, thereby reducing the risks related to counterfeiting and unauthorized modification. Stakeholder confidence has been enhanced by this increased security, which also highlights a dedication to quality and accessibility. Essentially, this implementation's effects go beyond simple technological progress; rather, they usher in a new era of industrial operations that are efficient, precise, and secure. They also unlock potential that has been unrealized and boost innovation and competitiveness in an increasingly automated environment.

11. CONCLUSION

In conclusion, the system is used to detect the keyway slot on the components with fully automation. When we place the component on the fixture the next procedure is fully automatic from detecting the keyway slot with the help of photoelectric sensors and it gives signal to the PLC with the controller and controller gives out signal to the laser printer and stored the components count on the computer to get all the combined data which is useful to record and store for future communication.

The exploration of an automated system for printing QR codes on components has revealed a groundbreaking solution that amalgamates PLC laser technology with machine learning algorithms. Through meticulous methodology, this system orchestrates various components synergistically to achieve unparalleled accuracy, speed, and reliability in QR code printing. The essence of this system lies in its holistic approach, where a PLC acts as the central command hub, orchestrating the intricacies of the laser marking process with precision timing. The laser marking system, guided by the PLC, delicately etches QR codes onto components, while a photoelectric sensor ensures meticulous alignment. Moreover, the integration of machine learning injects a layer of intelligence, empowering the system to discern between well-printed QR codes and defective ones, thereby enhancing quality control.

This frontier system provides numerous benefits. This automated approach results in a major decrease in production time and labour costs by improving the process efficiency & throughput significantly. Additionally, the laser marking system provides a highly precise mechanism that eliminates discrepancies and reduces accuracy-related defects in the component along with improving its operational efficiency. Built in scalability and flexibility ensure smooth incorporation into any production environment, regardless of size or setup. The nature of QR codes being inherently data-rich make them one of the most preferred choices for a robust tracking and tracing solution as it prevents safety rules to thwart fakes.

The road map for this system, in turn and looking ahead, is filled with potential for continued migration. Deeper machine learning integration may offer even smarter and more flexible quality control checks in the future. Standardization can enable data encoding and formatting to be simplified, promoting interoperability between manufacturing ecosystems, leading the way towards a fully connected state. In an age of digital threats, QR codes must be maintained against tampering and unauthorized access, ensuring the integrity of the manufacturing process categories security enhancements.

In short, this machine learning-enabled PLC laser marking system marks the QR code on parts by automation differently request Body being done. Its multi-faceted advantages efficiency, accuracy and security have put it under an indispensable category of. in present-day manufacturing. A technology that has and will continue to evolve with the upcoming of further advancements leading the global digitalisation revolution in manufacturing, propelling a new era of efficiency and innovation.

REFERENCES

Al Ayyubi, M. C., Mahmudah, H., Saleh, A., & Rachmadi, R. R. (2020, July). Implementation of poka-yoke system to prevent human error in material preparation for industry. In 2020 International Seminar on Intelligent Technology and Its Applications (ISITIA) (pp. 273-278). IEEE. DOI: 10.1109/ISITIA49792.2020.9163707

Ayvaz, S., & Alpay, K. (2021). Predictive maintenance system for production lines in manufacturing: A machine learning approach using IoT data in real-time. *Expert Systems with Applications*, *173*, 114598. DOI: 10.1016/j.eswa.2021.114598

Bamshad, A., & Cho, H. J. (2021). Laserjet printed micro/nano sensors and microfluidic systems: A simple and facile digital platform for inexpensive, flexible, and low-volume devices. *Advanced Materials Technologies*, *6*(12), 2100401. DOI: 10.1002/admt.202100401

Cakir, M., Guvenc, M. A., & Mistikoglu, S. (2021). The experimental application of popular machine learning algorithms on predictive maintenance and the design of IIoT based condition monitoring system. *Computers & Industrial Engineering*, *151*, 106948. DOI: 10.1016/j.cie.2020.106948

Chen, M. F., Chen, Y. P., & Hsiao, W. T. (2009). Correction of field distortion of laser marking systems using surface compensation function. *Optics and Lasers in Engineering*, *47*(1), 84–89. DOI: 10.1016/j.optlaseng.2008.07.017

De Simone, V., Di Pasquale, V., & Miranda, S. (2023). An overview on the use of AI/ML in Manufacturing MSMEs: Solved issues, limits, and challenges. *Procedia Computer Science*, *217*, 1820–1829. DOI: 10.1016/j.procs.2022.12.382

Diao, Q., Zeng, Y., & Chen, J. (2024). The Applications and Latest Progress of Ceramic 3D Printing. Additive Manufacturing Frontiers, 200113.

Durna, A., Fries, J., Hrabovsky, L., Sliva, A., & Zarnovsky, J. (2020). Research and development of laser engraving and material cutting machine from 3D printer. *Management Systems in Production Engineering*, *28*(1), 47–52. DOI: 10.2478/mspe-2020-0008

Guo, C. Y., Hsieh, T. L., Chang, C. C., & Perng, J. W. (2023). A novel smart photoelectric lock system: Speech transmitted by laser and speech to text. *Heliyon*, *9*(3), e14510. DOI: 10.1016/j.heliyon.2023.e14510 PMID: 36967961

Jandyal, A., Chaturvedi, I., Wazir, I., Raina, A., & Haq, M. I. U. (2022). 3D printing–A review of processes, materials and applications in industry 4.0. *Sustainable Operations and Computers*, *3*, 33–42. DOI: 10.1016/j.susoc.2021.09.004

Kaidarova, A., Alsharif, N., Oliveira, B. N. M., Marengo, M., Geraldi, N. R., Duarte, C. M., & Kosel, J. (2020). Laser-printed, flexible graphene pressure sensors. *Global Challenges (Hoboken, NJ)*, *4*(4), 2000001. DOI: 10.1002/gch2.202000001 PMID: 32257383

Kaidarova, A., & Kosel, J. (2020). Physical sensors based on laser-induced graphene: A review. *IEEE Sensors Journal*, *21*(11), 12426–12443. DOI: 10.1109/JSEN.2020.3034845

Kumar, V., Verma, R., Kango, S., & Sharma, V. S. (2021). Recent progresses and applications in laser-based surface texturing systems. *Materials Today. Communications*, *26*, 101736. DOI: 10.1016/j.mtcomm.2020.101736

Maddipatla, D., Narakathu, B. B., & Atashbar, M. (2020). Recent progress in manufacturing techniques of printed and flexible sensors: A review. *Biosensors (Basel)*, *10*(12), 199. DOI: 10.3390/bios10120199 PMID: 33287324

Mellado, J., & Núñez, F. (2022). Design of an IoT-PLC: A containerized programmable logical controller for the industry 4.0. *Journal of Industrial Information Integration*, *25*, 100250. DOI: 10.1016/j.jii.2021.100250

Monteiro, G., Camelo, L., Aquino, G., Fernandes, R. D. A., Gomes, R., Printes, A., Torné, I., Silva, H., Oliveira, J., & Figueiredo, C. (2023). A Comprehensive Framework for Industrial Sticker Information Recognition Using Advanced OCR and Object Detection Techniques. *Applied Sciences (Basel, Switzerland)*, *13*(12), 7320. DOI: 10.3390/app13127320

Moshayedi, A. J., Roy, A. S., Liao, L., Lan, H., Gheisari, M., Abbasi, A., & Bamakan, S. M. (2021). Automation attendance systems approaches: A practical review. *BOHR Int. J. Internet Things Artif. Intell. Mach. Learn*, *1*(1), 23–31. DOI: 10.54646/BIJIOTR.003

. Patel, K., Jogani, V., Bhajiyawala, P., Modi, A., & Panchal, P. (2021). Technological revolution in Industrial Automation using PLC.

Sehr, M. A., Lohstroh, M., Weber, M., Ugalde, I., Witte, M., Neidig, J., Hoeme, S., Niknami, M., & Lee, E. A. (2020). Programmable logic controllers in the context of industry 4.0. *IEEE Transactions on Industrial Informatics*, *17*(5), 3523–3533. DOI: 10.1109/TII.2020.3007764

Sethy, S., & Dadhich, M. (2023). *Adoption of QR Code Technology for Easy Access to Library and Management Information Services: A Comprehensive Analysis*. Library Philosophy & Practice.

Sweeney, D., Nair, S., & Cormican, K. (2023). Scaling AI-based industry 4.0 projects in the medical device industry: An exploratory analysis. *Procedia Computer Science*, *219*, 759–766. DOI: 10.1016/j.procs.2023.01.349

KEY TERMS AND DEFINITIONS

Industrial Internet of Things (IIoT): An industrial setting's network of physically connected equipment, sensors, and devices that gather and share data is referred to as the "Industrial Internet of Things." IIoT makes industrial processes better by facilitating real-time analytics and decision-making, automation, and predictive maintenance.

Automation: The implementation of information technology and control systems (computers, robots) into the factories to manage processes and machinery less human help Automation – The use of technology for controlling processes reducing human intervention. For example, in manufacturing facilities, automation improves the consistency and speed of production that also leads to a safer work environment.

Poka-Yoke: Poka-yoke is a Japanese phrase that means "mistake-proofing" or "inadvertent error prevention. It refers to devices or processes designed to lead users through tasks precisely or ensure them that missteps cannot take place, in an effort to control human errors in the production process.

PLC Controller (Programmable Logic Controller): The role of a PLC controller an industrial digital computer which has been ruggedised and adapted for the control of manufacturing processes such as assembly lines, robotic devices or any activity that needs high reliability control and ease of programming. PLCs are known for their longevity and performance in difficult industrial conditions and used to automate anything from standard functions to more complex processes.

Photoelectric Sensors: Photoelectric sensors detect objects, changes in surface conditions, or changes in the surrounding environment using a light transmitter and photoelectric receiver. Industrial Automation Systems used for measuring the distance, presence/absence of an Object this sensoric is always the first choice.

Chapter 13
Analyzing and Comparing Trajectory Planning Algorithms for Optimal Robotic Manipulator Motion: A Comprehensive Study

Vijay Kumar Dalla
https://orcid.org/0000-0002-0807-910X
National Institute of Technology, Jamshedpur, India

ABSTRACT

Robots and automatic machines are increasingly being designed to operate at ever increasing speeds to attain petite production times. The primary goal is to create a path trajectory from the initial to the desired position that meets critical objectives such as limiting jerk, trajectory path completion time, optimal energy requirement, or obstacle avoidance while also gratifying the manipulator's joint dynamics. As a result, special care must be taken to design a trajectory which could be implemented at higher speed while remaining safe for the manipulator in terms of evading redundant actuator accelerations and mechanical structure vibrations. As a result of this situation, trajectory planning algorithms are becoming increasingly important in automation. This work offers a systematic analysis, thorough understanding, and investigation of various trajectory planning algorithms for the trajectory selection of robotic manipulator which are primarily in Cartesian or Joint space.

DOI: 10.4018/979-8-3693-4365-4.ch013

1. INTRODUCTION

In trajectory path planning, the manipulator's position changes over time (Zha, 2002). Appropriate motion planning seems to be one of the most significant aspect of the design and advancement of a robotic manipulator. In motion planning, parameters, such as optimum time, low power consumption, obstacle avoidance are considered in optimization technique (Gasparetto & Zanotto, 2007). The optimization procedure is typically carried out with an objective function that has constraints which determines the formulation of the problem. The dynamic system is made up of constraints such as inertia, torque, and applied force. Many researchers stated that there are two kinds of constraints: System constraint and task-oriented constraint. It is known that system constraints represent the constraints in joint space, and task-oriented constraints means the constraints in Cartesian space. The researcher's main task is to plot the appropriate path trajectory from the initial target to the desired target while keeping the manipulator's base stabilized. According to (Jin et al., 2021) and (Ma et al., 2021), time optimal path could be produced in Joint space and Cartesian space. During motion planning in joint space, the motion of each joint is taken into account. While optimizing robot tip motion, the starting and end positions of the joints are taken into account. Cartesian trajectory planning is simpler to calculate, while Joint space trajectory planning takes into account Jacobian singularity, which is difficult to compute (Tsai, Kohli, & Arnold, n.d.). Hence, dynamic singularity should be avoided and individual trajectories should not interfere with each other.

Trajectories should be smooth and continuous, with enough space for the robot tip to change velocity or achieve retardation (Kala, 2018). Some of the techniques that can be utilized to generate path trajectories in Joint space as well as Cartesian space are polynomial profile, cubic spline, B-spline, quartic splines, 5th order polynomial, and 4th order quintic polynomial (Guan et al., 2005) (Gasparetto & Zanotto, 2010) (Liu et al., 2013). Tian and Collins (Tian & Collins, 2004) used a polynomial blend composed of Hermite cubic interpolation to evaluate the optimal trajectories in task space (Janjoš et al., 2017). Kucuk (2017) proposed time-optimal motion planning scheme for parallel manipulator subjected to kinematic and dynamic constraints. For this purpose, quartic and quintic polynomial splines were used in Cartesian space. Hence, the proposed technique is applicable to any parallel robot subjected to dynamic constraints. All of the above listed trajectory planning schemes present new challenges and necessitate advanced robotics research. The key problem in trajectory planning is that manipulator must travel in a time-efficient manner with smoother trajectories. Several researchers have suggested several motion planning approaches for mobile robots and land vehicles working on earth since 1970.

The purpose of trajectory path planning is to generate the reference input required for space robot motion control to achieve the desired pose. To plot trajectory path in joint space, primary extract a categorization of via-points through the intended robot tip path, afterwards conduct kinematic inversion to obtain the appropriate values of the manipulator joints. The trajectory is at that time created in Joint space using linear interpolation polynomial functions while accounting for the kinematic and dynamic constraints placed on the manipulator joints (in terms of Joint position, velocity, acceleration, and jerk). Typically, this method of trajectory planning may also avoid the challenges associated with stirring adjoining singular configurations and therefore could effectively cope through the existence of redundant DOF (Wie, 2005).

In CNC machining, interpolation and motion planning for parameterized curves are critical concerns. The main concept is to minimize the chord error limitation to an angular acceleration constraint, resulting in a velocity bound arc known as the chord error velocity limit curve (CEVLC) (Fan, Lee, & Chen, 2015). The trajectory planning process then seeks the time-optimized velocity curve defined through tangential acceleration constraint. In velocity planning, two main categories of acceleration patterns can be used: tangential acceleration as well as multi-axial acceleration, in which individual axis accelerates through control point (Fan et al., 2012). Sun et al. (2016) utilized a homogeneous parametric federate that did not account for chord error. In certain cases, Dong et al. Dong et al. (2014) employed a chord error constraint to adjust the feed-rate, whereas in others, they used a steady federate. Nevertheless, the system acceleration abilities were not taken into account. By determining optimum transit velocities at critical bends, Katrakazas et al. (2015) proposed a velocity planning approach that is based on tangential acceleration constraints. Huypens (2020) presented an interpolation approach for tangential ACC/DCC feed-rate optimization based on distinct velocity field. Numerous researchers (Beudaert et al., 2012) (Benko Loknar et al., 2021) (Wang et al., 2017) studied velocity planning algorithms with tangential acceleration and jerk constraints. Li et al. (2019) introduced velocity planning algorithms with limited acceleration, jerk, and chord error. Zhang et al. (2012) proposed a greedy approach for multi-axis jerk velocity planning. These ideal approaches employ the "Bang-Bang" control tactic, which ensures that as a minimum single axis must satisfy it for minimizing chord error.

This research paper serves two major objectives. First, to investigate the various optimization algorithms for manipulator's trajectory selection, highlighting major challenges. Secondly, this paper aspires to be a comprehensive review of proposed manipulator's motion planning schemes. This paper's key contribution is to analyze numerous optimization techniques for identifying the appropriate motion planning approach for robotic manipulator.

The paper is arranged as follows. In Section 2, trajectory path planning scheme in joint space is given. Section 3 presents trajectory path planning scheme in Cartesian space. Section 4 depicts classification of various trajectory planning methodologies. Section 5 represents results and discussions. Section 6 signifies current issues and future challenges in the area of motion planning. At last, conclusion is discussed.

2. TRAJECTORY PATH PLANNING SCHEME IN JOINT SPACE

Trajectories are given for each independent joint in joint space trajectories. Only at the initial and desired positions is the absolute Cartesian position of the end-effector specified. The velocity Jacobian is used to calculate joint motion. Trajectories are chosen to be fairly smooth in order to give the manipulator enough time to accelerate and decelerate. The trajectory path generated by the path controller is typically fed into the computational algorithm for motion planning, where kinematic and dynamic constraints such as resultant forces and internal torques are satisfied (Misra & Bai, 2017) (Xu & Chen, 2000) (Jin, Rocco, & Geng, 2021). As a result, the robot's arm tip generates path trajectories in the form of position, velocity, and acceleration variations with respect to time.

The variation of polynomial profile with time subjected to position, velocity, and acceleration is depicted in Fig.1. Here, the position profile appears to be smooth, whereas velocity and acceleration profile shows uneven fluctuations due to the abrupt change of joint velocity. Fig. 1 is obtained by calculating the coefficients of cubic polynomial such as $b_0 = \theta_i$, $b_1 = 0$, $b_2 = \frac{3}{t_f^2}(\theta_d - \theta_i)$, $b_3 = \frac{2}{t_f^3}(\theta_d - \theta_i)$, that connects any initial joint angle position with any desired final position. Then, put values of $\theta(t), \dot{\theta}(t), \ddot{\theta}(t)$, respectively. For $\theta(t), \dot{\theta}(t), \ddot{\theta}(t)$, the initial values are [10, 30, 12]°, respectively, whereas, the desired values are [60, 70, 75]°, respectively.

The procedure for calculating coefficients of cubic polynomial and values of $\theta(t), \dot{\theta}(t), \ddot{\theta}(t)$are expressed as:

For smoother joint trajectories, more than four constraints of θ_t are required. The choice of initial and final values imposes two constraints on the function values are:

$\theta(0) = \theta_i, \theta(t_f) = \theta_d$, (1) The function should have continual velocity, which implies that the start and terminal velocity are both zero in this situation:

$\dot{\theta}(0) = 0$

,

$$\dot{\theta}(t_f) = 0 \quad (2)$$

A cubic polynomial has four coefficients, so it can be made to satisfy the four constraints given by Equations. (1)-(2)

Eq. 1 resembles the trajectory path planning equation having 3rd degree polynomial, which can be represented as:

$$\theta(t) = b_0 + b_1 t + b_2 t^2 + b_3 t^3 \tag{3}$$

So, the joint velocity and acceleration along this path are clearly depicted as:

$$\dot{\theta}(t) = b_1 + 2b_2 t + 3b_3 t^2$$

,

$$\ddot{\theta}(t) = 2b_2 + 6b_3 t \tag{4}$$

Combining Eqs. (3)-(4) through four desired restraints vintages four equations as:

$$\theta_i = b_0$$

,

$$\theta_d = b_0 + b_1 t_f + b_2 t_f^2 + b_3 t_f^3$$

,

$$0 = b_1$$

,

$$0 = b_1 + 2b_2 t_f + 3b_3 t_f^2 \tag{5}$$

Elucidating these equations for b_i, we acquire

$$b_0 = \theta_i$$

,

$$b_1 = 0$$

,

$$b_2 = \frac{3}{t_f^2}(\theta_d - \theta_i)$$

,

$$b_3 = \frac{2}{t_f^3}(\theta_d - \theta_i) \quad (6)$$

where, $\theta(t)$represents angular displacement of the joint, $t_f$denotes travelling time, and, $(\theta_d - \theta_i)$ represents the difference between angular position at final and initial position of the joint.

Figure 1. Variation of polynomial profile with time subjected to position, velocity, and acceleration

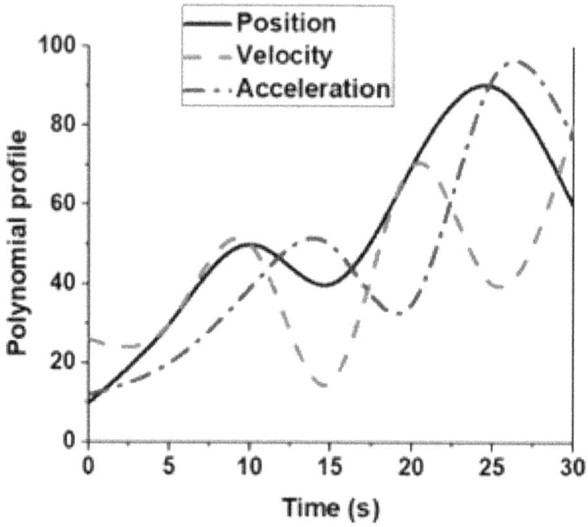

3. TRAJECTORY PATH PLANNING SCHEME IN CARTESIAN SPACE

In Cartesian space, the trajectory path planning involves primarily executing a set of data variables which satisfy the pose that the robot tip must achieve over each interval of time (Seereeram & Wen, 1995). By interacting with the trajectory path points through polynomial bends such as Bezier curves and cubic spline curves, the generated trajectory in Cartesian space could be presented in exact (original) or inexact (approximate) form (Xu et al., 2014). In Cartesian space, the kinematic inversion is requisite to convert the robot tip pose (position and orientation) into joint values of the manipulator (Feng et al., 2014). Lampariello et al. (2011) investigated the method for producing path trajectories in joint space by extracting intermediate points from the preferred robot tip position. They also implemented kinematic inversion to determine manipulator joint values. Following that, the interpolation function (Sundaralingam & Hermans, 2019) is used to compute a path trajectory in joint space by taking into account the kinematics and dynamics constraints imposed on manipulator's joints. Furthermore, joint space path planning will substantially reduce dynamic singularity while it will intelligently deal with joint instability (Dubowsky & Papadopoulos, 1993). Path planning in joint space, on the other hand, has the disadvantage of causing non-linearity in the manipulator's joints as a result of direct kinematics. Fig. 2 shows variation of linear interpolation with time subjected to position, velocity, and acceleration. Here, position curve tends to follow the constant straight-line path, whereas the velocity curve and acceleration curve follows linear path and parabolic path, respectively. Linear interpolation through flattening and rectilinear function with parabolic blends LSPB is used to obtain Fig. 2. The initial conditions for LSPB are stated as:

$$\theta_d^i(t)\begin{cases}\theta_i + \frac{\ddot{\theta}}{2}t^2 & 0 \le t \le t_a \\ \frac{\theta_d + \theta_i - \dot{\theta} t_f}{2} & t_a < t \le t_f - t_a \\ \theta_d - \frac{\ddot{\theta} t_f^2}{2} + \ddot{\theta} t_f t - \frac{\ddot{\theta} t_f^2}{2} & t_f - t_a < t < t_f \end{cases} \tag{7}$$

where, t_a is expressed as

$$t_a = \frac{\theta_i - \theta_d + \dot{\theta} t_f}{\dot{\theta}}$$

Then, put values of t_f from 1 to 20, to obtain the value of $\theta_d^i(t)$ (linear interpolation). For $\theta(t)$,$\dot{\theta}(t)$,$\ddot{\theta}(t)$, the initial values are [0, 0, 30]°, respectively, whereas, the desired values are [0, -10, 70]°, respectively.

Figure 2. Variation of linear interpolation with time subjected to position, velocity, and acceleration

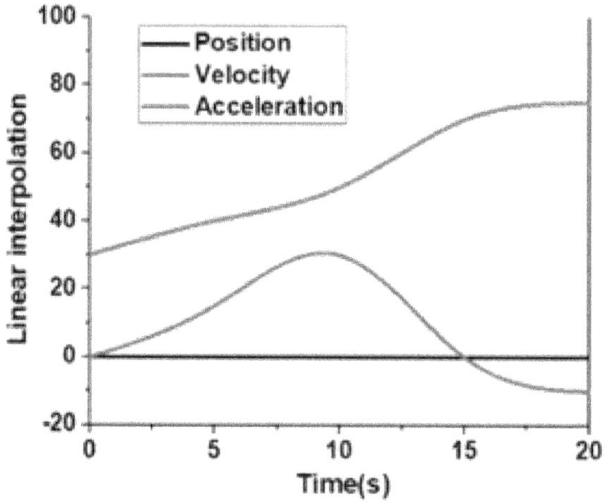

Huang et al. (2018) stated that acceleration at the manipulator's joints should be constant to minimize the impact of a jerk in the joint. Gasparetto and Zanotto (2008) concluded that restricting jerk is critical for reducing the effect of vibration caused in the robot's mechanical structure. According to Fang et al. (2020), vibrations induced by irregular joint trajectories could deteriorate the actuator and reduce the manipulator's trajectory tracking efficiency. Suriano et al. (2020) proposed a technique for solving real-time inverse kinematics and trajectory tracking issues for redundant robots in uncertain environments by integrating the Jacobian-based mathematical model with a customized vector field. Zhao et al. (2021) proposed a generic technique for the redundant planar manipulator's trajectory planning. Mishra et al. (2020) stated a technique for finding the minimum path in a cabledriven parallel robot's joint space for point-to-point movements This approach is based on the joint space domain, which means that the cord length and the shortest path in joint space between two spots are found. Zhou et al. (2021) developed an approach for optimum trajectory planning to decrease energy consumption and travelling time while assuring the smooth operation of a cooperative welding robot with several manipulators. In this approach, trajectory is modified using cubic B-spline for each joint space as well as particle swarm optimization (PSO) technique.

4. CLASSIFICATION OF VARIOUS TRAJECTORY PLANNING ALGORITHMS

The trajectory is usually planned following the optimality criteria as have been specified. The following are the most appropriate optimality criteria:

i. Reduced time seeking algorithms
ii. Reduced energy (actuator effort) algorithms
iii. Reduced jerk algorithms

4.1 Reduced Time Seeking Algorithms

In trajectory planning, determining the time optimum path for manipulators is a critical task. Many scholars have been interested in this topic, particularly in the previous two decades. One of the key objectives for adopting robots is to increase production through faster robots. A more practical method would be to reduce the amount of time needed to finish a task while being bound by physical limits. The minimal time criteria, which demands the robot to walk along a specific path in the least amount of time feasible, is the most often utilized criterion in industrial applications. It is critical to decrease manufacturing time while boosting profit.

The position-velocity phase plane is used to define the algorithms outlined in (Sahar & Hollerbach, 1986) (Nikolos et al., 2003). The application of dynamic programming techniques is an alternative method to shortest time trajectory planning (Shin & McKay, 1986). The main concept is to discretize the state space by forming a grid of points (known as state points). Based on the velocity, acceleration, and jerk constraints, it is possible to attach a set of the following permissible state points to each point and calculate the cost of each conceivable solution by considering the time required for the motion. This cost is calculated by assuming that each step has a constant acceleration value. Ultimately, a dynamic programming technique generates the shortest time trajectory. Dynamic programming methods, in contrast to phase plane approaches, do not require path parameterization and allow for the use of any performance index. As a result, such algorithms could be used as a general strategy for trajectory optimization. The phase plane approach, on the other hand, is quite efficient in terms of computation complexity; additionally, it can be used for on-line trajectory planning (Mettin et al., 2009). A model-based strategy to maximize the speed of industrial robots is used to obtain the minimum-time trajectories that satisfy various criteria commonly offered in industrial robot applications (Xiao et al., 2012).

Since the dynamic models employed presume the robot links to be rigid and thus do not consider the actuator dynamics, the mentioned methods earlier yield trajectories with abrupt accelerations and joint torques. Blatantly ignoring link flexibility and actuator dynamics almost always leads to unfavorable results. First, manipulator actuators cannot generate constant torques, resulting in joint motion being lagged relative to the reference trajectory. As a result, the precision of the desired trajectory is greatly decreased, and the tracking controller must be actuated often throughout trajectory execution. Additionally, each actuator switching may result in chatter, which is defined as high-frequency oscillations that create vibrations in the manipulator's mechanical structure. As a result, mechanical components deteriorate and trajectory following accuracy suffers. Another downside of an inaccurate model is that, because time-optimal control necessitates the activation of at most one robotic actuator at each time instant, a controller is unable to fix tracking errors induced by disturbances or mistakes in the model (Farid & Ruggiero, 2022).

The phase plane approach, combined with an actuator jerk constraint, is used in these works as a possible solution to these challenges (Consolini et al., 2021). The pseudo-jerk, which is described as the third derivative of the curved abscissa, is used as the control variable in the proposed technique, resulting in a dynamic equation of 3rd order. The experimental findings by Van Dijk (2006) indicate that time-optimal trajectories may be constructed by simply establishing an upper constraint on the pseudo-jerk with a typical PID controller. This graph depicts the relationship between trajectory following accuracy and low jerk values.

In addition to the execution time, another way to minimize torque fluctuations is to incorporate an energy contribution in the objective function. Even when typical PD controllers are utilized, the experimental results presented in (Chen et al., 2019) reveal that an increase in overall motion time is matched by increased trajectory following accuracy. As a consequence, actuator stresses are minimized, which has an obvious advantage in total lifespan for the manipulator's electro-mechanical components.

The most fundamental contrast is between constraint types. It may be used to characterize the difference between kinematic and dynamic trajectory planning for any trajectory planning approach (Goerzen et al., 2010). As input, kinematic trajectory planning methods employ upper and lower constraints on velocity, acceleration, and jerk. These boundaries are usually assumed to be constant. Staicu (2009) developed a method built on a parallel robot's inverse dynamic. The optimal spline function-based trajectory is found using an evolutionary method in this work, which addresses a multi-objective optimization challenge.

Splines are thus employed as trajectory primitives to maintain continuous acceleration. Another example is given by Zhang et al. (2022), which deals with the nonlinear optimization problem of finding the value of time intervals between via-points in order to reduce the overall execution time of a trajectory with kinematic constraints. The method is based on a combination of an unrestricted optimization technique known as flexible polyhedron search (FPS) and an algorithm known as feasible solution converter (FSC), which translates solutions that are not physically feasible (i.e., do not conform to the dynamical constraints) into viable ones by applying appropriate time scaling to the FPS method's trajectory. Held and Kasser (2014) uses the same optimization method as Piazzi and Visioli (1998), but instead of cubic Bezier curves, cubic B-splines are utilized as motion primitives.

Other techniques to minimum-time trajectory planning yield a global optimum, but the algorithms outlined above yield a local optimum. Piazzi and Visioli utilize interval analysis to generate a minimum-time trajectory with kinematic restrictions at the joints. The simulations given by Victoire et al. (2004) has showed an 18% reduction in overall execution time when compared to the results obtained utilizing a local optimization strategy. Bartz et al. and Talatahari et al. describes a global optimization method that blends a stochastic method, such as an evolutionary algorithm, with a deterministic interval strategy (Bartz-Beielstein & Zaefferer, 2017) (Talatahari et al., 2012). The suggested approach may be used to a variety of global optimization problems with semi-infinite constraints. The trajectories, given by cubic splines, are bound by maximal actuator torques as well as linear and rotational velocities of the robot tip in the operating space, as described by Fang et al. (2020). Rout et al. (2020) employed a mixture of polynomial functions of various orders to establish jerk continuity over a path built from a set of pre-defined via-points, resulting in a global lowest time solution. The fastest way for creating robot manipulator trajectory was proposed by Zanotto et al. (2011). In this approach, the objective function is comprised of two terms: the first includes the quadratic values of the optimization variables, and the secondary uses the average of the square's accelerations attained at the via-points.

Wen and Pagilla (2021) provided a technique for calculating time-optimal path-constrained trajectories with velocity, acceleration, and jerk constraints applied to the robot actuators. Medeiros et al. (2020) used a hybrid optimization technique to resolve the optimization problem, which takes the trajectory description, the robot's kinematic model, and the user's constraints into account. In regards to time, the resultant trajectories are optimum, but not in respect of smoothness. In the work of Huang et al. (2018), a mixture of spline functions up to the seventh order is used to identify least time solutions for velocities, acceleration, and jerk bounds. Additional research works done through kinematically constrained minimum-time algorithms can be found in Lin et al. (2018), Yuan et al. (2019), and Paing et al. (2021). Atiyah et al. (2021) provided a technique for addressing the minimum-time trajectory issue based on kinematic and dynamic constraints such as torque, jerk, and energy, while taking into consideration both robot dynamics and the existence of obstacles.

4.1.1 Kinematic Approach

The primary aim of the kinematics approach is to find a time sequence that reduces the overall amount of time spent while shifting from any two subsequent points in space (Sahar & Hollerbach, 1986). The important aspect of this strategy is that when determining fixed constraints such as position, velocity, and jerk, the strength of the robotic arm is ignored. Aside from that, most of the techniques necessitate the use of a series of knot points to define the trajectory path (Nikolos et al., 2003). Many researchers (Dinçer & Çevik, 2019) (Bharathi & Dong, 2016) (Thompson & Patel, 1987) (Sorek et al., 2017) concluded that the cubic polynomial functions are usually used for creating smoother trajectories whereas cubic splines incorporate piecewise continuous b^{th} order geometric polynomial which represents incessant derivatives in boundation of (b-2), depending on the problem formulation. The selection of an appropriate time-optimal path trajectory for the robotic manipulator can be encapsulated as follows:

Motion is parameterized in the spatial dimension by assuming that the cubic-spline polynomial is utilized to illustrate the Cartesian magnitude of robot's end tip. The parametric motion in x spatial dimension is expressed in Eq. 8, where v_j denotes the area specified path vertices and $b_j(a)$ represents the piecewise-cubic-B-spline basis function.

$$x(a)=\sum_{j=-1}^{m+1} v_j b_j(a) \quad 0 \le a \le m \tag{8}$$

The optimal trajectory path planning strategy could be described as minimizing the entire path transversed with time is given in Eq. 9.

$$t_f = \int_0^m \frac{da}{\dot{a}} \tag{9}$$

Polynomials are easy to compute and have the required homogeneity and boundary requirements. A quadratic (5th) order polynomial is commonly used.

$$r(t)= c_1 t^5 + c_2 t^4 + c_3 t^3 + c_4 t^2 + c_5 t + c_0 \tag{10}$$

where, t is the time $[0, T]$

Smooth polynomials can be found in the first and second derivatives.

$$\dot{r}(t)= 5c_1 t^4 + 4c_2 t^3 + 3c_3 t^2 + 2c_4 t + c_5 \tag{11}$$

$$\ddot{r}(t) = 20c_1t^3 + 12c_2t^2 + 6c_3t + 2c_4 \tag{12}$$

The position, velocity, and acceleration boundary conditions are all characterized, and the velocity and acceleration boundary conditions are generally set to zero. Writing Eq. 10 to Eq. 12 for both $t = 0$ and $t = T$ produces six equations that may be represented in matrix form as:

$$\begin{bmatrix} r_0 \\ r_T \\ \dot{r}_0 \\ \dot{r}_T \\ \ddot{r}_0 \\ \ddot{r}_T \end{bmatrix} = \begin{bmatrix} 0 & 0 & 0 & 0 & 0 & 1 \\ T^5 & T^4 & T^3 & T^2 & T & 1 \\ 0 & 0 & 0 & 0 & 0 & 1 \\ 5T^4 & 4T^3 & 3T^2 & 2T & 1 & 0 \\ 0 & 0 & 0 & 2 & 0 & 0 \\ 20T^3 & 12T^2 & 6T & 2 & 0 & 0 \end{bmatrix} \begin{bmatrix} c_1 \\ c_2 \\ c_3 \\ c_4 \\ c_5 \\ c_0 \end{bmatrix} \tag{13}$$

The Robotics Toolbox function "tpoly" constructs a quintic polynomial trajectory as specified in Eq (10). The following procedure is applied to obtain Fig. 3:

(i) By using function >> s = tpoly (0, 30, 30); it yields a 30× 1 column vector having values changing smoothly from 0 to 30 in 30-time steps that could be plotted by >> plot(s).
(ii) Optional output arguments can be used to obtain the appropriate velocity and acceleration as >> [s, sd, sdd] = tpoly (0, 30, 30);
(iii) We can see in Fig. 3 that the beginning and ultimate position, velocity and acceleration are all zero. The starting and ultimate velocities can both be non-zero. The function>> s = tpoly (0, 30, 30, 0.5, 0); is used to represent quintic polynomial trajectory.

From Fig. 3, it is observed that the starting velocity, final velocity, joint acceleration tends to be a default value i.e., 0. where acceleration peak is observed at ***t* = *15***, which indicates that for prolonged time $V_{actual} < V_{maximum}$. Hence, the value of mean acceleration tends to be 52% of the peak value.

Figure 3. The representation of a quintic polynomial trajectory with a zero-velocity boundary condition

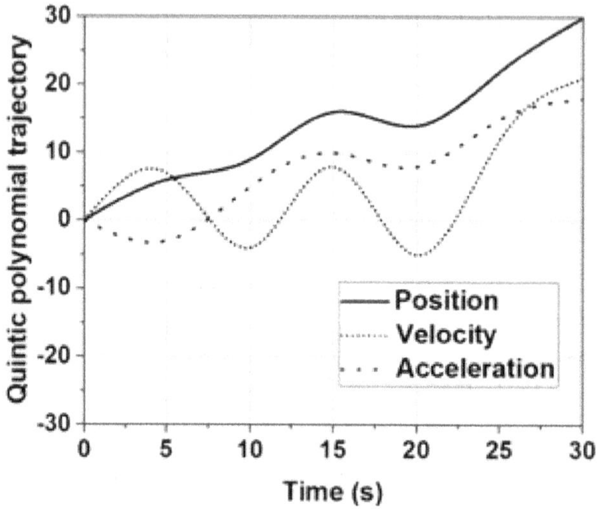

The "lspb" function in the Robotics Toolbox creates a linear segment (steady velocity) having parabolic blends. The word blend is typically used to describe a trajectory segment that combines linear segments smoothly. The following procedure is applied to obtain Fig. 4:

(i) By using function >> s = lspb (0, 30, 30); it yields a 30 × 1 column vector having values changing smoothly from 0 to 30 in 30-time steps that could be plotted by >> plot(s).

(ii) Optional output arguments can be used to obtain the appropriate velocity and acceleration as >> [s, sd, sdd] = lspb (0, 30, 30);

(iii) We can see in Fig. 4 that the beginning and ultimate position, velocity and acceleration are all zero. The starting and ultimate velocities can both be non-zero. The function >> s = lspb (0, 30, 50, 20); >> s = lspb (0, 30, 30, 25); is used to obtain linear segment parabolic blend (LSPB) path trajectory.

Figure 4. Representation of linear segment parabolic blend (LSPB) path trajectory

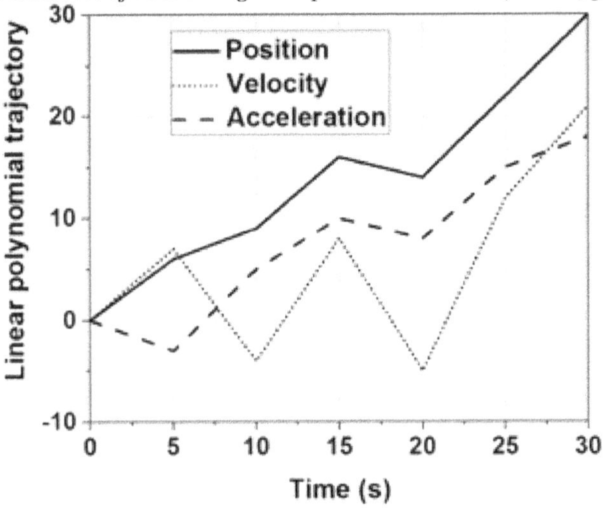

In robotics, it is common to travel along a path without stopping at any of the more intermediate or via points. It might be to avoid workplace risks or to finish a task that necessitates following a piecewise continuous trajectory, such as placing a coating bead in an industrial field. The Robotics Toolbox function "mtraj" is used to plot multisegmented trajectories from coordinates (0, 30) to (0, -25) in 30 steps as shown in Fig. 5. The function used is:

```
>> X = mtraj (@tpoly, [0 30], [0 -25], 30); >> X

            = mtraj(@lspb, [0 30], [0 -25], 30);
```

Fig. 5 shows the multi-dimensional motion, where X_1 differs from $0 \rightarrow 0$, whereas X_2 differs from $30 \rightarrow -25$.

Figure 5. Illustration of multi-dimensional motion approach

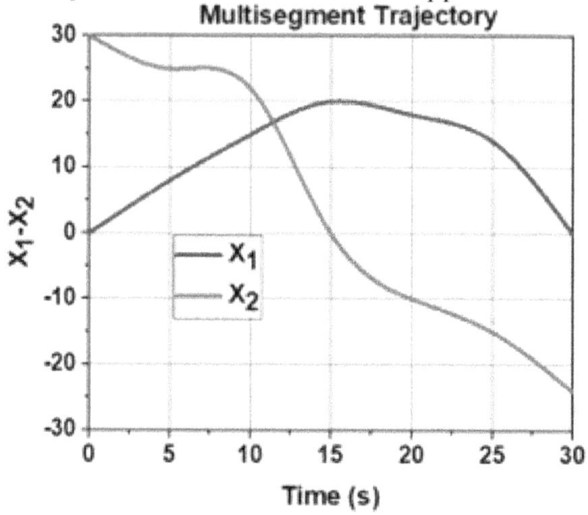

Fraichard (1998) explored the use of a novel concept to solve problems involving moving obstacles and dynamic constraints in three-dimensional space. The concept of state-time space as a tool for formulating dynamic trajectory planning problems is a novel concept. A robot's state-time space is its state space plus the time dimension. It permits a unified investigation of the many features of dynamic trajectory planning. As a result, both static forbidden regions of state-time space can be used to represent the constraints imposed by moving obstacles and dynamic constraints. Dynamic trajectory planning, in addition to a trajectory map to a curve in state-time space, simply entails finding a curve in state-time space. Then, this concept to find a time-optimal trajectory for a carlike robot moving along a given path on a dynamic planar workspace while subject to dynamic constraints. This presents an approximate method for finding the solution trajectory among a small number of canonical trajectories. These canonical trajectories do seem to have piecewise constant acceleration which varies only at particular moments. Apart from that, the acceleration may be set to minimum, zero, or maximum. The challenge of discovering the time-optimal canonical trajectory may be simplified to calculating the shortest path across a graph embedded in state-time space under these assumptions.

Xu (2008) stated that in order to formulate the robot's arm dynamic equation in parameterized form, then curvilinear abscissa of the trajectory path can be used as a self-dependent parameter. The quadratic abscissa (parameter) and its gradient pseudo-velocity are utilized to understand the state of a system, whereas the second derivative (i.e., pseudo-acceleration) is designated as the control variant. The primary aim of the kinematics method is to find a time sequence that reduces the overall amount of time spent while shifting from any two subsequent points in space

(Sahar & Hollerbach, 1986). Aside from that, most of the techniques necessitate the use of a series of knot points to define the trajectory path (Nikolos et al., 2003). The variational approach proposed by Seereeram and Wen (1995) is a global optimal path planning control strategy for dextrous robotic arms. There are two optimal path planning problems to contemplate: first, find the optimum paths of the joints based on the robot tip trajectory; and second, find the appropriate optimized paths for the robot tip as well as the joints based on the end effector's constraints. Agrawal (2008) and Baleanu et al. presented a numerical scheme for forward integration.

4.1.2 Dynamic Approach

Various dynamic programming techniques, including those listed in (Hansen et al., 2012) (Kim et al., 2010) (Lengyel et al., 1990) (Huang et al., 2018) (Abu-Dakka et al., 2017), represents novel approach to calculate the time-optimal path trajectory. The general idea, according to Gordon et al. (1993), is to linearize the search space by creating a grid of state points (Chiddarwar & Babu, 2012). Eventually, the shortest time trajectory is generated using a dynamic programming algorithm. Neglecting link-flexibility and actuator dynamics usually results in some unfavourable results. In reality, according to Kong et al. (2009), with reference to the target position, robot actuators are unable to generate the required torque, allowing joint motion to be distorted. As a result, trajectory tracking accuracy suffers significantly, and the trajectory tracking controller must be activated repeatedly during trajectory path execution.

According to Zhang and Wei (2017), another inevitable result of an inaccurate model is that the controller appears to be unable to correct tracking errors caused by fluctuations or model errors. (Shin & McKay, 1986) (Boscariol & Gasparetto, 2013) addressed a possible solution to these problems: the phase plane technique is used in these works, along with a torque variation constraint (actuator jerks). Codourey (1998) proposed a fascinating illustration of an iterative method based on the inverse dynamic of a multi-link manipulator. To find the best spline function-based trajectory, a multi-objective optimization problem is formulated. As a result, spline functions can now be used as trajectory primitives, ensuring that the acceleration stays constant (Fang et al., 2019).

The aforementioned algorithms produce local optimum solutions, whereas time optimal motion planning algorithms generates global optimum solutions. Piazzi and Visioli (1998) used time interval method to examine a time optimal trajectory restricted by dynamic restraints at subsequent robot's joints (Liu et al., 2018). Alvarez et al. (2008) and Jiang et al. (2007) suggested a global optimization approach that combines a probabilistic method, such as an evolutionary algorithm, and a deterministic interval analysis procedure. The proposed method could solve a wide range of linear differential equations with semi-infinite restraints. This algorithm is utilized by Pekarovskiy et al. (2017) to address the problem of kinematic and dynamic restraints in time optimized motion planning: the trajectories, which are represented by cubic splines, are constrained by optimal actuator rotational torque as well as translational, rotational velocities of an end-effector within the operating space.

The maximum speed of the first axis is considerably higher than that of the second axis is illustrated in Fig. 6(a). Because the slowest axis controls segment time, both axes begin moving at the same pace during the final segment. Fig. 6(b) represents that the path trajectory becomes much more curved when the acceleration increases. To acquire jerk uniformity across a path planned from a number of predefined via-points, Zalzala and Morris (n.d.) & Zhang et al. (n.d.) used a combination of polynomial functions of varying orders, which results in time optimal global solutions.

The Robotics Toolbox Function "mstraj" is used to obtain Fig. 6. This function generates a multi-segment multi-axis trajectory from a matrix of through points. In Fig. 6(a), the two-axis motion representing velocity and acceleration is generated by:

>> q = mstraj (via, [2,8; 8,8; 10,4; 20,3]); The function "mstraj" provides a matrix including one row for each time step and columns for the axes. The trajectory gets circular as the acceleration time increases, as seen in Fig. 6(b). The function used is >> q = mstraj (via, [2 10], [9 9], [14 7], [19, 2]). In case of multisegmented trajectories: when, $t_{acc} = 0$, the acceleration blend time is zero and when $t_{acc} = 0$, the acceleration blend time is non-zero, where, t_{acc} represents blend time.

Figure 6. Representation of mult-segment trajectories. a) When $t_{acc} = 0$. b) $t_{acc} = 1$

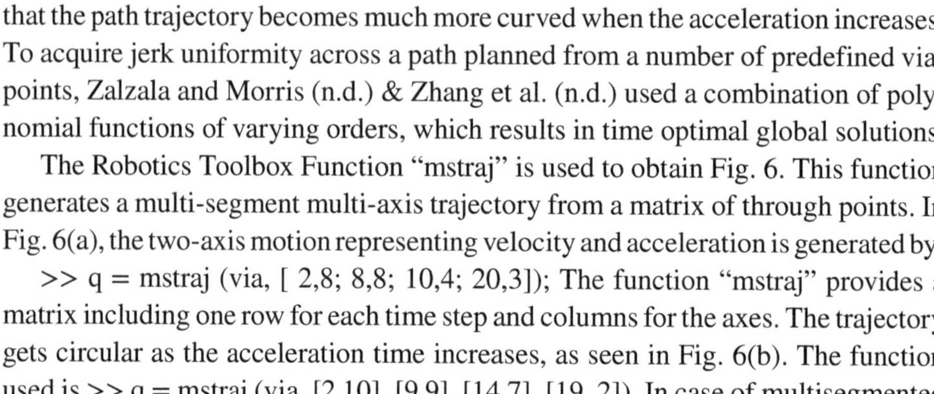

(a) (b)

Another description of robotic manipulator having lowest possible time trajectory could be depicted in (Chettibi, 2019). Rubio et al. (2012) and Van Dijk (2007) identified a method for determining the shortest time optimized path trajectory, while considering kinematic constraints such as joint velocity, joint acceleration, and jerk. (Fang et al., 2019) (Siddiqi et al., 2012) (Sahar & Hollerbach, 1986) (Hashemian et al., 2017) (Feddema, 1996) (Consolini et al., 2017) (Yuan et al., 2019) (Abu-Dakka et al., 2017) (Rubio et al., 2016) provide additional examples of kinematically constrained time-optimal algorithms.

In the last decade, some researchers (Kazem et al., 2008) (Tuncer & Yildirim, 2012) (Abo-Hammour et al., 2011) used the GA to determine the shortest possible time trajectory for robotic manipulator. A GA is an evolutionary algorithm that uses inherent genetics and selective breeding to optimize non-linear functions. When physical constraints on joint velocities, maximum acceleration, and joint jerks are considered, then GA could be used to coordinate the elapsed time between every set of identical configurations in such a manner that total travel time is minimized, according to Fonseca and Fleming (1995). The GA searching process, on the other hand, is subjected to enhanced probabilistic crossover as well as a normalized and validated measure of performance (Kazakovtsev et al., 2015) (Chen, 1991) (Saltzman & Kelso, 1987). Yano and Tooda (1999) used GA to calculate the pose of the robot tip of two robotic arms (Parker et al., 1989). Lin et al. (1983) clarified that in order to generate robot tip trajectory, the fitness function was specified in Cartesian and joint spaces before being combined. While several time-optimal trajectory planning techniques for continuous nonlinear systems have been devised, there has been little research on the issue for sporadic nonlinear dynamic systems (Foehn et al., 2021) (Tika et al., 2020). Fan et al. (2020) suggested a method for finding the time suboptimal trajectory of discrete nonlinear systems. Using his method, the control inputs in terms of time are segmented into nonlinear fractional functions. The fractional functions and time step intervals used in the system's discretization are then sought using a general-purpose non-linear programming optimization technique.

In an earlier approach to fix the trajectory selection problem, researchers (Li & Du, 2021) (Malyuta et al., 2021) typically linearized the dynamics in order to apply fundamental linear optimal control theory approaches to the solution. Suboptimal control inputs have been generated and explained on the basis of switching contours predicated on linearized differential equations (He et al., 2020). The above method, however, is just beneficial whenever the system mobility is constrained to a narrow area close the endpoint layout where the linearity assumption is applicable. Trajectory planning might be problematic due to inter-joint nonlinearities and the coupling properties of the robot manipulator (Dietrich et al., 2021). Only via identification can a robot arm's exquisite dynamics be attained. Nonetheless, coupled nonlinear dynamics are inapplicable to a class of robotic systems with high reduction (gear) ratios (Leahy & Valavanis, 2020).

4.2 Reduced Energy (Actuator Effort) Algorithms

For optimal motion planning of wheeled robots, several researchers used the minimal power consumption criterion (Kelly & Nagy, 2003) (Webb & Van Den Berg, 2013) (Aristidou et al., 2018) (Mei et al., 2006) (Dietmair & Verl, 2009) (Garg & Kumar, 2002). The dynamic programming approach was used to generate trajectory path for minimal energy consumption, and the results are compared with a map of null geometry described in (Chou & Song, n.d.) (Ma, 1995) (Garcia et al., 2018) (Connolly et al., 2017).

Luo et al. (2015) suggested a non-linear dynamic programming approach for generating lower power consumption joint trajectories subjected to time and actuator torque constraints. Chettibi et al. (2004) proposed a solution to the optimum trajectory planning problem, which was limited by joint torques, joint velocity, and the task requirement. He used the same method (Zlajpah, 1996), but provided extra constraints such as confined joint velocities and specified task requirement. Hirakawa and Kawamura (1997) proposed a discretization approach inside the B-spline function for limiting electrical energy consumption. In this approach, it is not required to compute the Jacobian inverse matrix. Therefore, reduction in deviation of trajectory state error is achieved while consuming the least amount of electrical power (Ata, 2007).

Shiller (n.d) presented a solution to the challenge of navigating a robot across a defined geometric path while using the least amount of actuator effort. The ideal travel time and the actuator's minimum mechanical energy were combined to form a multi-objective function, with the results influenced by the weighting factor. The mathematical analysis incorporates the manipulator's non-linear coupled equations of motion, and the optimization problem is constrained by physical parameters. Numerical results for robotic arms with 3 to 6 DOF were suggested for verification.

(Radmanesh et al., 2018) (Betts, 1998) proposed two-offline and three-dimensional optimal trajectory planning strategies for robotic systems in the presence of static obstacles. Verscheure et al. (2009) used a non-linear variable transformation to convert the time-energy optimum trajectory planning issue into a concave control issue with a single state variable. Korayem et al. (2009) proposed the technique for identifying the best joint path trajectory for stiff redundant manipulator. Boryga (2020) proposed and verified three trajectories based on the minimal energy consumption represented as: cubic profile, sinusoidal profile, and Gaussian profile. These three profiles depict a typical pick and place operation. They all have specified beginning and ending points, but the rate at which velocity is increased varies with time. They discovered that the Gaussian profile has the least amount of energy, followed by sinusoidal and cubic polynomial curves.

Humanoid robots employ GA to determine which proposed task requires the least amount of energy (Ezair et al., 2014). Shojaedini et al. (2019) suggested an adaptive algorithm, depending on evolutionary algorithm for a consistent, time-optimal trajectory planning. Sivcev et al. (2018) stated yet another interesting application of GA for underwater manipulators. Vezvari et al. (2021) presented for the robot's flawless torque compensation in point-to-point motions. The proposed method is a closed loop optimum control issue. The problem is stated as ideal trajectory planning for a planar 5R parallel robot with configurable design parameters to counteract applied torques. Scalera et al. (2020) presented an analytical model and experimental data for minimum-energy trajectories applicable to robotic axes. Using trapezoidal and cycloidal motion profiles, the dynamical and electro-mechanical models of a horizontal axis are integrated and used to the planning of ideal point-to-point trajectories. Carabin et al. (2021) developed a method for long-term 3D printing with a Linear Delta Robot equipped with elastic parts is given and tested. Energy is saved by installing and optimizing linear springs on the robot in different configurations, such as orthogonal to the prismatic joints or directly connecting the robot tip to the reference frame. The flexible components allow for the conversion of potential to kinetic energy during periodic motions such as solid layer generation in rapid prototyping, leading to energy savings for actuators.

4.3 Reduced Jerk Algorithms

It has been already mentioned that it is significant to generate joint trajectories that do not enforce actuator torque discontinuities at the robot's joints. However, this method necessitates the computation of the robot's 3rd order dynamics. The torque variations are determined by the predominant term of the inertia matrix augmented by the joint jerk vector (Macfarlane & Croft, 2003). Jerk reduction has many advantages such as: reduction in error mostly during path planning stage, reduction in resonance frequency stimulation, reduction in stresses caused in the robot's mechanical structure. Berti et al. (2016) suggested an algorithm based on interval analysis. Flash et al. (2013) presented an analytical solution to a point-to-point path trajectory planning problem that employs a minimum jerk optimization criterion. The optimization incorporates two objective functions such as: mean value of jerk (minimax approach), time integral of the mean squared jerk (Pontryagin's principle (Muenchhof & Singh, 2002)). Hogan (1984) minimized the squared jerk's integral along the enacted trajectory (Wada & Kawato, 2004). Qian et al. (2020) (Shin & McKay, 1986) (Lu et al., 2020) described another technique in which the via-points are interpolated using a trigonometric B-spline curve, ensuring that the joint jerk remains static. The domain of local area is one of the benefits of utilizing trigonometric splines to linear interpolation the trajectory via points: If a via-point is altered, only the two splines associated with a via-point is computed, rather than the whole path trajectory (Letizia et al., 2021) (Sun et al., 2018).

Ma et al. (2021) offered a new convex optimization (CO) technique to time-optimal trajectory planning (TOTP) that takes both torque and jerk restrictions into account. The approach's essential idea is that non-convex jerk restrictions are converted to linear acceleration constraints and then inserted into CO as linear acceleration constraints. The convexity of CO will not be violated, and the number of state variables will not rise, resulting in a fast-computing speed for the technique. By articulating an optimal control problem for time and jerk, Paing et al. (2021) proposed a kinematically limited reparameterization strategy for creating the trajectory. B-splines use both geometric path creation and trajectory reparameterization to achieve trajectory smoothness. To satisfy the axial kinematic restrictions, inequality constraints are suggested, as well as equality constraints to assure zero velocity and acceleration at the origin and destination of the trajectory.

4.4 On-Line Trajectory Planning Scheme

In robotics operations such as injection molding, laser flame cutting, and arc welding, on-line continuous path (CP) algorithms are in emerging stage (Landers et al., 2020). If the appropriate knot points are identified, Sahay et al. (2006) devised an on-line algorithm that can create a rough path trajectory required for any optimized "S" curve in Cartesian space. Bestaoui (1992) and Martins et al. (2003) presented a novel joint reference trajectory algorithm that took manipulator dynamics into account and was found suitable to be exposed by the position and velocity sensors. Elbanhawi et al. (2015) proposed an online system for designing continuous-path motion in Cartesian space with time optimal criterion. The time necessary to calculate the trajectory path was segmented to m-segments, as well as the polynomial coefficients were computed in mathematical recursion form at every segment. A set of variables in Cartesian coordinate system might be used to specify the trajectory path (Kewlani et al., 2009). Turissi et al. (2021) proposed planning and directing motions of underactuated robots with stochastic dynamics using an iterative technique. A learning process is at the heart of the system, and it assesses the perturbations caused by model errors on the active and passive degrees of freedom. The learnt data is used in both the planning phase, which is focused on optimization, and the control phase, wherein partial feedback equation of the active dofs is conducted on the model modified on-line, by the generalized iteration of the algorithm.

4.5 Hybrid Optimization Trajectory Planning Scheme

Hybrid time-jerk algorithms, include optimization of jerk and time with respect to time (Canali et al., 2014) (Oliveira & Barreto, 2020). These algorithms are improved by taking into consideration the energy consumption of the actuators as well as the physical limitations of the joints (Kota et al., 1999). (Nieuwenhuisen & Behnke, 2015) (Kapania, 2016) demonstrated a technique to plan low-effort trajectories using model-based techniques. The work (Ur-Rehman et al., 2010) (Khoukhi et al., 2009) (Hollerbach & Suh, 1987) (Bosscher et al., 2007) takes into account a set of points, including minimizing electrical energy, kinematic energy, minimizing time, all of which are accomplished by solving a single optimization problem. Any stiff link robot with serial or parallel kinematics may use any these aforementioned trajectory planning methods.

To maximize the computational efficiency, a hybrid scheme combining: PSO algorithm and the simulated annealing algorithm are used in motion planning of manipulator (Pott, 2018). Chen et al. (2012) also suggested a constrained multi-objective genetic algorithm (MOGA) based approach for a parallel kinematic manipulator. The planning process involves determining a motion that ensures the completion of the assigned task while reducing traverse time and energy consumption (Chen & Pham, 2012) (Brito, n.d.). Kashyap et al. (2020) proposed the hybridization of the Dynamic Window Approach (DWA) and the Teaching–Learning-Based Optimization (TLBO) techniques has been discussed, as well as its deployment on the NAO humanoid robot for locomotion. The target and barrier locations are used to generate the input. The variables are fed into the DWA algorithm, which determines the best speed. The intermediary result is fed into the TLBO approach, which works in two phases: instructor and learner. This hybridization offers the best angle for turning and avoids obstructions while it reaches the target.

4.6 Implementation of Genetic Algorithm (GA) For Obstacle Avoidance

Obstacle avoidance, while considering entire manipulator dynamics is challenging due to the nonlinearity of manipulator joints and the presence of obstacles in the trajectory path (Dubowsky et al., 1986). To move along such paths, a robot's tip motion has to come to a complete halt at each corner of obstacle (Duindam et al., 2008). The motion of each manipulator's arm joint as a time dependent was parameterized by Liang and Lee (2015). The joint movements were then optimized while guaranteeing that no obstructions were encountered throughout the robot's motion planning (Ichter et al., 2018) (Ratliff et al., 2009). Ozaki and Lin (1996) proposed a motion planning approach for describing the joint path trajectories of a space robot with B-spline curves. For collision-free maneuvering, this approach takes into account all dynamic and kinematic constraints and does not require the gradient to be evaluated during the optimization process. It can be used with space robots having n DOF.

Over the last twenty years, genetic algorithms (GA) are being effectively implemented to alleviate the obstacle avoidance issue in trajectory path planning problem of robotic manipulator (Abo-Hammour et al., 2011). GA is robust optimization tool for complex search problems based on natural evolution (Sivanandam & Deepa, 2008). It can handle discontinuities, non-linearities, and even noisy data. Since the performance of a GA is dependent on the coding, the coding of trajectory path is critical. GA compares natural evolution to global optimization in a simple way. It implies that an entity can be represented by a set of parameters as a potential solution to a problem. These parameters are chromosome genes, and they can be structured using a concatenated string of values. The way of representation of variables is determined by the encoding scheme. Fig. 7 depicts implementation of GA algorithm in path optimization problem through flowchart.

Figure 7. Implementation of the GA algorithm through flowchart

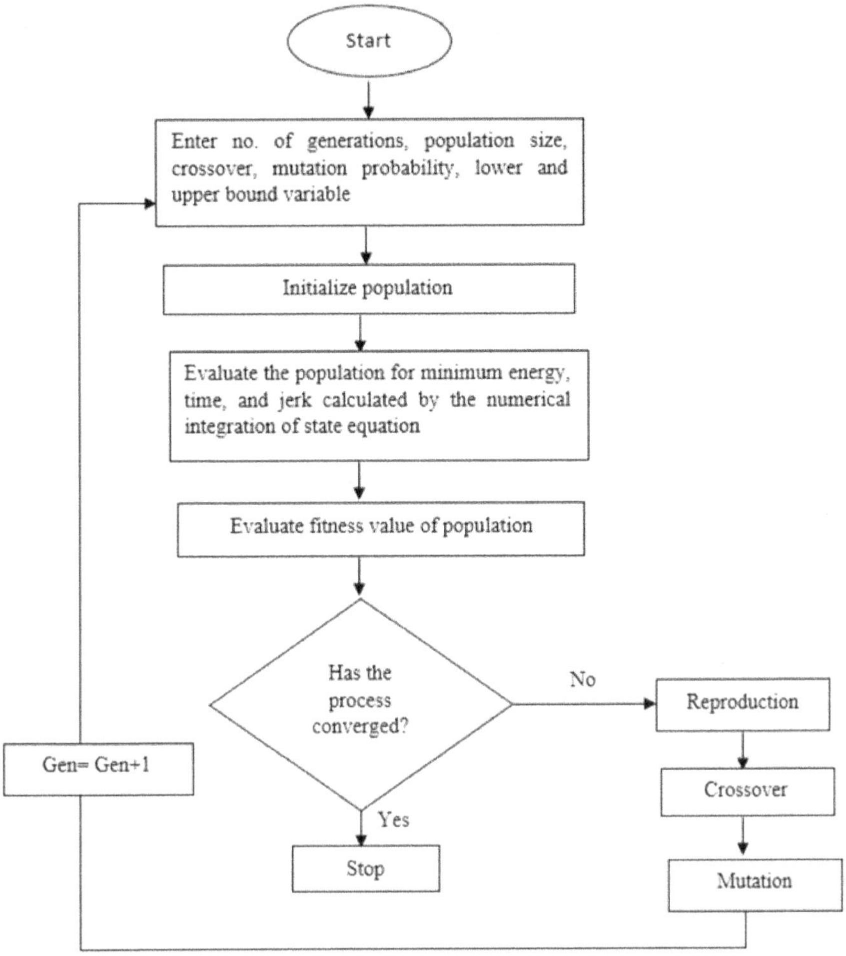

The GA technique is implemented by several researchers to resolve the major issue of obstacle avoidance during trajectory path planning of manipulator. Alezo et al. (2015) projected a method for obstacle avoidance that combined GA with the potential gradient algorithm. The proposed method is essentially a gradient-based iteration improving algorithm. Only B-spline control points are used to represent the spatial path. Kim et al. (2010) depicted that the control points can estimate the minimum travel time whenever a boundary condition and dynamical restraints of the trajectory path planning of robot is specified. Rubio et al. (2016) proposed a method for solving the optimal trajectory problem without obstacles in the trajectory path. Ali et al. (2005) implemented a trajectory path planning control strategy that employs GA to investigate the specified configuration space of manipulator for the best suitable path trajectory. The GA produces path solutions of higher accuracy, but it isn't reliable. The intelligent manipulator avoids collisions with objects in the space environment while minimizing the length of the path trajectory. Numerical simulations involving different robotic configurations and joint trajectories in the workplace was used to validate the proposed approach (Hwangbo et al., 2019).

Both online and offline, Campbell et al. (2012) and Softman et al. (2006) continued to work on GA for obstacle avoidance, and their proposed algorithms were found to be efficient. Salomon (1996) claimed that by integrating GA with any type of synchronized random search algorithm would enhance the system's characteristics. This viable approach is being used for obstacle avoidance with use of an external electromagnetic field. Vadakkepat et al. (2000) used a potential field algorithm to extend the application of GA in avoiding a collision between two manipulators. Each manipulator is treated as a moving obstacle in this approach, preventing collisions between them. Koker (2013) proposed adaptive GA to find the best end-effector trajectory path for a redundant manipulator by using a floating-point representation. A new evaluation function was devised, which was based on a number of factors such as joint displacement and the homogeneity of Cartesian and joint space velocities. To validate their technique, they simulate in free space and in a workplace with impediments. Han et al. (1997) described a GA-based strategy for avoiding point barriers in configuration space by looking for a suitable trajectory. Masehein and Katebi (2007) developed a method for designing a time-optimal, collision-free motion for multi-link manipulators. In joint space, the path is represented as a series of intermediate points connected by a B-spline curve.

5. RESULTS AND DISCUSSION

In this work, a total of 184 articles are reviewed, covering an adequate depth of work in the field of manipulator motion planning from (2000 to 2020). 200 articles are considered in order to make noteworthy comparisons between these motion planning approaches. Simultaneously, the researchers attempted to put together significant implementations of classical and heuristic algorithms in the literature to draw conclusions regarding the scope and direction of study in the motion planning approach of manipulator. Fig. 8 depicts comparison between classical and heuristic algorithms.

Figure 8. A comparison of classical and heuristic algorithms

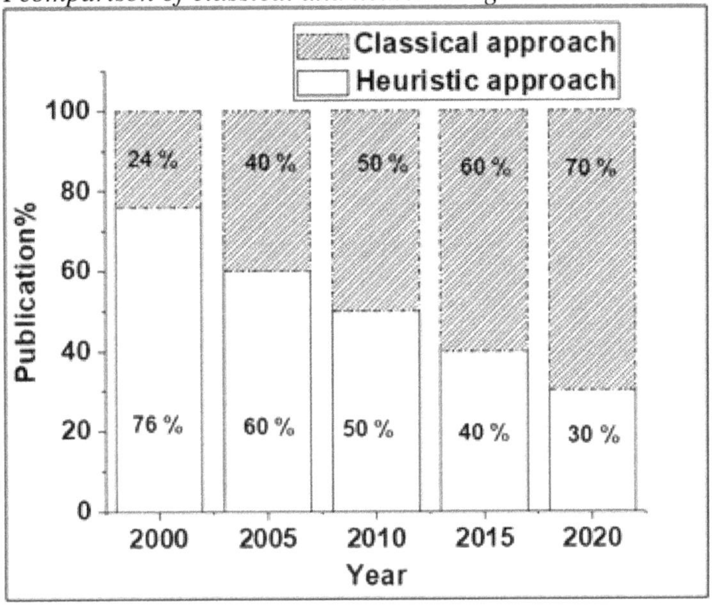

Some analytical methods were implemented to develop the trajectory planning algorithms, and after a while, some more sophisticated methods were also developed. In category of heuristic algorithm, the usage of Particle swarm optimization (PSO), Genetic algorithm (GA), Artificial neural network (ANN), Fuzzy logic, BUG, Potential field (PF) algorithm is increased in (%) in comparison to 10 algorithms compared in Year 2020 shown in Fig. 9. The last ten years (2010-2020) is the apex phase for heuristic methods as well as the breakthrough period for hybrid and meta-heuristic methods.

Tables 1 and Fig. 9 depicts the percentages of each heuristic motion planning algorithms (200 research papers) and compare them to each other. From Table 1, it is depicted that mostly GA, PSO, ANN, and fuzzy logic algorithms were implemented to solve motion planning problems of manipulator, whereas simulation annealing, tabu search, and BUG algorithms are used with limitations.

Table 1. A comparison of major heuristic motion planning algorithms

S. No.	Approach	10 Year time period		
		2000 (%)	2010 (%)	2020 (%)
1	PSO	8.09	10.98	11.43
2	GA	7.67	8.90	9.52
3	SA	1.89	4.89	5.71
4	ACO	2.59	5.12	7.62
5	Tabu search	1.12	3.12	7.43
6	Artificial Neural Network	9.12	10.80	13.33
7	Mathematical Programming	2.50	6.12	7.62
8	Potentials Field	7.12	8.12	9.52
9	Fuzzy Logic	11.90	12.23	14.48
10	BUG algorithms	8.22	11.10	13.00

Figure 9. The portion of each heuristic algorithms in (2020)

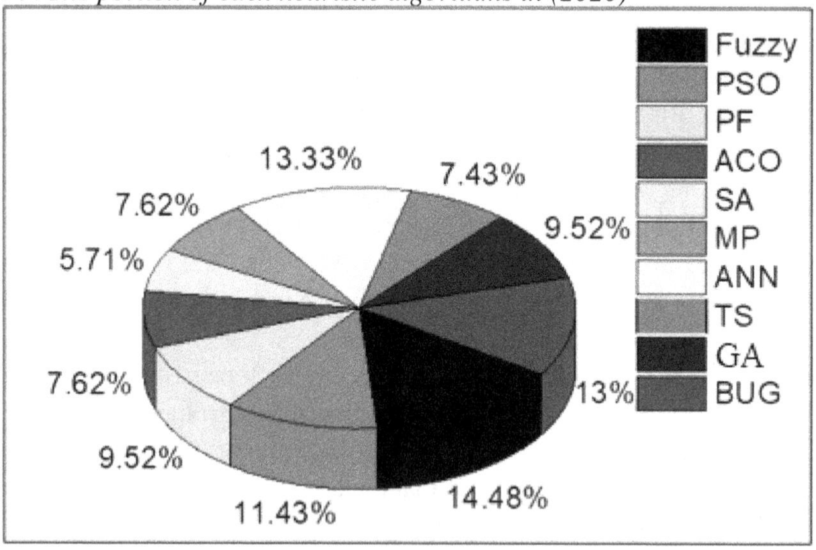

It is worth noting that fuzzy logic is the most thoroughly implemented motion planning approach for robotic manipulator. Fuzzy logic, as a statistical tool for dealing with uncertainties, is utilized to model non-linear dynamic systems.

6. CURRENT ISSUES AND FUTURE CHALLENGES IN THE AREA OF MOTION PLANNING

According to the articles discussed in this paper and the distinctions made between the various motion planning algorithms of manipulator, it is apparent that heuristic algorithms are more efficient than conventional algorithms. Heuristic algorithms such as: fuzzy logic, artificial neural network, genetic algorithms, and particle swarm optimization, illustrated in Tables 1. Following, an analysis of current research in the field of motion planning, the foregoing novel intellectual concerns, implementations, and evolving topics in motion planning of manipulator are recommended:

(i) Vision and motion planning of manipulator can be merged through online sensor-based motion planning.
(ii) Micro-scale robots, a recent emerging area, will provide conductive environment and novel problems for motion planning of manipulator. Non-equilibrium laws must be considered at certain microscopic scales.

An algorithm's effectiveness can be measured using a variety of parameters, including path traversal time, velocity of manipulator links or joints, energy consumption, actuator forces, and obstacle proximity. A statistical comparison of some of the major algorithms: computation time and path length are presented in Table 2.

Table 2. Merit and demerits of various heuristic motion planning algorithms

SI. No.	Approach	Merit	Demerit
1.	Genetic algorithm	• GA is robust with respect to local minima/maxima. • GA uses objective function information, not derivatives. • GA works well on discrete/ continuous motion planning problems. • GA supports multi-objective optimization.	• GA is computationally expensive i.e., time consuming. • The random heuristics sometimes does not find the optimum solution. • It is not a complete algorithm (not always the algorithm finds a suitable solution). Sometimes, it can get stuck with a local maximum problem.
2.	Particle swarm optimization	• It has higher probability and efficiency in calculating global optimum solution. • It has short computational time. • It can solve motion planning problems for complex mathematical models. • It can converge fast.	• It can be difficult to develop initial design parameters of manipulator. • It can converge prematurely and can be trapped into local minima in complex problems.
3.	Simulated annealing	• It can deal with arbitrary systems and cost functions. • It is applicable to large data sets	• To get a global optimal solution, a significant number of iterations may be necessary.
4.	Artificial neural network	• It can deal with non-linear data and has strong curve fitting ability. • It is capable of capturing non-linearity among predictors and outcomes.	• Heavy computation time; executes semi-free obstacle path
5.	Fuzzy logic	• It uses simple mathematics for non-linear, integrated, and complex systems.	• Lack of real response time, and restricted usage of input variables
6.	Ant colony optimization	• It can search among population in parallel.	• Have uncertain time to convergence

7. CONCLUSIONS

In this paper, after reviewing more than 211 research articles, the fundamental issues of motion planning in the field of robotics are identified and categorized. Following that, the dilemma of determining an optimal path trajectory is discussed, as well as the most significant trajectory planning algorithms are described. The kinematic approach is complicated to execute in reality due to inertia and torque constraints. However, in terms of combining torque restraints and physical joint constraints, the dynamic approach appears to be more realistic. Non-linearity and un-modeled dynamics can also be a challenge when it comes in completing the entire modelling approach, particularly for multilink redundant manipulators. Many researchers implemented minimum energy consumption criterion to select the appropriate path trajectory. Another method of acquiring smoother actuator torque profile is to minimize the jerk which is defined as the rate of change of acceleration with respect to time. In order to optimize the jerk and path trajectories with respect to time, hybrid time-jerk optimized algorithms are proposed in this paper. In the coming years, it appears that GA, PSO, ANN, fuzzy logic algorithms will have more applications in motion planning of manipulator. A GA, diversely, may find it intricate to converge to a global optimum solution. In terms of solution consistency and time consumption, the GA has some disadvantages such as delay in convergence rate during searching of optimal parameters in trajectory planning.

DECLARATIONS

Funding: There was no financial assistance for the author(s)' research, writing, or publishing of this work.

Conflict of interest/ Competing interest: The author declares that they have no known competing financial interests or personal relationships that could have influenced the work reported in this paper.

Ethics approval: Not applicable

Consent to participate: Not applicable

Consent for publication: Not applicable

Availability of data and material: Not applicable

REFERENCES

Abo-Hammour, Z. E., Alsmadi, O. M., Bataineh, S. I., Al-Omari, M. A., & Affach, N. (2011, December 1). Continuous genetic algorithms for collision-free cartesian path planning of robot manipulators. *International Journal of Advanced Robotic Systems*, *8*(6), 74. DOI: 10.5772/50902

Abu-Dakka, F. J., Assad, I. F., Alkhdour, R. M., & Abderahim, M. (2017, March 1). Statistical evaluation of an evolutionary algorithm for minimum time trajectory planning problem for industrial robots. *International Journal of Advanced Manufacturing Technology*, *89*(1-4), 389–406. DOI: 10.1007/s00170-016-9050-1

Agrawal, O. P. (2008, January 1). A quadratic numerical scheme for fractional optimal control problems. *Journal of Dynamic Systems, Measurement, and Control*, *130*(1), 011010. DOI: 10.1115/1.2814055

Alejo, D., Cobano, J. A., Heredia, G., & Ollero, A. (2015, June). Collision-free trajectory planning based on maneuver selection-particle swarm optimization. In 2015 international conference on unmanned aircraft systems (ICUAS) (pp. 72-81). IEEE.

Ali, M. A., Babu, N. R., & Varghese, K. (2005, April). Collision free path planning of cooperative crane manipulators using genetic algorithm. *Journal of Computing in Civil Engineering*, *19*(2), 182–193. DOI: 10.1061/(ASCE)0887-3801(2005)19:2(182)

Alvarez, V. H., Larico, R., Ianos, Y., & Aznar, M. (2008, June). Parameter estimation for VLE calculation by global minimization: The genetic algorithm. *Brazilian Journal of Chemical Engineering*, *25*(2), 409–418. DOI: 10.1590/S0104-66322008000200018

Aristidou, A., Lasenby, J., Chrysanthou, Y., & Shamir, A. (2018, September). Inverse kinematics techniques in computer graphics: A survey. *Computer Graphics Forum*, *37*(6), 35–58.

Ata, A. A. (2007). Optimal trajectory planning of manipulators: a review. Journal of Engineering Science and technology, 2(1), 32-54.

Atiyah, A. N., Adzhar, N., & Jaini, N. I. (2021, July). An overview: On path planning optimization criteria and mobile robot navigation. [). IOP Publishing.]. *Journal of Physics: Conference Series*, *1988*(1), 012036.

Banks, A., Vincent, J., & Anyakoha, C. (2008, March 1). A review of particle swarm optimization. Part II: Hybridisation, combinatorial, multicriteria and constrained optimization, and indicative applications. *Natural Computing*, *7*(1), 109–124. DOI: 10.1007/s11047-007-9050-z

Bartz-Beielstein, T., & Zaefferer, M. (2017, June 1). Model-based methods for continuous and discrete global optimization. *Applied Soft Computing*, *55*, 154–167. DOI: 10.1016/j.asoc.2017.01.039

Benko Loknar, M., Blažič, S., & Klančar, G. (2021, October 19). Minimum-time velocity profile planning for planar motion considering velocity, acceleration and jerk constraints. *International Journal of Control*, •••, 1–5.

Berti, A., Merlet, J. P., & Carricato, M. (2016, May). Solving the direct geometrico-static problem of underconstrained cable-driven parallel robots by interval analysis. *The International Journal of Robotics Research*, *35*(6), 723–739. DOI: 10.1177/0278364915595277

Bestaoui, Y. (1992, February). On-line reference trajectory definition with joint torque and velocity constraints. *The International Journal of Robotics Research*, *11*(1), 75–85. DOI: 10.1177/027836499201100106

Betts, J. T. (1998, March). Survey of numerical methods for trajectory optimization. *Journal of Guidance, Control, and Dynamics*, *21*(2), 193–207. DOI: 10.2514/2.4231

Beudaert, X., Lavernhe, S., & Tournier, C. (2012, June 1). Feedrate interpolation with axis jerk constraints on 5- axis NURBS and G1 tool path. *International Journal of Machine Tools & Manufacture*, *57*, 73–82. DOI: 10.1016/j.ijmachtools.2012.02.005

Bharathi, A., & Dong, J. (2016, February 1). Feedrate optimization for smooth minimum-time trajectory generation with higher order constraints. *International Journal of Advanced Manufacturing Technology*, *82*(5-8), 1029–1040. DOI: 10.1007/s00170-015-7447-x

Boryga, M. (2020). The use of higher-degree polynomials for trajectory planning with jerk, acceleration and velocity constraints. *International Journal of Computer Applications in Technology*, *63*(4), 337–347. DOI: 10.1504/IJCAT.2020.110414

Boscariol, P., & Gasparetto, A. (2013, August 1). Model-based trajectory planning for flexible-link mechanisms with bounded jerk. *Robotics and Computer-integrated Manufacturing*, *29*(4), 90–99. DOI: 10.1016/j.rcim.2012.11.003

Bosscher, P., Williams, R. L.II, Bryson, L. S., & Castro-Lacouture, D. (2007, November 1). Cable-suspended robotic contour crafting system. *Automation in Construction*, *17*(1), 45–55. DOI: 10.1016/j.autcon.2007.02.011

Brito, T. Intelligent collision avoidance system for industrial manipulators, Doctoral Dissertation), http://hdl.handle.net/10198/19319

Campbell, S., Naeem, W., & Irwin, G. W. (2012, December 1). A review on improving the autonomy of unmanned surface vehicles through intelligent collision avoidance manoeuvres. *Annual Reviews in Control*, *36*(2), 267–283. DOI: 10.1016/j.arcontrol.2012.09.008

Canali, F., Guarino Lo Bianco, C., & Locatelli, M. (2014, June 3). Minimum-jerk online planning by a mathematical programming approach. *Engineering Optimization*, *46*(6), 763–783. DOI: 10.1080/0305215X.2013.806916

Carabin, G., Scalera, L., Wongratanaphisan, T., & Vidoni, R. (2021, February 1). An energy-efficient approach for 3D printing with a Linear Delta Robot equipped with optimal springs. *Robotics and Computer-integrated Manufacturing*, *67*, 102045. DOI: 10.1016/j.rcim.2020.102045

Chen, C. T., & Pham, H. V. (2012, January). Trajectory planning in parallel kinematic manipulators using a constrained multi-objective evolutionary algorithm. *Nonlinear Dynamics*, *67*(2), 1669–1681. DOI: 10.1007/s11071-011-0095-2

Chen, Y., Agostini, L., Moretti, G., Fontana, M., & Vertechy, R. (2019, October 1). Dielectric elastomer materials for large-strain actuation and energy harvesting: A comparison between styrenic rubber, natural rubber and acrylic elastomer. *Smart Materials and Structures*, *28*(11), 114001. DOI: 10.1088/1361-665X/ab3b32

Chen, Y. C. (1991, October). Solving robot trajectory planning problems with uniform cubic B- splines. *Optimal Control Applications & Methods*, *12*(4), 247–262. DOI: 10.1002/oca.4660120404

Chettibi, T. (2019, March 1). Smooth point-to-point trajectory planning for robot manipulators by using radial basis functions. *Robotica*, *37*(3), 539–559. DOI: 10.1017/S0263574718001169

Chettibi, T., Lehtihet, H. E., Haddad, M., & Hanchi, S. (2004, July 1). Minimum cost trajectory planning for industrial robots. *European Journal of Mechanics. A, Solids*, *23*(4), 703–715. DOI: 10.1016/j.euromechsol.2004.02.006

Chiddarwar, S. S., & Babu, N. R. (2012, January 1). Optimal trajectory planning for industrial robot along a specified path with payload constraint using trigonometric splines. *International Journal of Automation and Control.*, *6*(1), 39–65. DOI: 10.1504/IJAAC.2012.045439

Chou LS, Song SM. Geometric work of manipulators and path planning based on minimum energy consumption.

Codourey, A. (1998, December). Dynamic modeling of parallel robots for computed-torque control implementation. *The International Journal of Robotics Research*, *17*(12), 1325–1336. DOI: 10.1177/027836499801701205

Connolly, F., Walsh, C. J., & Bertoldi, K. (2017, January 3). Automatic design of fiber-reinforced soft actuators for trajectory matching. *Proceedings of the National Academy of Sciences of the United States of America*, *114*(1), 51–56. DOI: 10.1073/pnas.1615140114 PMID: 27994133

Consolini, L., Locatelli, M., & Minari, A. (2021). A sequential approach for speed planning under jerk constraints. arXiv preprint arXiv:2105.15095.

Consolini, L., Locatelli, M., Minari, A., & Piazzi, A. (2017, May 1). An optimal complexity algorithm for minimumtime velocity planning. *Systems & Control Letters*, *103*, 50–57. DOI: 10.1016/j.sysconle.2017.02.001

Dietmair, A., & Verl, A. (2009, June 1). A generic energy consumption model for decision making and energy efficiency optimisation in manufacturing. *International Journal of Sustainable Engineering*, *2*(2), 123–133. DOI: 10.1080/19397030902947041

Dietrich, A., Wu, X., Bussmann, K., Harder, M., Iskandar, M., Englsberger, J., Ott, C., & Albu-Schäffer, A. (2021, September 1). Albu- chäffer A. Practical consequences of inertia shaping for interaction and tracking in robot control. *Control Engineering Practice*, *114*, 104875. DOI: 10.1016/j.conengprac.2021.104875

Dinçer, U., & Çevik, M. (2019, February 1). Improved trajectory planning of an industrial parallel mechanism by a composite polynomial consisting of Bezier curves and cubic polynomials. *Mechanism and Machine Theory*, *132*, 248–263. DOI: 10.1016/j.mechmachtheory.2018.11.009

Dong, J., Wang, T., Li, B., & Ding, Y. (2014, January 1). Smooth feedrate planning for continuous short line tool path with contour error constraint. *International Journal of Machine Tools & Manufacture*, *76*, 1–2. DOI: 10.1016/j.ijmachtools.2013.09.009

Dubowsky, S., Norris, M., & Shiller, Z. (1986, April). Time optimal trajectory planning for robotic manipulators with obstacle avoidance: A CAD approach. In Proceedings. 1986 [). IEEE.]. *IEEE International Conference on Robotics and Automation*, *3*, 1906–1912.

Dubowsky, S., & Papadopoulos, E. (1993, October). The kinematics, dynamics, and control of free-flying and freefloating space robotic systems. *IEEE Transactions on Robotics and Automation*, *9*(5), 531–543. DOI: 10.1109/70.258046

Duindam, V., Alterovitz, R., Sastry, S., & Goldberg, K. (2008, May). Screw-based motion planning for bevel-tip flexible needles in 3D environments with obstacles. In 2008 IEEE international conference on robotics and automation (pp. 2483-2488). IEEE.

Elbanhawi, M., Simic, M., & Jazar, R. N. (2015, December 1). Continuous path smoothing for car-like robots using Bspline curves. *Journal of Intelligent & Robotic Systems*, *80*(1), 23–56. DOI: 10.1007/s10846-014-0172-0

Ezair, B., Tassa, T., & Shiller, Z. (2014, May). Planning high order trajectories with general initial and final conditions and asymmetric bounds. *The International Journal of Robotics Research*, *33*(6), 898–916. DOI: 10.1177/0278364913517148

Fan, C., Miller, K., & Mitra, S. (2020, July). Fast and guaranteed safe controller synthesis for nonlinear vehicle models. In *International Conference on Computer Aided Verification* (pp. 629-652). Cham: Springer International Publishing.

Fan, W., Gao, X. S., Yan, W., & Yuan, C. M. (2012, September). Interpolation of parametric CNC machining path under confined jounce. *International Journal of Advanced Manufacturing Technology*, *62*(5), 719–739. DOI: 10.1007/s00170-011-3842-0

Fan, W., Lee, C. H., & Chen, J. H. (2015, September 1). A realtime curvature-smooth interpolation scheme and motion planning for CNC machining of short line segments. *International Journal of Machine Tools & Manufacture*, *96*, 27–46. DOI: 10.1016/j.ijmachtools.2015.04.009

Fang, Y., Hu, J., Liu, W., Shao, Q., Qi, J., & Peng, Y. (2019, July 1). Smooth and time-optimal S-curve trajectory planning for automated robots and machines. *Mechanism and Machine Theory*, *137*, 127–153. DOI: 10.1016/j.mechmachtheory.2019.03.019

Fang, Y., Qi, J., Hu, J., Wang, W., & Peng, Y. (2020, November 1). An approach for jerk-continuous trajectory generation of robotic manipulators with kinematical constraints. *Mechanism and Machine Theory*, *153*, 103957. DOI: 10.1016/j.mechmachtheory.2020.103957

Farid, Y., & Ruggiero, F. (2022, January 1). Finite-time extended state observer and fractional-order sliding mode controller for impulsive hybrid port-Hamiltonian systems with input delay and actuators saturation: Application to ball-juggler robots. *Mechanism and Machine Theory*, *167*, 04577. DOI: 10.1016/j.mechmachtheory.2021.104577

Feddema, J. T. (1996, April). Kinematically optimal robot placement for minimum time coordinated motion. In proceedings of IEEE international conference on robotics and automation (Vol. 4, pp. 3395-3400). IEEE.

Feng, Y., Yao-nan, W., & Yi-min, Y. (2014, September 18). Inverse kinematics solution for robot manipulator based on neural network under joint subspace. *International Journal of Computers, Communications & Control*, *7*(3), 459–472. DOI: 10.15837/ijccc.2012.3.1387

Flash, T., Meirovitch, Y., & Barliya, A. (2013, April 1). Models of human movement: Trajectory planning and inverse kinematics studies. *Robotics and Autonomous Systems*, *61*(4), 330–339. DOI: 10.1016/j.robot.2012.09.020

Foehn, P., Romero, A., & Scaramuzza, D. (2021, July 21). Time-optimal planning for quadrotor waypoint flight. *Science Robotics*, *6*(56), eabh1221. DOI: 10.1126/scirobotics.abh1221 PMID: 34290102

Fonseca, C. M., & Fleming, P. J. (1995, March). An overview of evolutionary algorithms in multiobjective optimization. *Evolutionary Computation*, *3*(1), 1–6. DOI: 10.1162/evco.1995.3.1.1

Fraichard, T. (1998, January 1). Trajectory planning in a dynamic workspace: A′ state-time space' approach. *Advanced Robotics*, *13*(1), 75–94. DOI: 10.1163/156855399X00928

Garcia, R. R., Bittencourt, A. C., & Villani, E. (2018, September). Relevant factors for the energy consumption of industrial robots. *Journal of the Brazilian Society of Mechanical Sciences and Engineering*, *40*(9), 1–5. DOI: 10.1007/s40430-018-1376-1

Garg, D. P., & Kumar, M. (2002, June 1). Optimization techniques applied to multiple manipulators for path planning and torque minimization. *Engineering Applications of Artificial Intelligence*, *15*(3-4), 241–252. DOI: 10.1016/S0952-1976(02)00067-2

Gasparetto, A., & Zanotto, V. (2007, April 1). A new method for smooth trajectory planning of robot manipulators. *Mechanism and Machine Theory*, *42*(4), 455–471. DOI: 10.1016/j.mechmachtheory.2006.04.002

Gasparetto, A., & Zanotto, V. (2008, June 1). A technique for time-jerk optimal planning of robot trajectories. *Robotics and Computer-integrated Manufacturing*, *24*(3), 415–426. DOI: 10.1016/j.rcim.2007.04.001

Gasparetto, A., & Zanotto, V. (2010, April 1). Optimal trajectory planning for industrial robots. *Advances in Engineering Software*, *41*(4), 548–556. DOI: 10.1016/j.advengsoft.2009.11.001

Goerzen, C., Kong, Z., & Mettler, B. (2010, January). A survey of motion planning algorithms from the perspective of autonomous UAV guidance. *Journal of Intelligent & Robotic Systems*, *57*(1), 65–100. DOI: 10.1007/s10846-009-9383-1

Gordon, N. J., Salmond, D. J., & Smith, A. F. (1993, April). Novel approach to nonlinear/non-Gaussian Bayesian state estimation. In IEE proceedings F (radar and signal processing) (Vol. 140, No. 2, pp. 107-113). IEE.

Guan, Y., Yokoi, K., Stasse, O., & Kheddar, A. (2005, July). On robotic trajectory planning using polynomial interpolations. In 2005 IEEE international conference on robotics and biomimetics-ROBIO (pp. 111-116). IEEE.

Han, W. G., Baek, S. M., & Kuc, T. Y. (1997, October). Genetic algorithm based path planning and dynamic obstacle avoidance of mobile robots. In 1997 IEEE International Conference on Systems, Man, and Cybernetics. Computational Cybernetics and Simulation (Vol. 3, pp. 2747-2751). IEEE.

Hansen, C., Öltjen, J., Meike, D., & Ortmaier, T. (2012, August). Enhanced approach for energy-efficient trajectory generation of industrial robots. In 2012 [IEEE.]. *IEEE International Conference on Automation Science and Engineering (CASE)*, •••, 1–7.

Hashemian, A., Hosseini, S. F., & Nabavi, S. N. (2017, December 17). Kinematically smoothing trajectories by NURBS reparameterization–an innovative approach. *Advanced Robotics*, *31*(23-24), 1296–1312. DOI: 10.1080/01691864.2017.1396923

He, S., Hu, C., Zhu, Y., & Tomizuka, M. (2020, December 1). Time optimal control of triple integrator with input saturation and full state constraints. *Automatica*, *122*, 109240. DOI: 10.1016/j.automatica.2020.109240

Held, M., & Kaaser, D. (2014, March 4). C2 approximation of planar curvilinear profiles by cubic B-splines. *Computer-Aided Design and Applications*, *11*(2), 206–219. DOI: 10.1080/16864360.2014.846092

Hirakawa, A. R., & Kawamura, A. (1997, April). Trajectory planning of redundant manipulators for minimum energy consumption without matrix inversion. In *Proceedings of international conference on robotics and automation* (Vol. 3, pp. 2415-2420). IEEE.

Hogan, N. (1984, November 1). An organizing principle for a class of voluntary movements. *The Journal of Neuroscience : The Official Journal of the Society for Neuroscience*, *4*(11), 2745–2754. DOI: 10.1523/JNEUROSCI.04-11-02745.1984 PMID: 6502203

Hollerbach, J. O., & Suh, K. (1987, August). Redundancy resolution of manipulators through torque optimization. *IEEE Journal on Robotics and Automation*, *3*(4), 308–316. DOI: 10.1109/JRA.1987.1087111

Huang, J., Hu, P., Wu, K., & Zeng, M. (2018, March 1). Optimal time-jerk trajectory planning for industrial robots. *Mechanism and Machine Theory*, *121*, 530–544. DOI: 10.1016/j.mechmachtheory.2017.11.006

Huypens, V. (2020, November). Constant speed lines–curves—NURBS reference pulse IPOs (part I). *International Journal of Advanced Manufacturing Technology*, *111*(5), 1247–1275. DOI: 10.1007/s00170-020-05339-1

Hwangbo, J., Lee, J., Dosovitskiy, A., Bellicoso, D., Tsounis, V., Koltun, V., & Hutter, M. (2019, January 16). Learning agile and dynamic motor skills for legged robots. *Science Robotics*, *4*(26), eaau5872. DOI: 10.1126/scirobotics.aau5872 PMID: 33137755

Ichter, B., Harrison, J., & Pavone, M. (2018, May). Learning sampling distributions for robot motion planning. In *2018 IEEE International Conference on Robotics and Automation (ICRA)* (pp. 7087-7094). IEEE.

Janjoš, F., Reichart, R., & Niermeyer, P. (2017, July 1). Smooth path-generation around obstacles using quartic splines and rrts. *IFAC-PapersOnLine*, *50*(1), 9108–9113. DOI: 10.1016/j.ifacol.2017.08.1708

Jiang, C., Han, X., Guan, F. J., & Li, Y. H. (2007, November 1). An uncertain structural optimization method based on nonlinear interval number programming and interval analysis method. *Engineering Structures*, *29*(11), 3168–3177. DOI: 10.1016/j.engstruct.2007.01.020

Jin, R., Rocco, P., & Geng, Y. (2021, January 1). Cartesian trajectory planning of space robots using a multi-objective optimization. *Aerospace Science and Technology*, *108*, 106360. DOI: 10.1016/j.ast.2020.106360

Kala, R. (2018, January). On repelling robotic trajectories: Coordination in navigation of multiple mobile robots. *Intelligent Service Robotics*, *11*(1), 79–95. DOI: 10.1007/s11370-017-0238-5

Kapania, N. R. (2016). *Trajectory planning and control for an autonomous race vehicle*. Stanford Digital University.

Kashyap, A. K., Parhi, D. R., Muni, M. K., & Pandey, K. K. (2020, November 1). A hybrid technique for path planning of humanoid robot NAO in static and dynamic terrains. *Applied Soft Computing*, *96*, 106581. DOI: 10.1016/j.asoc.2020.106581

Katrakazas, C., Quddus, M., Chen, W. H., & Deka, L. (2015, November 1). Real-time motion planning methods for autonomous on-road driving: State-of-the-art and future research directions. *Transportation Research Part C, Emerging Technologies*, *60*, 416–442. DOI: 10.1016/j.trc.2015.09.011

Kazakovtsev, L. A., Orlov, V., Stupina, A. A., & Kazakovtsev, V. (2015, February 17). Modied Genetic Algorithm with Greedy Heuristic for Continuous and Discrete p-Median Problems. Facta Universitatis, Series. *Mathematics and Informatics.*, *30*(1), 89–106.

Kazem, B. I., Mahdi, A. I., & Oudah, A. T. (2008, September). Motion planning for a robot arm by using genetic algorithm. *Jjmie.*, *2*(3), 131–136.

Kelly, A., & Nagy, B. (2003, July). Reactive nonholonomic trajectory generation via parametric optimal control. *The International Journal of Robotics Research*, *22*(7-8), 583–601. DOI: 10.1177/02783649030227008

Kewlani, G., Ishigami, G., & Iagnemma, K. (2009, October). Stochastic mobility-based path planning in uncertain environments. In *2009 IEEE/RSJ International Conference on Intelligent Robots and Systems* (pp. 1183-1189). IEEE.

Khoukhi, A., Baron, L., & Balazinski, M. (2009, August 1). Constrained multi-objective trajectory planning of parallel kinematic machines. *Robotics and Computer-integrated Manufacturing*, *25*(4-5), 756–769. DOI: 10.1016/j.rcim.2008.09.002

Kim, J., Kim, S. R., Kim, S. J., & Kim, D. H. (2010, January 12). A practical approach for minimum- time trajectory planning for industrial robots. Industrial Robot. *International Journal (Toronto, Ont.).*

Kim, N., Cha, S., & Peng, H. (2010, August 23). Optimal control of hybrid electric vehicles based on Pontryagin's minimum principle. *IEEE Transactions on Control Systems Technology*, *19*(5), 1279–1287.

Köker, R.KöKer R. (2013, February 10). A genetic algorithm approach to a neural-network-based inverse kinematics solution of robotic manipulators based on error minimization. *Information Sciences*, *222*, 528–543. DOI: 10.1016/j.ins.2012.07.051

Kong, K., Bae, J., & Tomizuka, M. (2009, February 13). Control of rotary series elastic actuator for ideal force-mode actuation in human–robot interaction applications. *IEEE/ASME Transactions on Mechatronics*, *14*(1), 105–118. DOI: 10.1109/TMECH.2008.2004561

Korayem, M. H., Nikoobin, A., & Azimirad, V. (2009, October 1). Trajectory optimization of flexible link manipulators in point-to-point motion. *Robotica, 27*(6), 825–840. DOI: 10.1017/S0263574708005183

Kota, S., Hetrick, J., Li, Z., & Saggere, L. (1999, December). Tailoring unconventional actuators using compliant transmissions: Design methods and applications. *IEEE/ASME Transactions on Mechatronics*, *4*(4), 396–408. DOI: 10.1109/3516.809518

Kucuk, S. (2017, December 1). Optimal trajectory generation algorithm for serial and parallel manipulators. *Robotics and Computer-integrated Manufacturing*, *48*, 219–232. DOI: 10.1016/j.rcim.2017.04.006

Lampariello, R., Nguyen-Tuong, D., Castellini, C., Hirzinger, G., & Peters, J. (2011, May). Trajectory planning for optimal robot catching in real-time. In *2011 IEEE International Conference on Robotics and Automation* (pp. 3719-3726). IEEE.

Landers, R. G., Barton, K., Devasia, S., Kurfess, T., Pagilla, P., & Tomizuka, M. (2020, November 1). A Review of Manufacturing Process Control. *Journal of Manufacturing Science and Engineering*, *142*(11), 110814. DOI: 10.1115/1.4048111

Leahy, M. B., & Valavanis, K. P. (2020, August). Dynamics-based control of robotic manipulators. In *Intelligent Robotic Systems* (pp. 245–282). CRC Press.

Lengyel, J., Reichert, M., Donald, B. R., & Greenberg, D. P. (1990, September 1). Real-time robot motion planning using rasterizing computer graphics hardware. *Computer Graphics*, *24*(4), 327–335. DOI: 10.1145/97880.97915

Letizia, N. A., Salamat, B., & Tonello, A. M. (2021, February 22). A Novel Recursive Smooth Trajectory Generation Method for Unmanned Vehicles. *IEEE Transactions on Robotics*, *37*(5), 1792–1805. DOI: 10.1109/TRO.2021.3053649

Li, D., & Du, L. (2021, September). AUV Trajectory Tracking Models and Control Strategies: A Review. *Journal of Marine Science and Engineering*, *9*(9), 1020. DOI: 10.3390/jmse9091020

Li, H., Wu, W. J., Rastegar, J., & Guo, A. (2019, July). A real-time and look-ahead interpolation algorithm with axial jerk-smooth transition scheme for computer numerical control machining of micro-line segments. *Proceedings of the Institution of Mechanical Engineers. Part B, Journal of Engineering Manufacture*, *233*(9), 2007–2019. DOI: 10.1177/0954405418809768

Liang, J. H., & Lee, C. H. (2015, January 1). Efficient collision-free path-planning of multiple mobile robots system using efficient artificial bee colony algorithm. *Advances in Engineering Software*, *79*, 47–56. DOI: 10.1016/j.advengsoft.2014.09.006

Lin, C., Chang, P., & Luh, J. (1983, December). Formulation and optimization of cubic polynomial joint trajectories for industrial robots. *IEEE Transactions on Automatic Control*, *28*(12), 1066–1074. DOI: 10.1109/TAC.1983.1103181

Lin, J., Somani, N., Hu, B., Rickert, M., & Knoll, A. (2018, October). An efficient and time-optimal trajectory generation approach for waypoints under kinematic constraints and error bounds. In *2018 IEEE/RSJ International Conference on Intelligent Robots and Systems (IROS)* (pp. 5869-5876). IEEE.

Liu, H., Lai, X., & Wu, W. (2013, April 1). Time-optimal and jerk-continuous trajectory planning for robot manipulators with kinematic constraints. *Robotics and Computer-integrated Manufacturing*, *29*(2), 309–317. DOI: 10.1016/j.rcim.2012.08.002

Liu, Y., Xu, H., & Zhang, X. (2018, January 1). Dynamic Characteristics and Restraint Mechanism of Elastically Restrained Connecting Joints on the Flexible Robotic Manipulator. *Shock and Vibration*, *2018*(1), 2018. DOI: 10.1155/2018/3217106

Lu, S., Ding, B., & Li, Y. (2020, March). Minimum-jerk trajectory planning pertaining to a translational 3- degree of freedom parallel manipulator through piecewise quintic polynomials interpolation. *Advances in Mechanical Engineering*, *12*(3), 1687814020913667. DOI: 10.1177/1687814020913667

Luo, L. P., Yuan, C., Yan, R. J., Yuan, Q., Wu, J., Shin, K. S., & Han, C. S. (2015, May 1). Trajectory planning for energy minimization of industry robotic manipulators using the Lagrange interpolation method. *International Journal of Precision Engineering and Manufacturing*, *16*(5), 911–917. DOI: 10.1007/s12541-015-0119-9

Ma, J. W., Gao, S., Yan, H. T., Lv, Q., & Hu, G. Q. (2021, February 12). A new approach to time-optimal trajectory planning with torque and jerk limits for robot. *Robotics and Autonomous Systems*, *140*, 103744. DOI: 10.1016/j.robot.2021.103744

Ma, J. W., Gao, S., Yan, H. T., Lv, Q., & Hu, G. Q. (2021, June 1). A new approach to time-optimal trajectory planning with torque and jerk limits for robot. *Robotics and Autonomous Systems*, *140*, 103744. DOI: 10.1016/j.robot.2021.103744

Ma, S. (1995, November). Real-time algorithm for quasi-minimum energy control of robotic manipulators. In Proceedings of IECON'95-21st Annual Conference on IEEE Industrial Electronics (Vol. 2, pp. 1324-1329). IEEE.

Macfarlane, S., & Croft, E. A. (2003, February 19). Jerk-bounded manipulator trajectory planning: Design for real- time applications. *IEEE Transactions on Robotics and Automation*, *19*(1), 42–52. DOI: 10.1109/TRA.2002.807548

Malyuta, D., Yu, Y., Elango, P., & Açıkmeşe, B. (2021, November 2). Advances in trajectory optimization for space vehicle control. *Annual Reviews in Control, 52*, 282–315. DOI: 10.1016/j.arcontrol.2021.04.013

Martins, J. M., Mohamed, Z., Tokhi, M. O., Da Costa, J. S., & Botto, M. A. (2003, July 1). Approaches for dynamic modelling of flexible manipulator systems. *IEE Proceedings. Control Theory and Applications, 150*(4), 401–411. DOI: 10.1049/ip-cta:20030496

Masehian, E., & Katebi, Y. (2007, May 26). Robot motion planning in dynamic environments with moving obstacles and target. *International Journal of Mechanical Systems Science and Engineering., 1*(1), 20–25.

Medeiros, V. S., Jelavic, E., Bjelonic, M., Siegwart, R., Meggiolaro, M. A., & Hutter, M. (2020, April 27). Trajectory optimization for wheeled-legged quadrupedal robots driving in challenging terrain. *IEEE Robotics and Automation Letters, 5*(3), 4172–4179. DOI: 10.1109/LRA.2020.2990720

Mei, Y., Lu, Y. H., Hu, Y. C., & Lee, C. G. (2006, June 5). Deployment of mobile robots with energy and timing constraints. *IEEE Transactions on Robotics, 22*(3), 507–522. DOI: 10.1109/TRO.2006.875494

Mettin, U., La Hera, P. X., Morales, D. O., Shiriaev, A. S., Freidovich, L. B., & Westerberg, S. (2009, June). Trajectory planning and time-independent motion control for a kinematically redundant hydraulic manipulator. In *2009 International Conference on Advanced Robotics* (pp. 1-6). IEEE.

Mishra, U. A., Chawla, I., & Pathak, P. M. (2020, September). On determining shortest path in joint space of a cable-driven parallel robot for point-to-point motion. In 2020 28th Mediterranean Conference on Control and Automation (MED) (pp. 984-989). IEEE.

Misra, G., & Bai, X. (2017, November). Task-constrained trajectory planning of free-floating space-robotic systems using convex optimization. *Journal of Guidance, Control, and Dynamics, 40*(11), 2857–2870. DOI: 10.2514/1.G002405

Muenchhof, M., & Singh, T. (2002, May). Desensitized jerk limited-time optimal control of multi-input systems. *Journal of Guidance, Control, and Dynamics, 25*(3), 474–481. DOI: 10.2514/2.4933

Nieuwenhuisen, M., & Behnke, S. (2015, September). 3D planning and trajectory optimization for real-time generation of smooth MAV trajectories. In *2015 European Conference on Mobile Robots (ECMR)* (pp. 1-7). IEEE.

Nikolos, I. K., Valavanis, K. P., Tsourveloudis, N. C., & Kostaras, A. N. (2003, November 17). Evolutionary algorithm based offline/online path planner for UAV navigation. *IEEE Transactions on Systems, Man, and Cybernetics. Part B, Cybernetics, 33*(6), 898–912. DOI: 10.1109/TSMCB.2002.804370 PMID: 18238242

Oliveira, P. W., Barreto, G. A., & Thé, G. A. P. (2020, January 9). A General Framework for Optimal Tuning of PID-like Controllers for Minimum Jerk Robotic Trajectories. *Journal of Intelligent & Robotic Systems, 99*(3-4), 1–20. DOI: 10.1007/s10846-019-01121-y

Ozaki, H., & Lin, C. J. (1996, April). Optimal B-spline joint trajectory generation for collision-free movements of a manipulator under dynamic constraints. In *Proceedings of IEEE International Conference on Robotics and Automation* (Vol. 4, pp. 3592-3597). IEEE.

Paing, M. S., Nshama, E. W., & Uchiyama, N. (2021, July 9). A Kinematically Constrained Reparameterization Approach to Optimal Time and Jerk Motion of Industrial Machines. *IEEE Access : Practical Innovations, Open Solutions, 9*, 97843–97854. DOI: 10.1109/ACCESS.2021.3095847

Parker, J. K., Khoogar, A. R., & Goldberg, D. E. (1989, January). Inverse kinematics of redundant robots using genetic algorithms. In *1989 IEEE International Conference on Robotics and Automation* (pp. 271-272). IEEE Computer Society.

Pekarovskiy, A., Nierhoff, T., Hirche, S., & Buss, M. (2017, November 14). Dynamically consistent online adaptation of fast motions for robotic manipulators. *IEEE Transactions on Robotics, 34*(1), 166–182. DOI: 10.1109/TRO.2017.2765666

Piazzi, A., & Visioli, A. (1998, January 1). Global minimum-time trajectory planning of mechanical manipulators using interval analysis. *International Journal of Control, 71*(4), 631–652. DOI: 10.1080/002071798221713

Piazzi, A., & Visioli, A. (1998, January 1). Global minimum-time trajectory planning of mechanical manipulators using interval analysis. *International Journal of Control, 71*(4), 631–652. DOI: 10.1080/002071798221713

Pott, A. (2018). Cable-driven parallel robots: theory and application, New York, Springer; NY, USA: Springer; 120: 52-56.

Qian, S., Bao, K., Zi, B., & Zhu, W. D. (2020, July 1). Dynamic Trajectory Planning for a Three Degrees-of- Freedom Cable-Driven Parallel Robot Using Quintic B-Splines. *Journal of Mechanical Design, 142*(7), 073301. DOI: 10.1115/1.4045723

Radmanesh, M., Kumar, M., & Sarim, M. (2018, June 1). Grey wolf optimization based sense and avoid algorithm in a Bayesian framework for multiple UAV path planning in an uncertain environment. *Aerospace Science and Technology, 77*, 168–179. DOI: 10.1016/j.ast.2018.02.031

Ratliff, N., Zucker, M., Bagnell, J. A., & Srinivasa, S. (2009, May 12). CHOMP: Gradient optimization techniques for efficient motion planning. In2009 [IEEE.]. *IEEE International Conference on Robotics and Automation*, •••, 489–494. DOI: 10.1109/ROBOT.2009.5152817

Rout, A., Mahanta, G. B., Bbvl, D., & Biswal, B. B. (2020). Kinematic and dynamic optimal trajectory planning of industrial robot using improved multi-objective ant lion optimizer. Journal of The Institution of Engineers (India): Series C, 101(3), 559-569.

Rubio, F., Llopis-Albert, C., Valero, F., & Suñer, J. L. (2016, December 1). Industrial robot efficient trajectory generation without collision through the evolution of the optimal trajectory. *Robotics and Autonomous Systems, 86*, 106–112. DOI: 10.1016/j.robot.2016.09.008

Rubio, F., Valero, F., Sunyer, J., & Cuadrado, J. (2012, January 6). Optimal time trajectories for industrial robots with torque, power, jerk and energy consumed constraints. Industrial Robot. *International Journal (Toronto, Ont.)*.

Sahar, G., & Hollerbach, J. M. (1986, September). Planning of minimum-time trajectories for robot arms. *The International Journal of Robotics Research, 5*(3), 90–100. DOI: 10.1177/027836498600500305

Sahay, R., Kumar, A., Udai, A. D., & Sinha, A. N. (2006). *An imitated human motion optimal b-spline path of a biped robot using genetic algorithm. InProceedings of All India Seminar on Advances in Product Development (APD-2006)*. New Age International.

Salomon, R. (1996, January 1). Re-evaluating genetic algorithm performance under coordinate rotation of benchmark functions. A survey of some theoretical and practical aspects of genetic algorithms. *Bio Systems, 39*(3), 263–278. DOI: 10.1016/0303-2647(96)01621-8 PMID: 8894127

Saltzman, E., & Kelso, J. A. (1987, January). Skilled actions: A task-dynamic approach. *Psychological Review, 94*(1), 84–106. DOI: 10.1037/0033-295X.94.1.84 PMID: 3823306

Scalera, L., Carabin, G., Vidoni, R., & Gasparetto, A. (2020). Minimum-energy trajectory planning for industrial robotic applications: Analytical model and experimental results. In *Advances in Service and Industrial Robotics: Results of RAAD* (pp. 334–342). Springer International Publishing.

Seereeram, S., & Wen, J. T. (1995, February). A global approach to path planning for redundant manipulators. *IEEE Transactions on Robotics and Automation, 11*(1), 152–160. DOI: 10.1109/70.345948

Shiller Z. Time-energy optimal control of articulated systems with geometric path constraints.

Shin, K., & McKay, N. (1986, June). A dynamic programming approach to trajectory planning of robotic manipulators. *IEEE Transactions on Automatic Control, 31*(6), 491–500. DOI: 10.1109/TAC.1986.1104317

Shojaedini, E., Majd, M., & Safabakhsh, R. (2019, April 1). Novel adaptive genetic algorithm sample consensus. *Applied Soft Computing, 77*, 635–642. DOI: 10.1016/j.asoc.2019.01.052

Siddiqi, S. S., Akram, G., & Iftikhar, M. (2012). Solution of seventh order boundary value problem by differential transformation method. *World Applied Sciences Journal, 16*(11), 1521–1526.

Sivanandam, S. N., & Deepa, S. N. (2008). *Genetic algorithms. InIntroduction to genetic algorithms*. Springer.

Sivčev, S., Coleman, J., Omerdić, E., Dooly, G., & Toal, D. (2018, September 1). Underwater manipulators: A review. *Ocean Engineering, 163*, 431–450. DOI: 10.1016/j.oceaneng.2018.06.018

Sorek, N., Gildin, E., Boukouvala, F., Beykal, B., & Floudas, C. A. (2017, April 1). Dimensionality reduction for production optimization using polynomial approximations. *Computational Geosciences, 21*(2), 247–266. DOI: 10.1007/s10596-016-9610-3

Staicu, S. (2009, May 31). Inverse dynamics of the 3-PRR planar parallel robot. *Robotics and Autonomous Systems, 57*(5), 556–563. DOI: 10.1016/j.robot.2008.09.005

Sun, S., Lin, H., Zheng, L., Yu, J., & Hu, Y. (2016, May). A real-time and look-ahead interpolation methodology with dynamic B-spline transition scheme for CNC machining of short line segments. *International Journal of Advanced Manufacturing Technology, 84*(5), 1359–1370.

Sun, Y., Ho, K. C., & Wan, Q. (2018, November 5). Solution and analysis of TDOA localization of a near or distant source in closed form. *IEEE Transactions on Signal Processing*, *67*(2), 320–335. DOI: 10.1109/TSP.2018.2879622

Sundaralingam, B., & Hermans, T. (2019, February). Relaxed-rigidity constraints: Kinematic trajectory optimization and collision avoidance for in-grasp manipulation. *Autonomous Robots*, *43*(2), 469–483. DOI: 10.1007/s10514-018-9772-z

Suriano, L., Otero, A., Rodríguez, A., Sánchez-Renedo, M., & De La Torre, E. (2020, June 26). Exploiting multi- level parallelism for run-time adaptive inverse kinematics on heterogeneous masons. *IEEE Access : Practical Innovations, Open Solutions*, *8*, 118707–118724. DOI: 10.1109/ACCESS.2020.3005202

Talatahari, S., Azar, B. F., Sheikholeslami, R., & Gandomi, A. H. (2012, March 1). Imperialist competitive algorithm combined with chaos for global optimization. *Communications in Nonlinear Science and Numerical Simulation*, *17*(3), 1312–1319. DOI: 10.1016/j.cnsns.2011.08.021

Thompson, S. E., & Patel, R. V. (1987, May). Formulation of joint trajectories for industrial robots using B-splines. *IEEE Transactions on Industrial Electronics*, *IE-34*(2), 192–199. DOI: 10.1109/TIE.1987.350954

Tian, L., & Collins, C. (2004, June 1). An effective robot trajectory planning method using a genetic algorithm. *Mechatronics*, *14*(5), 455–470. DOI: 10.1016/j.mechatronics.2003.10.001

Tika, A., Gafur, N., Yfantis, V., & Bajcinca, N. (2020, January 1). Optimal scheduling and model predictive control for trajectory planning of cooperative robot manipulators. *IFAC-PapersOnLine*, *53*(2), 9080–9086. DOI: 10.1016/j.ifacol.2020.12.2136

Tsai KY, Kohli D, Arnold J. Trajectory planning in joint space for mechanical manipulators.

Tuncer, A., & Yildirim, M. (2012, November 1). Dynamic path planning of mobile robots with improved genetic algorithm. *Computers & Electrical Engineering*, *38*(6), 1564–1572. DOI: 10.1016/j.compeleceng.2012.06.016

Turrisi, G., Capotondi, M., Gaz, C. R., Modugno, V., Oriolo, G., & De Luca, A. (2021, November 13). On-Line Learning for Planning and Control of Underactuated Robots with Uncertain Dynamics. *IEEE Robotics and Automation Letters*.

Ur-Rehman, R., Caro, S., Chablat, D., & Wenger, P. (2010, August 1). Multi-objective path placement optimization of parallel kinematics machines based on energy consumption, shaking forces and maximum actuator torques: Application to the Orthoglide. *Mechanism and Machine Theory*, *45*(8), 1125–1141. DOI: 10.1016/j.mechmachtheory.2010.03.008

. Vadakkepat P, Tan KC, Ming-Liang W. Evolutionary artificial potential fields and their application in real time robot path planning. InProceedings of the 2000 congress on evolutionary computation. CEC00 (Cat. No. 00TH8512) 2000 Jul 16 (Vol. 1, pp. 256-263). IEEE.

van Dijk, N. J. (2006, June 13). Generic trajectory generation for industrial manipulators. Eindhoven University of Technology, Department of Mechanical Engineering. *Dynamics and Control.*

Verscheure, D., Demeulenaere, B., Swevers, J., De Schutter, J., & Diehl, M. (2009, September 22). Time-optimal path tracking for robots: A convex optimization approach. *IEEE Transactions on Automatic Control*, *54*(10), 2318–2327. DOI: 10.1109/TAC.2009.2028959

Vezvari, M. R., Nikoobin, A., & Ghoddosian, A. (2021, July). Perfect torque compensation of planar 5R parallel robot in point-to-point motions, optimal control approach. *Robotica*, *39*(7), 1163–1180. DOI: 10.1017/S0263574720000971

Victoire, T. A., & Jeyakumar, A. E. (2004, September 1). Hybrid PSO–SQP for economic dispatch with valve-point effect. *Electric Power Systems Research*, *71*(1), 51–59. DOI: 10.1016/j.epsr.2003.12.017

Wada, Y., & Kawato, M. (2004, April 1). A via-point time optimization algorithm for complex sequential trajectory formation. *Neural Networks*, *17*(3), 353–364. DOI: 10.1016/j.neunet.2003.11.009 PMID: 15037353

Wang, G., Shu, Q., Wang, J., & Li, L. (2017, July). Research on adaptive non-uniform rational B-spline real- time interpolation technology based on acceleration constraints. *International Journal of Advanced Manufacturing Technology*, *91*(5), 2089–2100. DOI: 10.1007/s00170-016-9914-4

Webb, D. J., Van Den Berg, J., & Kinodynamic, R. R. T. (2013). asymptotically optimal motion planning for robots with linear dynamics, 2013 IEEE Int. Conf. Rob. Autom.

Wen, Y., & Pagilla, P. R. (2021, September). Path-constrained optimal trajectory planning for robot manipulators with obstacle avoidance. In *2021 IEEE/RSJ International Conference on Intelligent Robots and Systems (IROS)* (pp. 1421-1426). IEEE.

Wie, B. (2005, September). Singularity escape/avoidance steering logic for control moment gyro systems. *Journal of Guidance, Control, and Dynamics, 28*(5), 948–956. DOI: 10.2514/1.10136

Xiao, Y., Du, Z., & Dong, W. (2012, March 2). Smooth and near time-optimal trajectory planning of industrial robots for online applications. Industrial Robot. *International Journal (Toronto, Ont.).*

Xu, H. (2008, June 17). Tracking Lagrangian trajectories in position–velocity space. *Measurement Science & Technology, 19*(7), 075105. DOI: 10.1088/0957-0233/19/7/075105

Xu, X., & Chen, Y. A method for trajectory planning of robot manipulators in Cartesian space. InProceedings of the 3rd World Congress on Intelligent Control and Automation (Cat. No. 00EX393) 2000 Jun 26 (Vol. 2, pp. 1220-1225). IEEE.

Xu, Z., Wei, S., Wang, N., & Zhang, X. Trajectory planning with Bezier curve in Cartesian space for industrial gluing robot. In *International Conference on Intelligent Robotics and Applications* 2014 Dec 17 (pp. 146-154). Springer, Cham. DOI: 10.1007/978-3-319-13963-0_15

Yano, F., & Toyoda, Y. (1999, August). Preferable movement of a multijoint robot arm using a genetic algorithm. In *Intelligent robots and computer vision XVIII: algorithms, techniques, and active vision* (Vol. 3837, pp. 80–88). SPIE.

Yuan, M., Chen, Z., Yao, B., & Liu, X. (2019, December 16). Fast and accurate motion tracking of a linear motor system under kinematic and dynamic constraints: An integrated planning and control approach. *IEEE Transactions on Control Systems Technology, 29*(2), 804–811. DOI: 10.1109/TCST.2019.2955658

Zalzala A, Morris AS. An Optimum Trajectory Planner for Robot Manipulators in Joint- pace and Under Physical Constraints.

Zanotto, V., Gasparetto, A., Lanzutti, A., Boscariol, P., & Vidoni, R. (2011, November). Experimental validation of minimum time-jerk algorithms for industrial robots. *Journal of Intelligent & Robotic Systems, 64*(2), 197–219. DOI: 10.1007/s10846-010-9533-5

Zha, X. F. (2002, October 1). Optimal pose trajectory planning for robot manipulators. *Mechanism and Machine Theory, 37*(10), 1063–1086. DOI: 10.1016/S0094-114X(02)00053-8

Zhang, D., & Wei, B. (Eds.). (2017 Feb 3). *Adaptive control for robotic manipulators*. CRC Press. DOI: 10.1201/9781315166056

Zhang, H., Qiu, H., Zhang, X., & Hu, P. (2020, November 1). Recent Advances on Manipulator Trajectory Planning Methods. *Recent Patents on Mechanical Engineering, 13*(4), 303–327. DOI: 10.2174/2212797613666200319151513

Zhang, K., Gao, X. S., Li, H. B., & Yuan, C. M. (2012). A greedy algorithm for feed-rate planning of CNC machines along curved tool paths with confined jerk for each axis. *Robotics and Computer-integrated Manufacturing, 28*(4), 472–483. DOI: 10.1016/j.rcim.2012.02.006

Zhang, X., & Shi, G. (2022, April). Multi-objective optimal trajectory planning for manipulators in the presence of obstacles. *Robotica, 40*(4), 888–906. DOI: 10.1017/S0263574721000886

Zhao, L., Jiang, Z., Sun, Y., Zhao, J., & Liu, H. (2021, February 28). Collision-free kinematics for hyper-redundant manipulators in dynamic scenes using optimal velocity obstacles. *International Journal of Advanced Robotic Systems, 18*(1), 1729881421996148. DOI: 10.1177/1729881421996148

Zhou, J., Cao, H., Jiang, P., Li, C., Yi, H., & Liu, M. (2021, March 24). Energy-Saving Trajectory Planning for Robotic High-Speed Milling of Sculptured Surfaces. *IEEE Transactions on Automation Science and Engineering*.

Zlajpah, L. (1996, April). On time optimal path control of manipulators with bounded joint velocities and torques. In *Proceedings of IEEE international conference on robotics and automation* (Vol. 2, pp. 1572-1577). IEEE.

Chapter 14

Advancements in Examination Systems: IP-Based Robo Invigilator Technique

Swati Chowdhuri
Institute of Engineering and Management, Kolkata, India

Arka Rajak
JIS College of Engineering, India

Raktim Koner
JIS College of Engineering, India

Supratim Chakraborty
JIS College of Engineering, India

Moumita Pal
JIS College of Engineering, India

Biswarup Neogi
JIS College of Engineering, India

Souvick Kumar Guha
JIS College of Engineering, India

ABSTRACT

This research delves into the operational dynamics of invigilator robots, particularly in exam halls. Traditionally, human invigilators have overseen exams to prevent cheating and maintain fairness. However, the emergence of invigilator robots equipped with advanced features has revolutionized this process. These robots utilize cameras and sensors to monitor exam rooms in real time, detecting any suspicious behavior such as accessing prohibited materials or communicating with others. Employing facial recognition and identity verification, they ensure only authorized individuals are present for the exam. Furthermore, they analyze behavioral patterns to identify potential instances of fraud, distinguishing between normal actions and those indicating unauthorized assistance. Their ability to observe consistently without distraction or fatigue ensures thorough monitoring throughout the test. In case of

DOI: 10.4018/979-8-3693-4365-4.ch014

anomalies, they promptly alert human administrators. Additionally, these robots generate comprehensive reports and visual evidence of any irregular activities.

1. INTRODUCTION

The increase in the number of research articles published over various fields has made it difficult for researchers, academics and reviewers to find relevant works. With so many to sift through, it makes sense that we would need better ways of reviewing and evaluating this information. Though traditional peer review processes are necessary, the volume of manuscript submissions may outpace a journal's ability to review them in full — leading to magazine publication delays and biases or oversights during the assessment process. Robots and automation in the review process. The above-mentioned challenges are an area that robotics and automation have been working on even now where they can make a lot of difference. AI, NLP, and Analytics in Robotics Further Support Streamlining of Peer Review. The reviewing process may improve efficiency, objectivity, and quality in this process-from pre-sorting reviews to content analysis in depth-by using robotics through AI, NLP, and analytics. Based on synthesizing available literature and case studies and even emerging trends in scholarly publishing, this review is aimed at an integrated understanding of the role that robotics plays in assessing review papers.

It will discuss a range of applications, challenges, and implications of introducing robotics into the reviewing process, such as potential benefits it can have in relation to workload reduction on reviewers, improvement in the assessment of paper manuscript quality, and accelerating the dissemination of scientific knowledge. Through structured review of the current state-of-the-art, the objective of this review is thus to shed light onto developments in assessment within robotics-enabled reviews of paper and define potential avenues for future research and development.

More importantly, it will stimulate critical discussion on ethical matters, prejudices, and limitations concerning robotic peer review and will be consistent with academic integrity. In summary, the review offers timely insight into this revolutionizing robotics-scholarly publishing interface, demonstrating how automation can revolutionize the reviewing processes to project future scholarly discourse. The purpose is to inform stakeholders-on equally researchers, publishers, and policymakers-of the transformative power of technology for advancing the integrity and efficiency of peer review. Following which the work is presented in the following manner. Section 2 presents the past related works carried out in this domain and gap analysis of the review work is describes in section 3. Section 4 represents the technology behind the design followed by the operation of the robot during examination process in section 5. proposed methodology which outlines the preprocessing, feature extraction and training process. Section 6 represents strength and weakness of the chapter in context of the book. The work concludes in section 7 describes in brief the limitation and future scope of the work.

2. CONSTRUCTIVE LITERATURE SURVEY

Clodic A. et.al. (2006) describe that Rackham is these paper talks about an interactive robot-guide that has been installed in a number of places and events and explains its design and the perspectives it has brought along with it. The objective of the project was to enhance the functionality and decision-making capabilities of the robot by studying public interaction over it. This will, therefore, not only test the navigation efficiency of the robot in dynamic environments but also aim at developing and validating a systematic approach toward skills integration for human-robot interaction. Although the robot design considerations and results related to public deployment are discussed, important lessons learned will guide future design cycles in order to be more efficient and have a better reception in the user community.

This paper presents a customized variant of the LAAS architecture specifically targeted for human-robot interaction in Figure 1. According to Arocena et.al. (2022), a supervisor is developed that can perform its tasks of interaction along with traditional navigation tasks, explaining the actions and communicating its intention to the individuals in the robot's vicinity during its mission. This supervisor was integrated along with functional components like localization, motion control with collision avoidance, and user interface, thus resulting in the required degree of resilience to work under difficult conditions. We have thus strengthened Rackham's foundation for the purpose of robust navigation and interaction with humans in real scenarios. Although Rackham has been very successfully used in scores of presentations, and has been received very well, it represents more of the beginning of our journey. The work that has continued is focused on enhancing Rackham's interaction and perceptual capabilities in order to systematically incorporate advanced context interpretation and decision-making processes for synthesizing and controlling interactive tasks at multiple levels.

Figure 1. The functional level of Rackham and its 16 modules

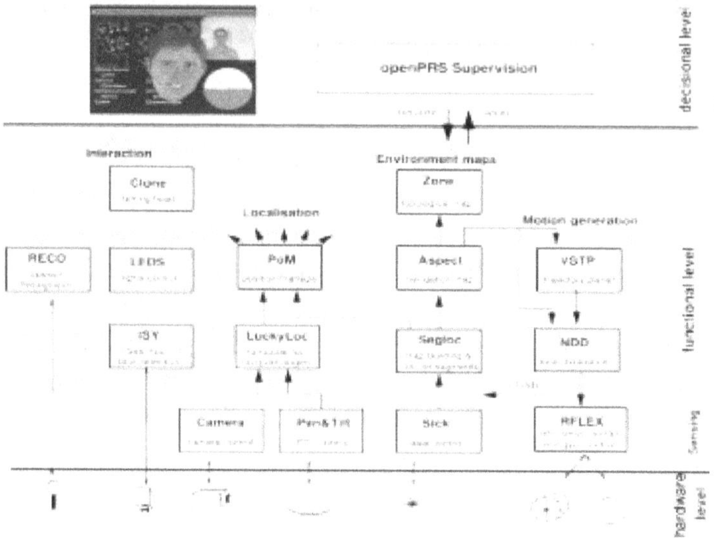

According to Erdogan R. et.al. (2023), the Covid-19 pandemic has impacted almost every aspect, which includes educational sectors. With a decision taken to shift the course contents into an online platform, delivering face-to-face instruction to students from remote locations presented several challenges to the teachers. Teachers using robotic coding tools are looking for ways to complete these activities remotely. Simulators can be used to animate real-world objects in a computer context. There are various commercial and free systems that simulate robotic coding tools. Tinkercad may be used to imitate Arduino and Micro: bit boards. Abdullahi et. Al (2015) and Alexenko et. Al. (2013) examines how instructors use the Tinkercad platform. Users are required to execute the actual tasks that they must conduct on the evaluated platform during usability testing. In this study, 12 Information Technology instructors attempted to accomplish 10 real assignments in the Tinkercad platform's Circuits area. While individuals were performing activities, their eye movements were captured and evaluated. During the exams, participants were given surveys, and data was collected using an observation form. As a result, the instructors who completed the usability tests indicated that the platform is acceptable for use by teachers and children above the age of ten, and that they may utilize the platform in their activities.

Figure 2. Steps of the research

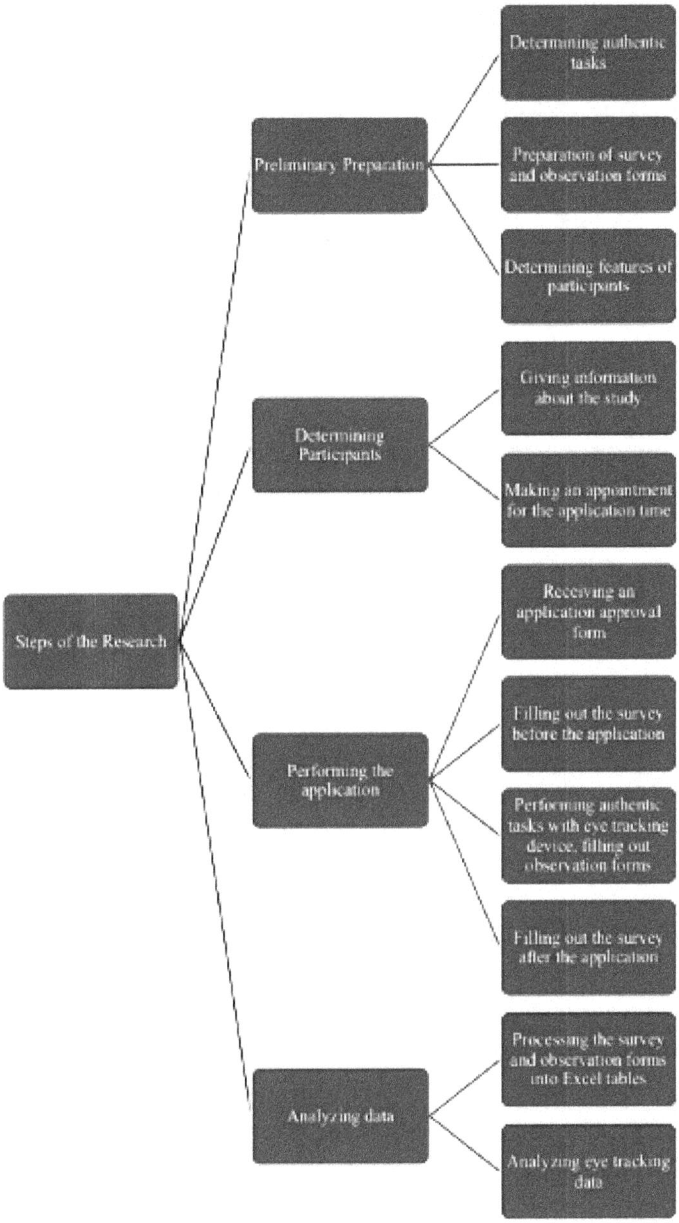

Kannan et al. (2015) and ÇAVAŞ et. Al. (2020) discovered that 11 participants who incorporated robotic coding studies using Arduino microcontroller cards into their courses and activities performed tasks to a high degree, despite having never used the Micro:bit microcontroller card previously. There were circuit components the students could easily find in the Tinkercad circuits category, and due to the block-based nature of the coding, they were good coders. Computer technology instructors found the Tinkercad website easy to use, with an average rate of completion at 89%. Attaran et. al. (2013) and Ayoobi et. Al. (2022) suggests that conducting independent usability tests with each user group such as teachers and students can be helpful for improving usability research in a program. The Arduino platform is flexible to work with in designing circuits and in the programming part, so it could be used in a variety of courses. According to the study, Tinkercad circuits might be improved by doing usability testing with youngsters above the age of 10. Teachers of science or physics that incorporate electrical and circuits concepts in their classes might benefit from similar applications. The Tinkercad platform also allows you to build 3D creations. It is favored not only in computer science classes, but also in other fields. (Tasti et al. (2015)) found that 3D learning artifacts are effective in embodying abstract concepts and sustaining learning. Tinkercad content is thought to pique kids' interest and help them develop their creative talents and creativity. In this regard, a usability test of the 3D design category, which will be used by other branch teachers in future studies, may be conducted, and the platform's excellent and negative points revealed. It is also important to identify the most common issues encountered during user testing using the Tinkercad platform. When a new circuit is added to the circuits section then the circuit elements added to the circuits creation area have to be deleted one by one. It might more conveniently clean the circuit region in a shorter time period. Furthermore, when the circuit is built, the circuit components are closed when writing the code blocks, and the code blocks section may be closed automatically when the simulation is run. The building blocks in the code blocks section are quite large, making them difficult to see in the coding area.

According to Gonzalez F. et.al. (2023), Robotics is a curriculum so sought after in many curricula of education, especially as one for technical replacement. Most of this course is focused on teaching the student how to program motion through an arm that features within a robot by setting the velocity of its numerous joint motors - this is oftentimes known as joint programming. Students should understand the writing algorithms to control the arm end effector, which requires changeable properties like instantaneous velocity of every joint motor. The students make use of most of the instances containing either physical or virtual robotic arms to help them learn to do so (Krishnan et.al. (2022)). Visual feedback of the movement of the arm is very essential to guide students in knowing the correctness of the methods that they have programmed for their joints. However, a problem arises in teaching to have the robotic arm move along a planned path using defined velocity, a subtopic known as differential movements. For them to master this, students would have to design and experiment algorithms for differential movements and validate their correctness. The human eye fails to distinguish between correct and incorrect movements of an end effector, especially if it entails tiny differences in velocities. A study found that simulating the spray painting on an artificial canvas may better assess the correctness of the differential movement algorithm itself than tracking the movement of arms. To address this, an existing virtual robotic arm educational tool was enhanced with a model of spray-painting equipment and a canvas. This upgrade was implemented in an Introduction to Robotics class at Florida Gulf Coast University in spring 2019 and spring 2020. While the spring 2019 class used the virtual arm without the spray-painting feature, the spring 2020 class utilized the newly added spray-painting function. Test results revealed a significant difference in performance: 59.4% of students who utilized the spray-painting feature scored at least 85% on the differential movements exam question, compared to only 5.6% of those who did not use the feature. The exam question required students to develop a differential movement algorithm that would move the arm with a set velocity along a straight line.

Figure 3. Manual control of the arm and simulation

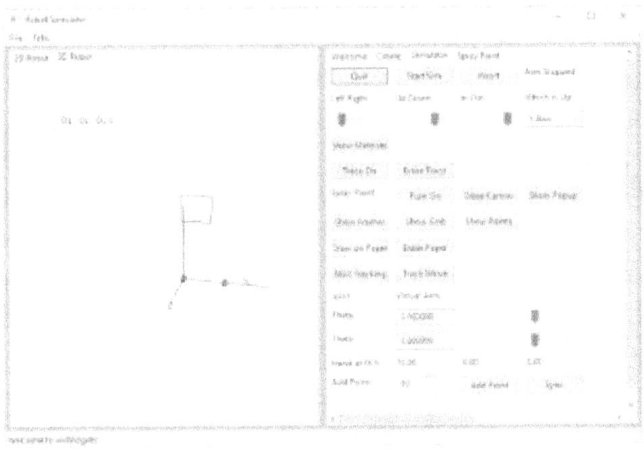

To make the theory of differential motion of the control of precise arm motions with a velocity understandable to students, it would be advisable to write and demonstrate algorithms on a real arm, most probably used for painting or welding. However, there are disadvantages associated with this: the expensive equipment, special requirements for the workshop, to say nothing about the danger. A virtual arm with an application model can overcome all these challenges. The suggested solution would be the addition of a virtual spray-painting application to an existing simulator that would help students test their differential movement algorithms on a robotic virtual arm by evaluating the final canvas after virtually painting. Those who used the virtual spray-painting arm to program a painting operation using differential motions performed better on the exam question. Specifically, 59.4% of students who utilized the virtual spray-painting arm scored at least 85% on the relevant exam question, demonstrating their ability to develop software for arm movement using distinctive gestures. In contrast, only 22.2% of students who didn't use the tool achieved a grade of at least 70%, and merely 5.6% attained a score of 85% or higher. A long-term objective is to enhance the virtual arm by incorporating features that aid in learning robot dynamics. As noted by Mahmood et. al. (2022), it would be changing power requirements with location, velocity, acceleration, and load on the end effector changes.

Bakker P. et.al. (1996) indicated that there are only two major ways of robot teaching currently: either to direct the robot in what to do directly (programming) or to let it come up with it for itself (reinforcement learning or genetic algorithms). We review a new method, such as imitation learning, which enables direct teaching of behaviors by presenting a demonstration of some others' behaviors to the robot. We outline the psychological roots of this approach, define what imitation is, and identify key challenges in using imitation in robotic systems. Based on this approach, we assess recent published work in this subject and describe an imitation effort now underway at Japan's Electrotechnical Laboratory.

Robot learning is a promising topic but beset with insurmountable challenges. Imitation learning is a unique method that promises a richer environment for the learning robot, provides an easier channel of communication between people and robots, and may eventually enable robots to induct innovative behaviors. The purpose of this study has been to examine both the potential and the challenges of this emerging new subject in robotics. While recent research is exciting, several fundamental difficulties in perception and representation must be addressed before broad imitation learning becomes a reality in robots. Recent work at ETL draws on concepts from developmental psychology to solve some of these difficulties in a practical way.

Zhang S. et.al. (2023) describe that currently, it is necessary to improve robots' ability to provide various educational services; nevertheless, while applying robots to educational scenes, examination scenes are commonly disregarded. Because the Mobile Educational Robot (MER) performing intelligent invigilation tasks frequently encounters unknown behaviors of examinees or different examination environments and must deal with complex uncertainties, this paper comprehensively employs various advanced technologies in the field of artificial intelligence (AI) to build a MER autonomous system that performs intelligent proctoring in which the three-dimensional simultaneous localization and mapping (3D-SLAM) is used

Figure 4. A structural network visualization of YOLOv5 in an examination setting

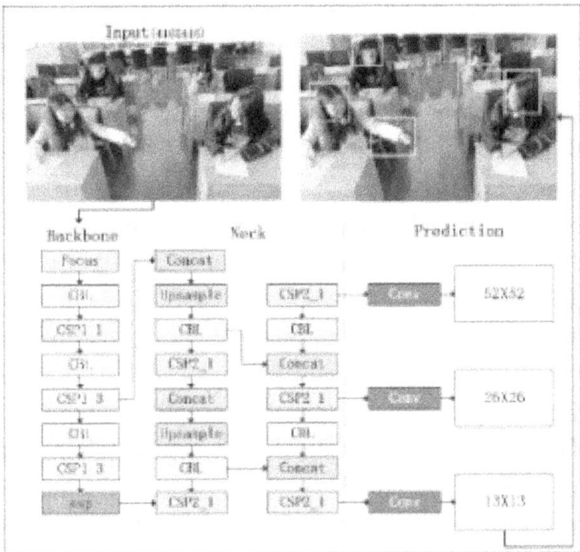

The research discusses an autonomous MER system that is capable of performing intelligent invigilation operations, integrating the latest AI technologies. Figure 4 illustrates a structural network visualization of YOLOv5 within an examination environment. This method allows for considerable autonomous invigilation on the part of the MER so that it can efficiently and intelligently observe examinations within uncertain environments. Leveraging the autonomy and interactive capabilities of robots to monitor exam rooms not only saves human resources and enhances flexibility in exam organization but also increases efficiency while ensuring the fairness and scientific integrity of the invigilation process, thus safeguarding candidates' rights and interests to some extent. In the future, two aspects of work are: one is to combine MER with clever monitoring; and carry out moral induction within students on a more specific age range related; the other can not only have double effect in acceleration Moral education also has an aspect (that pulls itself up by ties). Development of an independent outdoor navigation algorithm in complex interference environments (e.g., controlling an outdoor sport event) It is directed towards the development of the autonomy and adaptability of future educational robots in heterogeneous educational settings. Suman. et.al. (2012) have mentioned that in the present IT-enabled scenario, online examination has gained tremendous popularity. It has removed all the disadvantages of Offline Examination.

However, there is still a significant concern about the security of online testing. The verification of genuine candidates throughout the online test is a major issue. In this study, a new mechanism for online exam security using IP cameras is provided. We have, in this study, presented a reliable online examination system based on stringent online security and authentic candidacy. The centrally administered control room will monitor all activity in the approved centers through IP cameras installed in each one.

Online exam security based on wireless IP Cam is proposed to be the solution intended for worldwide certification exams, such as CCNA, CCNP, CCIE, MCSE, MCTS, MCITP, OCJP, and so on. The fake applicants have increased day after the other. This is the reason that we never come to discover some real-world hidden talents. Another disadvantage to the suggested model is the cost of implementation. If not, this model is pretty secure in terms of the testing.

Xiu I. et. al. (2023) elaborates that the digital age has taken a strong hold in our society, and its impact on education as many other sectors of human life is undeniable. This article engages in an exploration of the ways in which technology — particularly robots, as today's topic deals with them rather extensively — has altered our school curriculum over the past century. Using some key terms in searching yielded an extensive cover across three of the most relevant education databases used here, Education Resources Information center, Scopus, and Web of Science, because they are all relevant for education studies. Although interesting findings were produced, they show that, while there is promise for robotics in developing teamwork in the classroom, it is mostly used to improve individual skills.

During the last two decades, published works on socio-educational robots in education have been growing rapidly. This fact is known in the scientific world because of a clear outlook regarding the possibility of using robots as an effective educational tool and supporting their assimilation in education. The review indicates that most of the educational interventions are usually held as after-school programs and are just a negligible percentage happening in schools. Most are offered as extracurricular activities rather than school-based programs. Furthermore, most of these interventions are designed to be implemented into schools later on. This is careful but step-by-step toward the full implementation of robots into curricula. Regarding objectives, most of these interventions mainly aim at teaching the programming skills using the robots. Additionally, most the improvement interventions are centered along the STEM subjects; that includes science, technology, engineering, and mathematics. While some interventions embarked on using robotics for subjects other than STEM, the preliminary results seem promising.

It is also interesting that a huge percent of these educational interventions did not use physical robots but opted to use computer or tablet software that made it possible for the students to practice programming on their own. This may well be because the very high cost and maintenance requirements tied with commercial models placed the use of software as more viable for schools.

According to Valle M. et.al., (2008), the robotic devices become more and more integrated in life, which brings interfacing and learning issues for future robots. These problems are major for self-directed learning and safe interaction with the environment. Human beings communicate through five prime sense modalities: touch, vision, hearing, olfaction, and taste. The absence of any of these modalities leads in inadequate knowledge about the surroundings. Understanding their distinct contributions is critical for increasing sensing capabilities in future robotic systems. In robotics, tactile sensing is sometimes confused with force sensing, however this is not totally accurate. The chapter defines tactile sensing, its role in robotics, and the characteristics required for tactile sensors to accomplish tasks. This book addresses a wide range of technologies and transduction methods to enhance touch perception in robotic machinery. The slow pace with which tactile sensing developed has been attributed to the lack of tactile analogs of Complementary Metal Oxide Semiconductor (CMOS) or Charge Coupled Devices (CCD) optical arrays. The chapter describes the creation of tactile sensing arrays employing piezoelectric polymers and silicon micromachining, with the goal of creating tactile analogs of CMOS optical arrays. Section 5 discusses the first phase implementation of these arrays. The chapter finishes with a consideration of the present state of tactile sensing and the issues that need to be addressed.

Touch sensing is less commonly employed in robots than other sensory modalities like as vision and hearing, which limits their cognitive and real-world interaction capabilities. This is due in part to the complexity and distribution of tactile sensing, as well as the lack of acceptable tactile sensors. Tactile sensors can not be used so efficiently since they are still faced with physical issues such as the location, strength of tactile and wiring complexity. The notion of tactile sensing has been largely limited to measuring static interaction forces, even though real-world interactions are dynamic and not just static. Many of these sensors are designed to measure either static pressure or some kind of force thus making it hard to get information about friction, stickiness, roughness. hardness and elasticity; In practice, all contact characteristics need the use of various transduction mechanisms at the same time, such as capacitive and piezoelectric transducers. Furthermore, few studies on tactile sensing have considered system limits such as those imposed by other sensors or the robot controller, processing power, and so on. Huge numbers of touch sensors impose a strain on computing resources necessary to handle huge volumes of data, which can be alleviated or lowered by implementing distributed computing beginning at the transducer level. A systematic approach to tactile sensing can assist bridge the gap between tactile sensing and other sense modalities.

Grella F. et.al. (2021) describe that in this paper, we present a sensing device that enables safe and reactive human-robot interaction during collaborative tasks involving industrial heavy-duty robots. We designed and manufactured a handle with artificial skin that responds to the operator's touch. The pre-trained convolutional neural network is used by the tactile processing system for the discrimination between human voluntary activities and tactile images. We present the accuracy results of the fine-tuning training of the network with extra data from the handle. We also carry out the performance analysis of model inference on a board of Nvidia Jetson Nano, and compare the times of inference of two optimization models fp16 and fp32, to prove that fp16 performs better by time.

Figure 5. CAD representation of the device

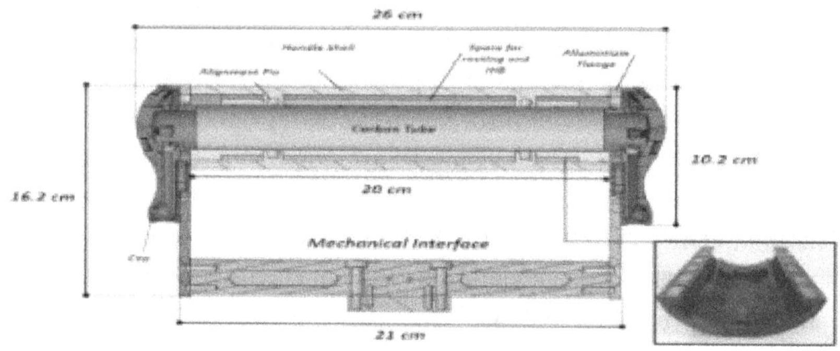

In this work, we presented a tactile sensing device that enables realistic physical human-robot interaction in industrial settings. We refined the architecture proposed to discriminate between intentional and involuntary interactions using distributed tactile sensing measurements. Finally, the classification architecture was implemented on the NVIDIA Jetson Nano, a small host PC that can be easily attached to the robot gripper. As previously indicated, the architecture will be tested against the CoLLaboratE project's use-case scenario. We are now performing additional research on sensor architecture. In addition to the force/torque sensor, we wish to provide motion orders to the robot via force data from the handles. Furthermore, researchers are working on building learning algorithms that can understand the semantics of complicated hand motions on handles.

Jolad B. et.al. (2017) describe that the study describes how to manage a robotic vehicle using voice instructions for remote operation. An ARM series microcontroller and an Android application are used to perform the needed task. The Android application communicates with the Robot's Bluetooth module (HC-05) via Bluetooth. Orders are transmitted to the robot via push buttons or voice instructions via the

Android application. At the receiving end, two DC servo motors are linked to the microcontroller and used to push the vehicle. The Bluetooth RF transmitter may take switch push or voice instructions, which are converted to encoded digital data at a suitable range (up to 100 meters) from the robot. The data is decoded by the receiver before being sent to another microcontroller, which then drives DC motors via motor driver IC to complete the necessary job. This technique outperforms RF technology in terms of communication range. The idea may be developed with IoT technology, allowing users to control the robot remotely.

Figure 6. Block schematic and description

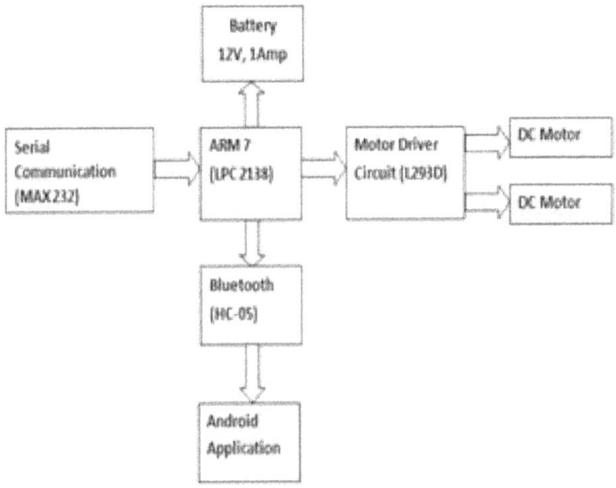

This development improves the performance of robotic cars as they can run without errors and carry out voice commands with ease. Soon, its applications can be extended to home security and military applications where the robot can receive commands from a distance while the camera systems use it for surveillance purposes. 1. Here, the objective of the project was the integration of the short-range Bluetooth module. Advanced technologies might support long-range transmission in later versions. 2. Power-saving attributes such as sleep and wake-up schedules might draw lower powers than needed. 3. Color and object detection might be supported by combining the image processing module with the robotic module. 4. Thermal camera attachment can be helpful in thermal patterns identification, hence useful when employed in military situations; it can trace enemy at the battle field. 5. Automatic Targeting System to make the robot automatically locate and shoot targets.

3. REVIEW GAP ANALYSIS

Human Invigilators have always provided maintenance to the aspects of examination supervision. However, the incorporation of robotic technologies with much advance sensors and cameras gives a new dimension to the level of examination, wherein real-time monitoring and intervention become possible.

However, existing literature does not provide a holistic understanding of characteristics, uses and even the impact of nemesis robots in the testing process. Specifically, there is a limitation on their difficult to observe behaviors such as determine the identity of participants being tested, as well as to monitor the testing process in real time. It is also clear that so many things are yet to be known about the ability and capacity of the invigilator robots about enhancing accountability and transparency through detailed reports and visual evidence of incidences. The lack of such literature goes forward to indicate why critical and empirical research should be conducted in order to understand the role of invigilator robots in changing examination environments of the future. Further research is thus pivotal to allow proper guidance of educators as well as policymakers on the impact of their activities.

New issues have arisen and deserve discussion because robots of this kind have become rather usual practice and the corresponding ethical concerns along with their negative implications have barely been discussed in the literature over the last two years. Although such technological advancements may help preserve the integrity of the exam process in many ways, the other concerns remain wide and far-reaching; these include issues like bias in algorithms, privacy breach, and lack of human presence in the entire exam process. In addition, effective implementation problems like high cost, technicalities involved, even non-readiness on part of education stakeholders might be pivotal.

And, at the same time, it is also necessary to fill up these research gaps, that the legislators, educators, and the developers can present their views based on the fruitful use of deployment of invigilator robots in the examination rooms. Such further research will evaluate merit and demerit in the adoption of the technology of invigilator robots into Olympic educational practices and then follow it subsequently.

The chapter "Advancements in Examination Systems: IP-based Robo Invigilator Technique" is related to the discussion, concerns, and issues of a wider context, which are given importance and attention in the book "Intelligent Robotic Process Automation: Development, Vulnerability, and Applications" in the following ways:

1. Automation of Processes: The book to an extent talks of impulsive tendences of applying intelligent systems in the automation of several processes including education. The chapter on IP-based Robo Invigilators is no exception to this trend as it demonstrates how the discovery of such robotic systems will help automate

invigilation work during examination thus reducing human participation which may increase efficiency rate.

2. The issue of security: Another very crucial headache when dealing with intelligent systems applying automation concerning the topics in the book is the stability of the systems. Concerning the IP-based Robo invigilator tackling the question of how the system could stand and/or counter any form of cheating, data leakage or hacking would be a relevant issue; hence, linking the chapter to the wider discussion about the vulnerabilities of the systems in the book.
3. Development of Robotic Technologies Chapter four would also fall under the development side of RPA and outline how imaging, text and communication technologies may be used in the construction of autonomous systems for invigilating examinations.

This is in agreement with the factors associated with technological progress and described in the book.

4.Usage across Different Sectors: The book most likely discusses the various uses of RPA in various sectors, and education sector is one which surely is used the most with such technologies. The Robo Invigilator method is a classic example where RPA has been implemented in the academic sector, and it strictly proves how intelligent robots can be very useful in all fields aside from production or business-related areas.

5. Ethical Issues and Human Effects: Replacing humans with robotic invigilators in the examination system generates, as in all cases of RPA application, ethical impacts that cannot be overlooked. Probably this is handled in this book on a substantive level with the description of the contexts of automation and employment and control by man.

The chapter does not work in stand-alone form but puts itself into the overall story line of the book since it not only deals with an issue that is specific, non-threatening, and security-related and developmentally relevant but also associates and ties it with other issues present within the book.

4. TECHNOLOGY BEHIND THE DESIGN

To enhance its efficiency in supervising an examination room, the invigilation robot must be of higher intelligence and autonomy. Teachers in an examination hall get bored and do not engage in light conversation that a robot would not face. The concept of "IP Cam based online exam security" in this investigation basically refers to a video-surveillance system. The video security sector is increasingly evolutionary with the merger of video technology and computer networking. New solutions that are best in video wiring are mostly computer-driven and rather easily embeddable on existing IP infrastructure. Internally, IP cameras combined with standalone video server devices generate images in the form of MPEG, MJPEG, or H.264. These images can then be accessed by any other networked device through a standard IP network interface. Fully software-based NVRs which work on normal PC hardware are capable enough to oversee and record footage. Advanced video analytics tools can detect suspicious activities or happenings automatically without requiring human eyes to keep an eye on multiple screens continuously.

4.1. Components Used

1. ***Arduino UNO***: It is a control mechanism of power for the invigilator robot, within which the charging monitoring tasks are given a human-like approach. Equipped with advanced sensors and programming, the robot can engage with the environment, identify anomalies, and uphold exam integrity. This ability to respond accurately under various conditions ensures safe and regulated exam environments while blending technological prowess with human-like interaction. In this sense, considering the fact that Adriano Uno is a robot of adaptability type, its lifelike interaction provided innovative solutions regarding modern needs of supervising, such as the invigilator robot.
2. ***Web Camera:*** The robot uses a human component by employing the use of a web camera, allowing the monitoring and supervision of a station to occur from afar with an accurate visual view. This technology allows the robot to keep efficient watch over places such as classrooms or examination halls to therefore enforce rules and dissuade misconduct among individuals. Adding the web camera to the robot elevates the accuracy and rapid response of the robot while giving that personal, physical presence that can be lost in remote operation.
3. ***Microphone:*** Another major equipment for human interaction in the invigilator robot is the microphone, as it captures spoken instructions or conversations and picks surrounding noises. It is through audio monitoring that the robot detects possible cheating or unauthorized behavior, which improves its ability to interact with them, answer their questions, and maintain a controlled examination environment.
4. ***Servo Motor:*** The servo motor is used to replicate the sense of human touch within the invigilator robot. It helps in interactions that are accurate and controlled, and servo motor technology is required for manipulation and gauging the intensity of touch because it replicates subtle nuanced movements required for such tasks. The servo motor thus responds correctly to different pressure

levels to ensure the robotic system's ability to carry out tasks which require human-like tactile sensitivity and interaction. Mainly, it is used to move the robot's mouth and handle examination papers.

5. ***Display Screen:*** The display screen is the integration feature of the invigilator robot used in order to interact with humans. The interface may be utilized by operators to control movements while viewing live video streams or retrieving relevant information about the robot. Using the interface through touch-based interaction allows the users to navigate menus, change settings, or view notifications. It is actually the tactile interface that enhances interaction and provides smooth collaboration between human and robot, especially in monitoring or surveillance tasks.
6. ***Wheel:*** It plays a very important role within the mobility of the invigilator robot and its human interaction. They allow the robot to move in a smooth manner within an allocated area, and with it, proactive monitoring will be made during exams. Using wheels will allow the robot to move freely, to reach destinations quickly, and to respond rapidly to any unusual activities. The robotics invigilator assisted in putting apprehension to rest while at the same time providing the much-needed exam integrity with its presence, which was human-like. This helped create a safe as well as controlled testing environment that would ensure maintenance of exam integrity.
7. ***Speaker:*** Through gentle yet accurate movement and soothing voice, the invigilator robot creates an environment for which test-takers are given care and oversight. Thus, there is created an impression of security and comfort, which eventually promotes a testing experience that is in concert and agreeable with all parties involved.

4.2. Examination Process

In the current examination procedure, a three-set paper is set before the examination. Among these three sets, one set is prepared by a moderation teacher while the moderation and finalizing of the paper occurs. These set papers are issued during the examination session. A third set of questions, for use as a backup, is also prepared as a set B. The three question-making teachers and the moderating teacher are given the complete syllabus at every stage of this process. But with the introduction of our invigilation robot, all these tasks are automated. The invigilator robot is given the complete syllabus and has the option of emailing it to the teachers. The compilation of the questions into one paper is sent back to the robot for moderation hence setting up the final paper. The whole process is carried out online as shown in Figure 7.

Figure 7. Process of Examination System using our proposed design

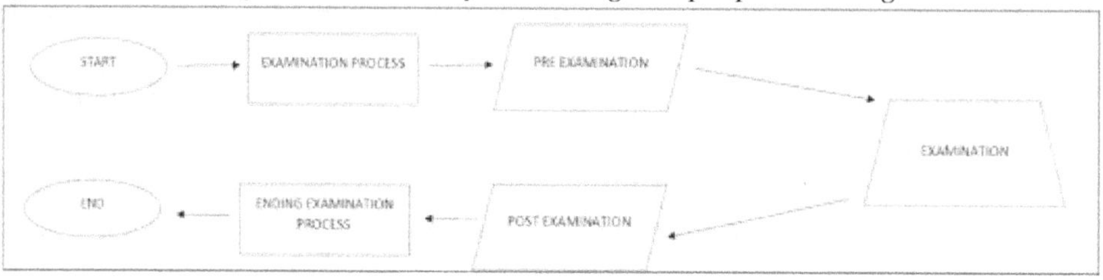

5. OPERATION DURING THE EXAMINATION PROCESS

5.1. Pre-Examination

The teacher and students arrive in class together twenty to twenty-five minutes before the exam, bringing in all the materials for the exam. Any instructions at the very least are handed out before the exam. Most of the time, there will be some late-comer students asking the teacher for their exam paper, and the teacher will often consent to giving it to them. However, it introduced a new protocol whereby the robot would dish out the exam paper 15 minutes before the time indicated to begin the exam. Dishing out continues to late comers at least 30 minutes past the time the exam had started. No one else gets a paper after that. One who arrives after this time will not get the exam paper.

To achieve this, the robot is provided with detailed information regarding the students. When a student reports to class, the robot captures the face of the student and subsequently flashes the student's name along with all other details on the monitor. If everything is correct, the student confirms and clicks the "OK" button. Subsequently, it prints all relevant details that are the student's name, roll number, and registration number, and then leaves space for the signature of the student.

5.2. During Examination

With respect to the said examination, the student puts his or her name, roll number, and registration number on the paper and, besides that, appends his or her signature. The design now under discussion would pre-print all the details on the paper. Hence, there is no need to write all the details to the paper at the student end. Rather the students should put their signature on the paper and write the roll number in the OMR portion. If a child requires to withdraw for water or a restroom break, they raise their hand, and the robot goes and meets them. The robot shows options on the screen as far as either access to water or the use of a restroom is concerned. If the child requires the former, permission is accorded; if the latter, then the face is scanned to verify, and after scanning, permission is granted. Every detail scanned is documented.

The invigilator robot patrols the room using its wheels and four-eyed vision system, making it easy to monitor for instances of cheating. In case the cheating is observed, the robot issues three warnings and then seizes the paper of the concerned student. If the student does not desist in cheating despite the warnings, his or her paper is canceled. Furthermore, the robot records the entire exam duration to be used as evidence.

Figure 8. Process of Examination System using our proposed design

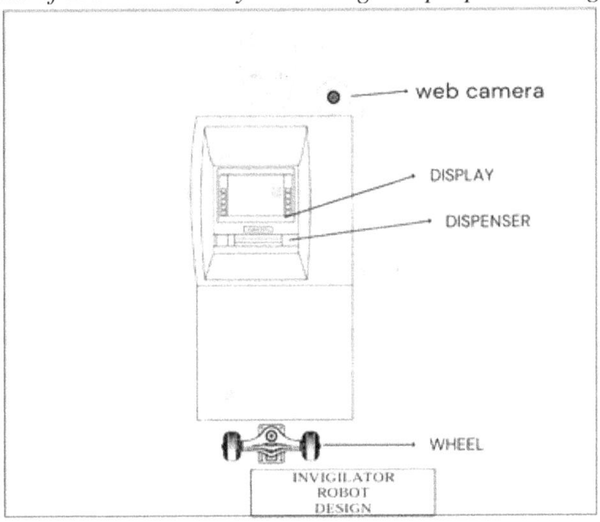

An artistic view of the invigilator robot in Figure 8, the head of the robot is at the top where a webcam captures live video, and any printed instructions and announcements are posted. A dispenser mechanism is used to dispense and collect the exam papers at the entrance. The bottom part of the robot is made with wheels for movement.

5.3. Post-Examination

Many students submit the paper before the end of the examination. So, the teacher started to take this before the examination, and when the exam was finished, they took all of the paper. And they count all the papers and arrange the papers of the students according to their roll number. Our robots also work like this. He collects the papers one by one as soon as the examination gets over according to their roll number. There is a roll displayed by the robot, and as per this roll, the students submit their papers. As per the student, if he doesn't hand them in on time, then he skips this paper of the student and goes to the next one. After all rolls, I again display the skipping students' rolls and gives them also ten seconds to present the paper. On the second time also he is showing the roll, and then if anybody skips, he doesn't take their paper.

5.4. Ending Examination Process

The papers are then forwarded to the examination cell after the exam wherein they sign it. The papers are then forwarded to the respective teachers for marking. The papers, after marking, get brought back to the examination cell. All the papers are scanned by the robots and electronically forwarded to the examination cell before being submitted physically. The paper then gets distributed to the teachers for marking purposes. After evaluation, the grade lists are sent back by the teachers to the robots. Finally, the papers are submitted to the examination cell again.

6. STRENGTH AND WEAKNESS OF THE CHAPTER

The Chapter Offers Several Strengths:

Increased Efficiency: The process of automation of invigilation increases the consistency of the monitoring of exams and reduces human staff requirements, mainly in large testing environments.

Reduced Human Error: Robots can reduce human error opportunity in exposing such cheating or suspicious activities by providing immediate analysis and detection of all that usually goes unnoticed by human invigilators: unapproved head movement and communication.

Scalability: All these systems are scalable and can be applied to any number of testing environments within the education and professional worlds. It could be used for multiple exam rooms, or even remote invigilation, with little human intervention.

Long-term Cost Saving: The major capital cost is going to be the negative factor, but one could drastically reduce invigilation costs over time, making the IP-based robo invigilator a cost-effective solution in the long run.

Real-Time Monitoring and Reporting: It can generate data and reports in real-time so that appropriate action could be undertaken by the authorities immediately after identifying unethical activities, thus upholding the integrity of an examination.

This chapter brings to the attention of the reader the manner in which robotic automation can address some of the challenges traditionally inherent in the control of an examination. There is great potential for a long-term solution with regard to educational institutions and professional testing centers in this modern, scalable approach.

Chapter “Advancements in Examination Systems: IP-based Robo Invigilator Technique” of the book Intelligent Robotic Process Automation: Development, Vulnerability and Applications describes the development of robotic systems for better exam invigilation with a special focus on the minimization of human error and manpower. The weaknesses in this approach are:

Technological Limitations: Advanced algorithms and hardware are significantly determining factors of the accuracy and efficiency of invigilation. The system may not catch faint cheating behavior or may be unable to function effectively within a dynamic exam environment when real-time human judgment is needed, such as in the case of an instance of a robotic invigilator.

Ethical and Privacy Considerations: The use of AI-based monitoring systems in exam halls brings up various ethical and privacy concerns about the students' privacy and the ethically binding nature of permanent surveillance.

High Up-front Costs and Maintenance: Although robotic invigilators may save human resources in the longer term, their high up-front costs plus maintenance costs and the complexity of integrating such systems in existing examination frameworks is a significant challenge.

These limitations indicate that although promising, robotic invigilation, as above, requires considerable refinement and balancing of technological, ethical, and financial considerations before it can replace human invigilators.

7. CONCLUSION

Robotic invigilators is the milestone technology and hence supervision of an examination. Its impact will include monitoring video analysis in real-time facial recognition that ensures control of examinations do not take place and assessment surroundings are not compromised. Preliminary results are promising for the reduction of bias as well as error often associated with traditional methods of invigilation. But then, it should be highlighted that there are drawbacks like privacy concerns of students, technical malfunctions, and lack of the psychological judgment of the human being. In order to achieve the optimum balance between efficiency and emotional intelligence, there should be an integration of an invigilator robot with its human supervisor. In summary, although the invigilator robots offer promise as valuable assets in exam settings, their implementation does require careful consideration of ethical concerns, rigorous testing, and ongoing human oversight to maximize potential benefits and mitigate adverse impacts.

ACKNOWLEDGEMENT

We would like to thank JIS College of Engineering AICTE IDEA Lab for making this research possible.

REFERENCES

Abdullahi, Z. H.. (2015). Mobile robot voice recognition in control movements. [IJCSEE]. *International Journal of Computer Science and Electronics Engineering*, *3*(1), 11–16.

Alexenko, T., Biondo, M., Banisakher, D., & Skubic, M. (2013, March). Android-based speech processing for eldercare robotics. In *Proceedings of the companion publication of the 2013 international conference on Intelligent user interfaces companion* (pp. 87-88).

Arocena, I., Huegun-Burgos, A., & Rekalde-Rodriguez, I. (2022). Robotics and Education: A Systematic Review. *TEM Journal*, *11*, 1.

Attaran, M. (2023). The impact of 5G on the evolution of intelligent automation and industry digitization. *Journal of Ambient Intelligence and Humanized Computing*, *14*(5), 5977–5993. DOI: 10.1007/s12652-020-02521-x PMID: 33643481

Ayoobi, H., Kasaei, H., Cao, M., Verbrugge, R., & Verheij, B. (2022). Local-HDP: Interactive open-ended 3D object category recognition in real-time robotic scenarios. *Robotics and Autonomous Systems*, *147*, 103911. DOI: 10.1016/j.robot.2021.103911

Bakker, P., & Kuniyoshi, Y. (1996, April). Robot see, robot do: An overview of robot imitation. In AISB96 Workshop on Learning in Robots and Animals (Vol. 5, pp. 3-11).

Clodic, A., Fleury, S., Alami, R., Chatila, R., Bailly, G., Brethes, L., . . . Montreuil, V. (2006, September). Rackham: An interactive robot-guide. In ROMAN 2006-the 15th IEEE International Symposium on Robot and Human Interactive Communication (pp. 502-509). IEEE.

Dahiya, R. S., & Valle, M. (2008). *Tactile sensing for robotic applications*. Sensors, Focus on Tactile, Force and Stress Sensors.

De Barrie, D., Pandya, M., Pandya, H., Hanheide, M., & Elgeneidy, K. (2021). A deep learning method for vision based force prediction of a soft fin ray gripper using simulation data. *Frontiers in Robotics and AI*, *8*, 631371. DOI: 10.3389/frobt.2021.631371 PMID: 34113655

Deepak, A.. (2024). Image Processing based Robotic Car for Agricultural Ploughing using Machine Learning Approach. *International Journal of Intelligent Systems and Applications in Engineering, 12*(2s), 718–724.

Erdogan, R., Saglam, Z., Cetintav, G., & Karaoglan Yilmaz, F. G. (2023). Examination of the usability of Tinkercad application in educational robotics teaching by eye tracking technique. *Smart Learning Environments, 10*(1), 27.

Gonzalez, F. (2023). Learning robot differential movements using a new educational robotics simulation tool. *Education and Information Technologies, 28*(9), 1–24. DOI: 10.1007/s10639-022-11433-6 PMID: 36846490

Grella, F., Baldini, G., Canale, R., Sagar, K., Wang, S. A., Albini, A., . . . Zoppi, M. (2021, December). A tactile sensor-based architecture for collaborative assembly tasks with heavy-duty robots. In 2021 20th International Conference on Advanced Robotics (ICAR) (pp. 1030-1035). IEEE.

Jolad, B., Arora, M., Ganu, R., & Bhatia, C. (2017). Voice controlled robotic vehicle. International Research Journal of Engineering and Technology (IRJET), 4(06).

Kannan, K., & Selvakumar, J. (2015). Arduino based voice controlled robot. [IRJET]. *International Research Journal of Engineering and Technology, 2*(01), 49–55.

Krishnan, M. G., & Sankar, A. (2022). Image space trajectory tracking of 6-DOF robot manipulator in assisting visual servoing. *Automatika (Zagreb), 63*(2), 199–215. DOI: 10.1080/00051144.2021.2022889

Mahmood, F., Arshad, J., Ben Othman, M. T., Hayat, M. F., Bhatti, N., Jaffery, M. H., Rehman, A. U., & Hamam, H. (2022). Implementation of an Intelligent Exam Supervision System Using Deep Learning Algorithms. *Sensors (Basel), 22*(17), 6389. DOI: 10.3390/s22176389 PMID: 36080848

Mehmet, Ç. A. V. A. Ş., & Emin, A. Ğ. R. A. L. İ. (2020). V-REP robotik simülatör ile robot kol simülasyonu. *Fırat Üniversitesi Mühendislik Bilimleri Dergisi, 32*(2), 435–444.

Rashid, H., Ahmed, I. U., Osman, S. B., Newaz, Q., Rasheduzzaman, M., & Reza, S. T. (2017, January). Design and implementation of a voice controlled robot with human interaction ability. In *Proceedings of the International Conference on Computer, Communication, Chemical, Materials and Electronic Engineering*, Rajshahi, Bangladesh (pp. 26-27).

Suman, M. C., & Prathyusha, K. (2012). Wireless camera based online examination security. [IJERA]. *International Journal of Engineering Research and Applications, 2*, 1432–1435.

Xie, Z., Liang, X., & Roberto, C. (2023). Learning-based robotic grasping: A review. *Frontiers in Robotics and AI*, *10*, 1038658. DOI: 10.3389/frobt.2023.1038658 PMID: 37082744

Xue, J., Wu, W., & Cheng, Q. (2023). Intelligent invigilator system based on target detection. *Multimedia Tools and Applications*, *82*(29), 44673–44695. DOI: 10.1007/s11042-023-15474-y PMID: 37362652

Zhang, S., & Luo, J. (2022, December). Construction of Autonomous System of Educational Robot Under the Background of Artificial Intelligence Taking the Mobile Educational Robots toward Performing Intelligent Invigilation as an Example. In 2022 3rd International Conference on Artificial Intelligence and Education (IC-ICAIE 2022) (pp. 810-816). Atlantis Press.

Chapter 15
Navigating the Future:
Challenges and Opportunities in Intelligent RPA Implementation

Mohan Kumar
https://orcid.org/0000-0002-5171-0892
SGT University, Gurugram, India

Manju Dagar
https://orcid.org/0009-0002-5677-7647
SGT University, Gurugram, India

Mohit Yadav
https://orcid.org/0000-0002-9341-2527
O.P. Jindal Global University, India

Tanupriya Choudhury
University of Petroleum and Energy Studies, India

Thi Minh Ngoc Luu
https://orcid.org/0000-0002-5972-7752
International School, Vietnam National University, Hanoi, Vietnam

ABSTRACT

The Chapter investigates the transition from RPA to IPA and its revolutionary impact on businesses in the digital age. The study carefully analyzes the key challenges and opportunities that have accompanied the adoption of IPA in organizational workflows. The report exploits a detailed literature review to establish major trends and business issues when using IPA systems. It addresses the organizational, technological, and human-centric challenges that relate to leadership matters, resistance to change management, and gaps in skills for handling automation technologies.

DOI: 10.4018/979-8-3693-4365-4.ch015

The findings suggest that although IPA can greatly improve operational efficiency through automated routine tasks, it poses challenges in role redefinition at work, in higher-order talent enhancement for employees, and in countering opposition to automation. However, for IPA adopting AI technologies such as machine learning and natural language processing will provide new ways to manage complex activities and streamline data-driven decision-making.

INTRODUCTION

Among the most revolutionary technologies to have emerged lately and enabled enterprises to automate dull, repetitive, rule-based operations that are predominantly human-centric is robotic process automation. RPA utilizes software bots to automate routine work, such as data entry, invoice processing, and customer support, by mimicking human-like interactions with digital platforms (Siderska et al., 2023This technology was adopted very fast because it would help reduce operational costs, increase accuracy, and free up human resources for other strategic and more beneficial tasks (Lopes et al., 2021Initially, RPA capabilities were limited to simple structured, and rule-based processes. Unstructured input could not be processed, nor could it work outside of the set parameters to make intelligent decisions. In order to overcome these, RPA has grown into Intelligent Process Automation, which is known today. IPA contains various sophisticated AI technologies such as machine learning, NLP, and DL, which enable far better decision making and automate difficult cognitive tasks (Siderska et al., 2023).

The evolution from RPA to IPA is a major technological shift where the robots are able to not only execute tasks, but also learn from data, adapt to new situations and predict outcomes with information already existing (Berruti et al., 2017). It lets companies automate a greater range of operations, which involves the use of discernment, structured data analytics, and human emotional interaction, such as customer service or even taking financial decisions (Richardson et al., 2020)Importance of Automation in Modern Business Landscape

Automations now form a very important feature of contemporary corporate activity, keeping up with the imperative for efficiency, precision, and scale in highly complicated environment s. In any case, RPA is definitely one of the most important technologies as they enable an organization to automate routine, rule-based tasks, releasing human employees to concentrate more on higher-level activities (Siderska et al., 2023). The penetration of AI in the RPA-based activity stream has strengthened automation in assuming duties increasingly complex and unstructured as well (Richardson et al., 2020This development has propelled automation not only as a way to gain operational efficiency but as an innovation and competitiveness driver.

One major advantage of automation is that it reduces the cost of operations while still improving the accuracy. Automating routine processes gives firms the potential to significantly lower the chances of human mistake, hence improved service quality and customer satisfaction (Kholiya & Kumar, 2023a)Automation also allows for operating on a 24/7 basis minus the constant reliance on humans-this is highly applicable to industries, such as banking, healthcare, and customer care, in which bulk work is prominently done (Aryawati et al., 2023)IPA enhances decision-making through AI technology, including predictive analysis and machine learning. These abilities make it possible to analyze massive sets of data in real time and provide knowledge that could be the basis for strategic recommendations as well as business process improvement (Süße et al., 2019). This is especially important nowadays in such fast-developing digital economies where an enterprise has to respond rapidly to changing market circumstances. Automation plays a key role in efforts of digital transformation. Organisations are embracing automation at a tremendous pace in pursuit of operational optimization, agility, and enhanced productivity to fuel their competition quest (Lievano-Martínez et al., 2022)Automation, particularly through Intelligent Process Automation (IPA), is acquiring an increasingly strong imperative for businesses looking to prosper and transform against this new path.

This review article examines the emerging RPA Landscape and its transformation to IPA. It brings to the fore the major challenges that include organizational, technological, and human-centric issues that those deploying intelligent RPA systems are faced with. The paper will give consideration to opportunities that AI integration opens up, specifically how advanced technologies such as ML and NLP can make companies more efficient and automated. This essay will attempt to provide a comprehensive understanding toward further research and practical implementations of intelligent RPA systems.

THEORETICAL FRAMEWORK

Historical Evolution of Robotic Process Automation (RPA): The history of Robotic Process Automation (RPA) reflects its evolution through clearly defined milestones in technological invention, from the original, basic process automation methods in the late 20th century to the sophisticated, intelligent systems that are prevalent today. This chapter outlines the key events in the history of RPA-from its infancy to its growth and transformation into an integral part of digital business strategy.

Early Automation: Origination of RPA: Concept of automation traces its origins back to the Industrial Revolution that began during the 18th century. However, automation of the process of business activity came into focus somewhat later in 1980s and 1990s with the help of workflow automation tools. The initial efforts of automation were geared towards manufacturing processes where machines were used for the purpose of automating physical operations (Siderska et al., 2023). In the business world, automation had begun from scripts and macros, simple software that involves automation of repetitive tasks across the software applications (L. P. Willcocks et al., 2015). By the late 1990s, corporations started gradually accepting software that would automate basic business operations, such as data entry and file maintenance. Among these technologies were screen scraping and workflow automation, which helped precede RPA. Screen scraping, a method that was invented in the 1990s, enabled (L. P. Willcocks et al., 2015)applications to read data off a display screen through simulation of human activity interacting with the user interface (Moreira et al., 2023). Such an attempt at mimicking human actions in digital environments marks the beginning of what later became known as RPA.

Emergence of RPA: Emerged as an independent technology during the early years 2000s. As such, the term "Robotic Process Automation" was used for referred to software which could automate rule-based jobs by performing human behaviors across multi applications and systems (L. P. Willcocks et al., 2015). The RPA software starts giving the flexibility to interact with many systems simultaneously to automatically perform operations like application login, data entry and file transfer (Geyer-Klingeberg et al., 2018). One of the prime features of the early RPA technologies was that they were non-intrusive. Unlike other automation solutions that require considerable changes in the underlying IT structure, RPA operates at the level of the user interface, making it a very adaptive solution. The first generation of RPA tools, starting from Blue Prism, UiPath, and Automation Anywhere, have been used as an enabler of its widespread adoption in various sectors.

Adoption and Growth: 2010s: RPA really gained popularity in the 2010s. Its adoption was initially driven by advancements in AI, ML, and cloud computing. At the later part of the decade, RPA expanded to include AI and cognitive computing capabilities in addition to basic job automation. Definitely, one of the stand-out developments is its capability to integrate with AI technologies that can be used to perform decision-making tasks. In reality, all organizations-even those in most industries-like banking and finance, healthcare, and insurance-have, on average, started to implement RPA for a number of time-consuming, paper-based tasks such as customer onboarding, claims processing, or invoice management (Lacity & Willcocks, 2018). RPA becomes the component of digital transformation plans that intend to increase savings in costs, increase accuracy, and provide efficiency in operations (K. K. H. Ng et al., 2021).

As RPA technologies continued to evolve, the front-office work scheduled with RPA started incorporating simple back-office work and expanded to customer services, support, etc. Incorporation of RPA with AI has led to Cognitive RPA, such as helping systems decipher unstructured data in emails and social media messages, thus portraying Intelligent Process Automation, which eventually is called.

Intelligent Process Automation: Late 2010s to Present: A new turn in the late 2010s and early 2020s started when NLP and machine learning technologies were integrated into the systems used in RPA, giving birth to Intelligent Process Automation or IPA (Aryawati et al., 2023). IPA is the innovation of RPA in practice with AI and other sophisticated technologies that enable automation systems to be able to handle increasingly complex tasks that would require reasoning, learning, and adaptation (Asadov, 2023). The shift from RPA to IPA has been compelled by the need for automation systems that can perform repetitive jobs, but also adapt to the change of events and make decisions after analyzing data. Cognitive automation, where AI will mimic human choice-making processes, was identified in some research as the latest factor of IPA-detectable by da Silva Costa et al., in 2022. Today, IPA systems could automate an incredibly vast array of operations - from the most basic, rule-based processes to intricate workflows that require interpretation and decision-making.

Global Adoption and Impact: Global adoption of RPA and IPA is changing the very fabric of industries since organizations around the world are adopting the technologies to boost efficiency in operations, reduce costs, and strengthen competitiveness. As a 2021 report by Deloitte reveals, more than 50% of worldwide organizations have implemented RPA, while thousands more were looking into how IPA can be adopted in more complicated AI-driven processes (Parsons et al., 2023). Expansion of the market proves an increase in the adoption of automation. According to industry analysts, the global RPA market was $2.7 billion in 2021 and is to rise to $13.7 billion by 2030. It is a rapid growth due to the resultant high demand for the optimization of operations and the emergence of advanced automation solutions on a growing scale in industries that span from banking and manufacturing to healthcare.

With the global expansion of RPA, the additional regulatory challenges take places. In sectors as highly regulated as finance and healthcare, where data privacy and security are particularly sensitive, there is now a call for governance frameworks (Sarker, 2022). Furthermore, RPA systems tend to deal with sensitive personal information; hence it is mandated that organizations adhere to data protection legislation such as GDPR from Europe and HIPAA in the United States. What all this really means is that there needs to be close supervision of such automated systems and upgrade frameworks as technologies advance. Since automation is now being adopted more frequently in corporate operations, there have always been debates over the ethical implications surrounding an increased use of AI. Among such debates are issues such as displacement in employment, issues of algorithmic biases, and the more enormous scope associated with labor market influences. This is the general policy and strategy being discussed today as far as the future of work is concerned

(Inam et al., 2024). For instance, though RPA can carry out routine jobs, there is an argument regarding the level of permanent job loss with automation increasing. And IPA systems, which often produce decisions that rely on high dimensions of information tend to feed in and exacerbate biases unless monitored well or trained on biased data (Wilson & Daugherty, 2018). This led to the demand for responsible AI practices and ethical standards on the implementation of automation technology (Binns, 2018).

CHALLENGES FOR INTELLIGENT RPA SYSTEMS

Organizational Challenges for Intelligent RPA Systems

Challenges in Implementation with Leadership and Management: Intelligent RPA systems require powerful leadership and proper management in their implementations. Many organizations experience management issues, that is, a defined automation strategy, aversion to change, and poor strategic alignment between IT and business goals (Agostinelli et al., 2020). However, when there is a lack of leadership buy-in, the IPA might be adopted in bits and pieces, and then many organizations find themselves applying them unevenly across different business units quite frequently (Kholiya & Kumar, 2023b). Thirdly, insufficient full understanding of IPA technical details would lead to poor decisions and resource allocation (Zeiser et al., 2022). Business settings are dynamic. In such fluid business settings, issues concerning leadership are also compounded because the development of competencies in IPA technologies calls for rather flexible strategic plans (Berruti et al., 2017).

Governance and Task Redesign: there is a need for effective governance to ensure the seamless performance of Intelligent RPA systems. Many organizations face difficulties in defining extensive governance frameworks that define roles, responsibilities, and workflows separately both for human staff and RPA bots (Siderska, 2020). Effective governance requires consistent policies, communication protocols, and decision-making frameworks to ensure homogenous task design and orchestration (Furman & Seamans, 2019). The redesign of work activities is problematic because standard procedures must be optimized for automation, often forcing the organization to revisit its frameworks of operations (Jurdi et al., 2022). Without proper governance in place, RPA projects can quickly become a mess, leading to waste, duplication, and missed opportunities for further improvement (Lievano-Martínez et al., 2022).

Compliance and IT Security Challenge: The considerations of compliance and IT security are a big challenge in implementing intelligent RPA systems. It being the case that IPA deals with confidential information, thereby the companies will have to comply with multiple legal requirements for implementation compliance, like GDPR and HIPAA, which creates complexity in the process of implementation (Veale & Zuiderveen Borgesius, 2021).Security issues that will be encountered include the security from hacking, unauthorized access, as well as ensuring secure integration of RPA technology with the available systems (Al-Slais & Ali, 2023). As companies fail to define specific security controls and auditing practices, they are at risk of vulnerability and increased chances of breaching data and compliance breaches as well (Murhekar et al., 2021). In this regard, risk mitigation and successful sustainability of IPA projects could be achieved by integrating robust IT security solutions with a compliance-based culture (Meegahapola et al., 2023).

Technological Risks for Intelligent RPA Solutions

One of the big technological hurdles Intelligent RPA has is inputting and relying on good quality data through machine learning. The major dependency for intelligent automation today is based on AI technologies, more so the aspect of machine learning; it guarantees clean, correct, and representative data use (Agostinelli et al., 2020). Low data quality may supply wrongly created models. Then, these often bring about unreliable automation results and subsequently defective decision-making processes (Geiger et al., 2020). In addition, implementation of machine learning in the robotic process automation system offers some problems in the training of models which requires huge amounts of data, high computational power, and continuous time-updates because the business environment changes over time (D. T. K. Ng et al., 2021).

Software and infrastructure integration is one of the significant challenges in the intelligent RPA systems. It is challenging for most organizations to integrate RPA technology with their IT infrastructures. RPA systems need to be interoperable with legacy systems and still to be able to adapt to new innovations (Kholiya & Kumar, 2023a). The non-standardization of the platform is one of the factors, which deepen complexity that resulted in integration problems that can delay deployment and increase costs ((Lievano-Martínez et al., 2022)). Another challenge is more pronounced in scalability and system security, mainly because organizations aim at deploying RPA more widely in different departments or geographically (Flechsig, 2021). Ensuring high IT security and data protection throughout the process of integrating RPA with enterprise infrastructure is critically challenging due to differing compliance standards from different regions (Kedziora & Hyrynsalmi, 2023).

Technological challenges are related mainly to issues with data quality and the integration of machine learning, software, and infrastructure along with their integration challenges. Deployment on a mass scale would, therefore require concerted efforts in data management, software standards, and cross-platform interaction.

Trends for Human-Centric Challenges for Intelligent RPA Systems: Redefining Job Roles, Employee Skills Development, and Resistance to Automation

Such employee-related issues come into play with the redesign of job functions, the development of staff competence, and the rejection of automation as the implementation of smart RPA introduces dramatic shifts in the character of job definitions and staff. As RPA technology increasingly cuts deeper into the businesses, people modify the traditional job function they carry out at work to more analytical and inspirational tasks. This change also creates a state of confusion since employees have to foster the development of high-order cognitive abilities and actualize decision-making skills to cater for their changing roles (Herm et al., 2021). Redefinition of job positions will form human workforces needed to attend to creativity and problem-solving jobs, and RPA systems basically undertake routine or rule-based jobs (Berruti et al., 2017). This role reinterpretation enhances the productivity of the organization and promotes innovation since human employees are relieved to focus on work that requires more strategic thinking.

With role reinterpretation, there is an emerging demand for people to learn skills so that they remain relevant in an increasingly automated world. Over and above Intelligent RPA, technical and digital skills must be developed to work effectively with such systems. Training and upskilling programs enable the employees with the capabilities to manage and optimize automations effectively (Kholiya & Kumar, 2023a). Research findings conclude that in this present age, constant skill building is required by a company for its workforce such that they can master data analytics, machine learning, and systems management (Schulte et al., 2024). Hybrid human-RPA systems, fast growing in integration, underscore the need for both artificial intelligence and automation technologies to enhance collaboration among humans and machines (Battling et al., 2022).

A vital limitation for RPA adoption is resistance to automation among employees. The majority of employees feel that automation risks their job security, which further propels anxiety about losing a job and moves towards technophobia over the progress of technology (Moulaï et al., 2022). Organizations should have effective open communication and let the people understand how automation will help and support the work of humans instead of replacing them. In fact, telling positive stories about automation, encouraging professionalism by up-skilling, and launching trust-building initiatives can be helpful in alleviating this resistance (Jha et al., 2022). The transition process and employee anxiety should be eased by implementing these methods, which leadership should pay special attention to (Kedziora & Hyrynsalmi, 2023).

PROSPECTS OF INTELLIGENT RPA SYSTEMS: AI, NLP, ML, DL, AND PREDICTIVE ANALYTICS

This integration of AI into the RPA system offers unprecedented opportunities to enterprises. Moving beyond IRPA systems, integrating RPA with AI technologies like Natural Language Processing, Machine Learning, Deep Learning, and predictive analytics have revolutionized automation by enabling the execution of systems that are capable of a more complexly cognitively taxing task. There are many opportunities here for increasing productivity efficiency and productivity in deciding improvement in the operation.

AI in RPA: In order to automate business processes beyond simple repetitive, rule-based operations, AI and RPA have been integrated to great success. The application of AI in RPA allows it to manage unstructured data, recognize patterns, and make real-time decisions, as noted by Kaur et al. in 2022. This capability enables businesses to implement processes that require high-order cognitive intelligence, such as answering service-related queries and fraud prevention, thus making the entire process much more streamlined (Manning et al., 2021) Additionally, AI-based RPA systems are capable of learning from data and improving functionality step by step, a significant difference compared to traditional RPA, which relies purely on the pre-defined rules (Manning et al., 2021)Natural Language Processing in RPA: Natural Language Processing, a subcategory of Artificial Intelligence, aids the robots in reading, understanding, and responding to human language efficiently. With NLP for instance, RPA systems can automate contact-intensive processes such as reaching out to customers, processing documents, and analyzing sentiments (Devlin, 2018). The integration of NLP with RPA allows organizations to automate the processing of unstructured data, such as emails, reports, and social media content, thereby improving customer services as well as internal processes (Chowdhury et al., 2021) This integration supports RPA bots in understanding textual data, classifying it, and developing high-utility insights, which is very important in dealing with client request types and paperwork (Rath et al., 2022).Machine Learning (ML) in RPA: Machine learning allows RPA systems to do more than just automate monotonous tasks as they are equipped with the ability to learn from data and make intelligent decisions. Machine learning algorithms enable them to study history and predict the outcomes; thus, the robotic process automation system becomes more responsive and intelligent (Sarker et al., 2021). With machine learning and robotic process automation, the bots can be taught automatically to identify anomalies, forecast future trends, and make specific suggestions; thus increasing the effectiveness of corporate activities (T. K. Nguyen et al., 2021). The ability of machine learning in self-learning enhances the malleability of RPA systems to make them learn new business scenarios with no significant opportunities in further retraining (Yapıcı & Akcayol, 2022).

Deep Learning (DL) in RPA: Deep learning is a kind of machine learning which focuses on building neural networks on the pattern of imitating and replicating the human brain. Such creation of neural networks enables the systems to identify patterns, images, or voices with great accuracy. The incorporation of deep learning into robotic process automation further opens up the horizon of automation to even more complex areas, such as image processing, voice processing, and predictive modeling. An example is in healthcare by incorporating robotic process automation systems that have deep learning capabilities for interpreting medical images to enhance the precision of diagnosis and alleviate some burden on healthcare providers (Esteva et al., 2021). Deep learning could help classify fraudulent transactions in the financial sector by detecting patterns that imply fraudulent activities (Goodfellow et al., 2020). Application of deep learning in RPA is ubiquitous, and it can allow for automation in a great deal of industries.

Predictive Analytics in RPA: Powered by artificial intelligence and machine learning, predictive analytics will enable the robotic process automation systems to analyze the previous data so that a smart forecasting can be done in the future. It is particularly useful in finance, retail, and manufacturing where predictive capabilities are essential for many organizations to make decisions (Shmueli & Koppius, 2011). Predictive analytics allows RPA to predict customer behavior, optimize supply chains, and update schedules for machinery maintenance in cutting down on operational costs and enhancing efficiency (Khan et al., 2020). The capacity to integrate predictive analytics with RPA enables firms to automate decision-making processes, thus ensuring that decisions are time-and-data-driven (Wong et al., 2022).

Future of Intelligent Robotic Process Automation Using AI, NLP, ML, DL, and Predictive Analytics Future Opportunities The adoption of AI, NLP, ML, DL, and predictive analytics in RPA systems has set a wide future scope. The intelligent RPA system can thus transform sectors as it can automate the activities that it feels are only possible by human intervention. The improvements in NLP enable the organization to automate jobs related to customer services. Also, improvements in ML and DL boost the decision-making abilities of a system as these algorithms enable systems to learn and improve with time (Duan et al., 2019). Predictive analytics enhances the intelligence of RPA systems, aiding businesses in taking decisions based on data-driven insights that are true-to-fact.

Intelligent in RPA systems may manage the inquiries of the client, process loan applications, and detect fraud in banking services, thus enhancing customer experience and operation efficiency (Zhao et al., 2020). In healthcare, RPA can automate entry and scheduling of patient data and assist with disease diagnosis through DL algorithms (Hamet & Tremblay, 2017). For the industrial firms, predictive analytics improves operations within the supply chain and repair equipment on time to avoid losses in both time and costs incurred (Lee, 2020). AI, NLP, ML, DL, and predictive analytics are major inclusions in RPA solutions, implying a great jump in the level of capabilities of automation. These technologies allow for smart RPA systems to execute complex activities, improve decision-making quality, and provide predictive insights-all creating new opportunities across industries. Companies can expect a yield in terms of increased efficiency, reduced operational costs, and improved customer support as the use and integration of smart RPA systems become more widespread. But with all these advantages come the difficulties associated with the introduction of such technologies, including the upskilling of labor and the abhorrence of automation. Future of RPA: Revolutionizing Potential of Intelligent Technologies

IMPROVEMENT OF BUSINESS PROCESSES

The application of Intelligent Robotic Process Automation can have the potential to make improvements in business operations across a multitude of fields. Intelligent RPA for the banking and healthcare industries and for manufacturing basically brings about an entirely new level of operational speed because it can maximize human capacity on other relevant activities and automate repetitive ones. In the financial sector, RPA has streamlined different processes, including data entry, invoice management, and compliance reporting, into faster transactions and minimal errors (Arif & MacNeil, 2022; Yao, 2024). By integrating RPA with other AI technologies like machine learning, the system allows the system to make decisions in real-time, thereby enhancing the overall effectiveness of financial operations (Arif & MacNeil, 2022).

Another sector where RPA has been impacting the business environment is in the manufacturing sector. Here, RPA automates various supply chain management, inventory tracking, and quality control processes in the manufacturing environment. If integrated with predictive analytics, RPA systems can predict demand patterns in advance and optimize schedules of production, reducing down time and overall enhancing productivity (Xia et al., 2021). In this case, human collaboration with RPA allows employees to focus on the creative facets of innovative problem-solving, while RPA will take care of mundane, rule-based activities such as data analysis and process optimization (Zhao et al., 2020). This way, it ensures human expertise is maximized and that machine efficiency is maximized, thus promoting innovation in manufacturing (Liu et al., 2021).

The integration of RPA into healthcare operations is revolutionary. It helps healthcare professionals focus more on proper care by streamlining administrative tasks like patient registration, billing, and appointments. RPA allows rapid data analysis, which has evolved as an essential tool for the management of patient records, compliance, and medical record coding optimization (Topol, 2020). This automation minimizes the possibility of human error and supports service delivery through fast yet more accurate handling of information (Topol, 2020).

The scope of collaboration between human and RPA for better productivity in many areas is quite vast. While human workers bring creativity, empathy, and complex decision capabilities to an organization, repetitive or a data-intensive job is something that RPA does well. In customer service, insurance, and retail industries, for example, human-RPA collaboration has increased customer satisfaction through quicker time-to-response and improved services (Damian et al., 2024). After all, while organizations implement hybrid human-RPA systems, the urge for the development of skills, particularly within control and monitoring of automated processes becomes progressively urgent, as highlighted in Moulin et al. (2023).

FUTURE RESEARCH DIRECTIONS

Strong advancement opportunities for IPA still lie in many areas, which indicate the most needed channels for future research: IPA implementation algorithms, AI algorithms for cognitive automation, the ethical and regulatory problems of these technologies, and IPA implementation mechanisms. Improvement in IPA implementation mechanisms should become the critical focus for future study. Despite the fact that the existing body of research has emphasized more on process automation in structured environments, as shown in Bui et al., 2023; Farhadi & Zhao, 2024, there is limited emphasis on the execution of IPA in more highly unstructured domains like creative sectors and healthcare areas, as discussed in Ella et al., 2021. More in-depth research, therefore, has to be done in understanding the interconnection of IPA with other emerging technologies, as recently brought to the fore by Nogueira-de-Almeida et al. This should make IPA applications revolutionary and consolidate workflows more complex and dynamic in the future, according to Ruiz et al.

IPA systems need further research on advanced algorithms in cognitive automation, and deep research is also needed in the field of increasing the flexibility of IPA systems along with further learning capabilities. Advanced machine learning algorithms and AI algorithms have restricted IPA's ability to learn from experience and continuously adapt its behavior against new, unexpected processes (T. Nguyen et al., 2021). Therefore, this is required: enhancement of IPA systems with real-time learning will help IPA systems process more complex tasks in terms of decision-making issues (Park et al., 2021). Another area of furthering research is in the

The effects of cognitive automation fusion with deep learning and NLP in IPA systems remain poorly explored today. One of the large domains to be explored in the near future is ethical and regulatory issues. The ethical impacts of IPA systems, in particular concerning the issue of worker replacement and accountability in the decision-making process, should be explored extensively in detail (Vlahovic et al., 2021). The current legal frameworks do not consider the complexity IPA presents to data privacy, security issues, and even the transparency of AI-generated decisions (Jackson et al., 2022). Future studies should discuss creating a set of global regulatory standards or guidelines that mitigate such risks (Patel & Kaur, 2021). Equally relevant to this is an examination of the ethical implications of IPA designs in that human welfare is placed at the forefront to guide the development and application of AI responsibly (Chakraborty et al., 2022).

Emerging Leading Technologies: The emerging technologies in the class include LCDP and Conversational AI. These technologies create potential challenges as well as opportunities faced by organizations in searching for the optimization of efficiency and creativity in a domain as fast-paced as Intelligent RPA systems. Low-Code Development Platforms are being found to enable non-technical people to develop applications and automate tasks with minimal coding know-how (Nurk et al., 2022). The pace of RPA system integration accelerates in the democratization of software development. With RPA, enterprises can quickly deploy automation solutions. The scope for integrating intelligent RPA with business applications enhances the possibilities for enterprises as LCDP matures in creating complex automated systems (Allgaier et al., 2023). The user-friendly feature of LCDP poses issues, like the maintenance of strong governance and the security of fast-built applications (Mendling et al., 2020). While developing LCDP, organizations face the trade-off between innovation with proper oversight (Moss et al., 2021).

On the other hand, advancements in Conversational AI have a critical impact on Intelligent RPA systems. Converse AI has roots in NLP and machine learning. It allows RPA systems to have many more conversations with humans than previous and pushes the horizon of automation past customer support, virtual assistants, and user-centric operations (Kim et al., 2021). Because it can deal with more dynamic, unstructured data, more interactivity strengthens this kind of automation beyond simple repeated rule-based operations, expanding the horizon (Sheth, 2020). The innovations of conversational agents in RPA processes have enabled companies to improve human-machine collaboration, which increases flexibility in the interaction between people and other integrated systems (Antoniadis et al., 2023). Still, other challenges like constraints in language understanding, biasness in AI models, and data privacy issues still need to be mitigated when deploying Conversational AI in the realms of RPA (Bender et al., 2021). A portfolio of Intelligent RPA combining LCDP and Conversational AI presents various opportunities and challenges. These technologies offer better capabilities in automation but raise concerns related to governance, security, and AI bias (Verhoef et al., 2021)(Verhoef et al., 2021). Continuous learning, strong governance frameworks, and ethics in AI practice should characterize organizational activities when using these new technologies (Verhoef et al., 2021). With time and development of RPA systems, the effects of how interaction between technology will take place to further encourage innovations going forward and resolve the challenges presented from their adoption (von Mehren et al., 2022).

CONCLUSION AND DISCUSSION

Intelligent RPA is being applied to the operations of companies with problems and prospects. As these systems grow in adoption, firms experience human-centered, technological, and organizational challenges that require strict control. However, Intelligent RPA has the potential to transform future digital transitions into increasing operational efficiency and optimizing workflows while raising overall productivity. Problems related to Intelligent RPA are divided into three areas: work-force adaptation, technological limitations, and resistance from organizations. One major issue is the disruption of work roles because automation takes over routine, mundane activities that are governed by rules. This forces human resources to move to more complex, analytical, and creative jobs (Herm et al., 2021). This associated uncertainty may cause discomfort, especially for those who lack the skills to adjust to a new context (Herm et al., 2021). A redefinition of tasks, therefore is an essential component for firms seeking to exploit Intelligent RPA without making employees irrelevant (Berruti et al., 2017).

Related to this is the challenge of continuous education and development of skills. Employees, with the development of RPA systems, need to gain skills in data analytics, machine learning, and management of multiple systems ((Kholiya & Kumar, 2023b). Such instances of lack of proper training may result in employees not achieving efficient interaction with automation technology which leads to growing job loss fears. Upskilling employees to interact with RPA technologies efficiently has been proven to require huge investments from firms (Jarrahi, 2018). In the immediate future, employees will have to possess a much better understanding of AI and RPA systems in order to manage hybrid human-machine collaboration (Battling et al., 2022).Resistance to automation is another major challenge. The majority of employees fear job displacement from the application of automation. It adds cultural and operational resistance that slows down the pace of digital change (Moulaï et al., 2022). To counter such resistance, organizations need to inculcate a culture of openness and trust. The members' minds should be filled with the benefits of automation and the progressive role of RPA. Key interventions will be led by leadership since they would have to promote positive stories about automation along with possibilities for career progression and personal growth through upskilling and reskilling initiatives (Hlásny et al., 2021).

Though these are some of the challenges posed, opportunities that Intelligent RPA presents are huge. RPA systems could greatly make operations more efficient by developing automation for repetitive activities, reducing the risk from human errors, and making it possible to concentrate human workers on other activities that are value-enhancing. Organizations can increase overall productivity and save money through the simplification of processes and optimization of resource utilization (Vial, 2019). At the same time, RPA helps enterprises become more agile in their operations, which make them respond rapidly to market needs and technical progress (Aguirre & Rodriguez, 2017). Intelligent RPA systems improve the experience of customers by automatic front-end operations that include data entry and order processing in a very short period while providing service precise and faster (Salviotti et al., 2019). The intelligent RPA opens new tracks of innovation. This will thereby allow human resources to move away from routine tasks and then direct the creative potential of their workforce to innovation and problem-solving activities (Lindner et al., 2020). This further creates a more vibrant work setting where employees can participate in strategy projects and drive digital transformation initiatives (Colbert et al., 2016). Besides, RPA also contributes to better data-driven decision making since it provides real-time insights into corporate operations, therefore making easier on leaders to make more informative decisions that are pro-active in nature as well (Schulte et al., 2024).

The potential of Intelligent RPA for the digital transformation is enormous. With enterprises heavily depending on automation technology, the role of Intelligent RPA will clearly determine how future work will look like ((L. Willcocks et al., 2017). Improvements from AI and ML would make them cover more complex cognitive tasks that give these systems an enhanced value proposition (Colombo et al., 2020). RPA will merge with AI to offer "hyper-automation," where sophisticated automation systems will be able to independently act in solving problems and responding to changes in circumstances (Lacity et al., 2021).Digital transformations later on will then enable businesses to develop significantly more adaptable and scalable business models (Marr, 2019).By implementing RPA technologies across various functions, organizations are planning to make connectivity between different departments and systems become remarkably easier to organize (Davenport & Ronanki, 2018).In doing so, efficiency in operations will improve while fostering innovation and business agility so that organizations can compete effectively in an increasingly digital global marketplace (Vial, 2021). In addition, Intelligent RPA will help firms to unlock the enormous sources of big data, promising real-time analytics and insights that could be used to update strategic planning and assure continuous improvement (Chen et al., 2020). As RPA continues to evolve, ethics questions will also grow in importance. Most of the issues that have arisen concerning the widespread adoption of automation of work related to employment loss, data privacy, and decision-making authority ((McAfee & Brynjolfsson, 2017).companies have to handle these issues effectively by creating ethical norms and following clear procedures that focus on employee welfare and societal implications (Jarrahi, 2018).Upon solving these issues, businesses can make fully effective use of Intelligent RPA capabilities while remaining perfectly balanced and sustainable in their quest for digital transformation.

After all, although the process of Intelligent RPA poses gigantic challenges, its outcome benefits far surpass these obstacles. Focus on upskilling, increased transparency and the culture of innovation can help unlock the full potential of Intelligent RPA from future digital revolutions for enterprises. Automation technologies are deeply going to change The future of work, and enterprises using these products and services in the right manner will find themselves favorably positioned to prosper in the age of digitization.

REFERENCES

Agostinelli, F., Doepke, M., Sorrenti, G., & Zilibotti, F. (2020). *It takes a village: The economics of parenting with neighborhood and peer effects*. National Bureau of Economic Research. DOI: 10.3386/w27050

Aguirre, S., & Rodriguez, A. (2017). Automation of a business process using robotic process automation (RPA): A case study. *Applied Computer Sciences in Engineering: 4th Workshop on Engineering Applications, WEA 2017, Cartagena, Colombia, September 27-29, 2017. Proceedings*, *4*, 65–71.

Ahmad, H., Hanandeh, R., Alazzawi, F., Al-Daradkah, A., ElDmrat, A., Ghaith, Y., & Darawsheh, S. (2023). The effects of big data, artificial intelligence, and business intelligence on e-learning and business performance: Evidence from Jordanian telecommunication firms. *International Journal of Data and Network Science*, *7*(1), 35–40. DOI: 10.5267/j.ijdns.2022.12.009

Allgaier, J., Mulansky, L., Draelos, R. L., & Pryss, R. (2023). How does the model make predictions? A systematic literature review on the explainability power of machine learning in healthcare. *Artificial Intelligence in Medicine*, *143*, 102616. DOI: 10.1016/j.artmed.2023.102616 PMID: 37673561

Antoniadis, J., Arumugam, P., Arumugam, S., Babak, S., Bagchi, M., Nielsen, A.-S. B., Bassa, C. G., Bathula, A., Berthereau, A., & Bonetti, M. (2023). The second data release from the European Pulsar Timing Array-III. Search for gravitational wave signals. *Astronomy & Astrophysics*, *678*, A50. DOI: 10.1051/0004-6361/202346844

Arif, S., & MacNeil, M. A. (2022). Predictive models aren't for causal inference. *Ecology Letters*, *25*(8), 1741–1745. DOI: 10.1111/ele.14033 PMID: 35672133

Aryawati, N. P. A., Harahap, T. K., Yanti, N. N. S. A., Mahardika, I. M. N. O., Widiniarsih, D. M., Ahmad, M. I. S., Mattunruang, A. A., & Amali, L. M. (2023). *Manajemen keuangan*. Penerbit Tahta Media.

Asadov, J. (2023). Azerbaijan's Multicultural Policy of Ethnic Diversity. *Państwa Postradzieckie w Procesie Modernizacji, April*.

Battling, S., Pastoors, J., Deitert, A., Götzen, T., Hartmann, L., Schröder, E., Yordanov, S., & Büchs, J. (2022). Development of a novel defined minimal medium for Gluconobacter oxydans 621H by systematic investigation of metabolic demands. *Journal of Biological Engineering*, *16*(1), 31. DOI: 10.1186/s13036-022-00310-y PMID: 36414992

Beithou, N., Mansour, M. A., Abdellatif, N., Alsaqoor, S., Tarawneh, S., Jaber, A. H., Andruszkiewicz, A., Alsqour, M., Borowski, G., & Alahmer, A. (2023). Effect of the Residential Photovoltaic Systems Evolution on Electricity and Thermal Energy Usage in Jordan. *Advances in Science and Technology. Research Journal, 17*(3).

Bender, E. M., Gebru, T., McMillan-Major, A., & Shmitchell, S. (2021). On the dangers of stochastic parrots: Can language models be too big? . *Proceedings of the 2021 ACM Conference on Fairness, Accountability, and Transparency*, 610–623. DOI: 10.1145/3442188.3445922

Berruti, A., Grisanti, S., Pulzer, A., Claps, M., Daffara, F., Loli, P., Mannelli, M., Boscaro, M., Arvat, E., Tiberio, G., Hahner, S., Zaggia, B., Porpiglia, F., Volante, M., Fassnacht, M., & Terzolo, M. (2017). Long-Term Outcomes of Adjuvant Mitotane Therapy in Patients With Radically Resected Adrenocortical Carcinoma. *The Journal of Clinical Endocrinology and Metabolism, 102*(4), 1358–1365. DOI: 10.1210/jc.2016-2894 PMID: 28324035

Binns, R. (2018). Fairness in machine learning: Lessons from political philosophy. *Conference on Fairness, Accountability and Transparency*, 149–159.

Bui, D. P., Bajema, K. L., Huang, Y., Yan, L., Li, Y., Rajeevan, N., Berry, K., Rowneki, M., Argraves, S., & Hynes, D. (2023). Development of a prediction model for 30-day COVID-19 hospitalization and death in a national cohort of Veterans Health Administration patients-March 2022-April 2023. MedRxiv, 2011–2023. DOI: 10.1101/2023.11.17.23298653

Chakraborty, R., Asthana, A., Singh, A. K., Jain, B., & Susan, A. B. H. (2022). Adsorption of heavy metal ions by various low-cost adsorbents: A review. *International Journal of Environmental Analytical Chemistry, 102*(2), 342–379. DOI: 10.1080/03067319.2020.1722811

Chen, N., Zhou, M., Dong, X., Qu, J., Gong, F., Han, Y., Qiu, Y., Wang, J., Liu, Y., Wei, Y., Xia, J., Yu, T., Zhang, X., & Zhang, L. (2020). Epidemiological and clinical characteristics of 99 cases of 2019 novel coronavirus pneumonia in Wuhan, China: A descriptive study. *Lancet, 395*(10223), 507–513. DOI: 10.1016/S0140-6736(20)30211-7 PMID: 32007143

Chowdhury, P., Paul, S. K., Kaisar, S., & Moktadir, M. A. (2021). COVID-19 pandemic related supply chain studies: A systematic review. *Transportation Research Part E, Logistics and Transportation Review, 148*, 102271. DOI: 10.1016/j.tre.2021.102271 PMID: 33613082

Colbert, A., Yee, N., & George, G. (2016). The digital workforce and the workplace of the future. [). Academy of Management Briarcliff Manor, NY.]. *Academy of Management Journal*, *59*(3), 731–739. DOI: 10.5465/amj.2016.4003

Colombo, C., Burgel, P.-R., Gartner, S., van Koningsbruggen-Rietschel, S., Naehrlich, L., Sermet-Gaudelus, I., & Southern, K. W. (2020). Impact of COVID-19 on people with cystic fibrosis. *The Lancet. Respiratory Medicine*, *8*(5), e35–e36. DOI: 10.1016/S2213-2600(20)30177-6 PMID: 32304639

Costa, D. A. D. S., Mamede, H. S., & Silva, M. M. D. (2022). Robotic Process Automation (RPA) adoption: a systematic literature review. Engineering Management in Production and Services, 14(2).

Damian, C. S., Devarajan, Y., & Jayabal, R. (2024). Biodiesel production in India: Prospects, challenges, and sustainable directions. *Biotechnology and Bioengineering*, *121*(3), 894–902. DOI: 10.1002/bit.28643 PMID: 38164703

Davenport, T. H., & Ronanki, R. (2018). Artificial intelligence for the real world. *Harvard Business Review*, *96*(1), 108–116.

Devlin, J. (2018). Bert: Pre-training of deep bidirectional transformers for language understanding. *ArXiv Preprint ArXiv:1810.04805*.

Duan, Y., Edwards, J. S., & Dwivedi, Y. K. (2019). Artificial intelligence for decision making in the era of Big Data–evolution, challenges and research agenda. *International Journal of Information Management*, *48*, 63–71. DOI: 10.1016/j.ijinfomgt.2019.01.021

Ella, R., Reddy, S., Blackwelder, W., Potdar, V., Yadav, P., Sarangi, V., Aileni, V. K., Kanungo, S., Rai, S., Reddy, P., Verma, S., Singh, C., Redkar, S., Mohapatra, S., Pandey, A., Ranganadin, P., Gumashta, R., Multani, M., Mohammad, S., & Waghmare, S. (2021). Efficacy, safety, and lot-to-lot immunogenicity of an inactivated SARS-CoV-2 vaccine (BBV152): Interim results of a randomised, double-blind, controlled, phase 3 trial. *Lancet*, *398*(10317), 2173–2184. DOI: 10.1016/S0140-6736(21)02000-6 PMID: 34774196

Ens, B., Bach, B., Cordeil, M., Engelke, U., Serrano, M., Willett, W., Prouzeau, A., Anthes, C., Büschel, W., & Dunne, C. (2021). Grand challenges in immersive analytics. *Proceedings of the 2021 CHI Conference on Human Factors in Computing Systems*, 1–17.

Esteva, A., Chou, K., Yeung, S., Naik, N., Madani, A., Mottaghi, A., Liu, Y., Topol, E., Dean, J., & Socher, R. (2021). Deep learning-enabled medical computer vision. *NPJ Digital Medicine*, *4*(1), 5. DOI: 10.1038/s41746-020-00376-2 PMID: 33420381

Farhadi, M., & Zhao, L. (2024). Exploring the Impact of Iran-China Trade on Environmental Sustainability. *Journal of Energy and Environmental Policy Options, 7*(1), 1–8.

Flechsig, C. (2021). The impact of intelligent process automation on purchasing and supply management–Initial insights from a multiple case study. *Logistics Management: Contributions of the Section Logistics of the German Academic Association for Business Research, 2021, Dresden, Germany*, 67–89.

Furman, J., & Seamans, R. (2019). AI and the Economy. *Innovation Policy and the Economy, 19*(1), 161–191. DOI: 10.1086/699936

Geiger, M., Jacot, A., Spigler, S., Gabriel, F., Sagun, L., d'Ascoli, S., Biroli, G., Hongler, C., & Wyart, M. (2020). Scaling description of generalization with number of parameters in deep learning. *Journal of Statistical Mechanics, 2020*(2), 23401. DOI: 10.1088/1742-5468/ab633c

Geyer-Klingeberg, J., Nakladal, J., Baldauf, F., & Veit, F. (2018). Process Mining and Robotic Process Automation: A Perfect Match. *BPM (Dissertation/Demos/Industry), 2196*, 124–131.

Goodfellow, I., Pouget-Abadie, J., Mirza, M., Xu, B., Warde-Farley, D., Ozair, S., Courville, A., & Bengio, Y. (2020). Generative adversarial networks. *Communications of the ACM, 63*(11), 139–144. DOI: 10.1145/3422622

Hamet, P., & Tremblay, J. (2017). Artificial intelligence in medicine. *Metabolism: Clinical and Experimental, 69*, S36–S40. https://doi.org/https://doi.org/10.1016/j.metabol.2017.01.011. DOI: 10.1016/j.metabol.2017.01.011 PMID: 28126242

Herm, L.-V., Janiesch, C., Reijers, H. A., & Seubert, F. (2021). From symbolic RPA to intelligent RPA: challenges for developing and operating intelligent software robots. *Business Process Management: 19th International Conference, BPM 2021, Rome, Italy, September 06–10, 2021. Proceedings, 19*, 289–305.

Hlásny, T., König, L., Krokene, P., Lindner, M., Montagné-Huck, C., Müller, J., Qin, H., Raffa, K. F., Schelhaas, M.-J., Svoboda, M., Viiri, H., & Seidl, R. (2021). Bark beetle outbreaks in Europe: State of knowledge and ways forward for management. *Current Forestry Reports, 7*(3), 138–165. DOI: 10.1007/s40725-021-00142-x

Inam, A., Joseph, S., & Chemingui, H. (2024). Beyond the Screen. *Hyperautomation in Business and Society*, 266.

Jackson, C. B., Farzan, M., Chen, B., & Choe, H. (2022). Mechanisms of SARS-CoV-2 entry into cells. *Nature Reviews. Molecular Cell Biology, 23*(1), 3–20. DOI: 10.1038/s41580-021-00418-x PMID: 34611326

Jarrahi, M. H. (2018). Artificial intelligence and the future of work: Human-AI symbiosis in organizational decision making. *Business Horizons*, *61*(4), 577–586. DOI: 10.1016/j.bushor.2018.03.007

Jha, P., Deshmukh, Y., Tumbe, C., Suraweera, W., Bhowmick, A., Sharma, S., Novosad, P., Fu, S. H., Newcombe, L., Gelband, H., & Brown, P. (2022). COVID mortality in India: National survey data and health facility deaths. *Science*, *375*(6581), 667–671. DOI: 10.1126/science.abm5154 PMID: 34990216

Jurdi, A. P. S., Queiróz, M., Feio, C. B. A. P. R., Barros, N. A., Cagnin, G. M., Iassia, A. M., Santos, L. C., & Pinheiro, A. C. (2022). PET–Saúde Interprofissionalidade CER II: Narrativas de um grupo sobre ações cotidianas de trabalho frente às alterações provocadas pela Covid-19. *Interface: Comunicacao, Saude, Educacao*, *26*, e210597. DOI: 10.1590/interface.210597

Kar, N., Kar, B., & Kar, S. (2021). Stress and coping during COVID-19 pandemic: Result of an online survey. *Psychiatry Research*, *295*, 113598. DOI: 10.1016/j.psychres.2020.113598 PMID: 33264677

Kaur, B., Kumar, S., & Kaushik, B. K. (2022). Recent advancements in optical biosensors for cancer detection. *Biosensors & Bioelectronics*, *197*, 113805. DOI: 10.1016/j.bios.2021.113805 PMID: 34801795

Kedziora, D., & Hyrynsalmi, S. (2023). Turning robotic process automation onto intelligent automation with machine learning. *Proceedings of the 11th International Conference on Communities and Technologies*, 1–5. DOI: 10.1145/3593743.3593746

Khan, S. A. R., Yu, Z., Belhadi, A., & Mardani, A. (2020). Investigating the effects of renewable energy on international trade and environmental quality. *Journal of Environmental Management*, *272*, 111089. DOI: 10.1016/j.jenvman.2020.111089 PMID: 32854892

Kholiya, N., & Kumar, A. (2023a). Phytotoxicity of Chromium (VI) on Germination. *Growth Attributes and Pigmentation in Cluster Bean.*, *19*(3), 94–101.

Kholiya, N., & Kumar, A. (2023b). Phytotoxicity of chromium (VI) on germination, growth attributes and pigmentation in cluster bean. *Journal of Stress Physiology & Biochemistry*, *19*(3), 94–101.

Kim, S., Chen, J., Cheng, T., Gindulyte, A., He, J., He, S., Li, Q., Shoemaker, B. A., Thiessen, P. A., Yu, B., Zaslavsky, L., Zhang, J., & Bolton, E. E. (2021). PubChem in 2021: New data content and improved web interfaces. *Nucleic Acids Research*, *49*(D1), D1388–D1395. DOI: 10.1093/nar/gkaa971 PMID: 33151290

Lacity, M., Willcocks, L., & Gozman, D. (2021). Influencing information systems practice: The action principles approach applied to robotic process and cognitive automation. *Journal of Information Technology, 36*(3), 216–240. DOI: 10.1177/0268396221990778

Lacity, M., & Willcocks, L. P. (2018). *Robotic process and cognitive automation: The next phase*. SB Publishing.

LeCun, Y., Bengio, Y., & Hinton, G. (2015). Deep learning. *Nature, 521*(7553), 436–444. DOI: 10.1038/nature14539 PMID: 26017442

Lee, S. A. (2020). How much "Thinking" about COVID-19 is clinically dysfunctional? *Brain, Behavior, and Immunity, 87*, 97–98. DOI: 10.1016/j.bbi.2020.04.067 PMID: 32353520

Lievano-Martínez, F. A., Fernández-Ledesma, J. D., Burgos, D., Branch-Bedoya, J. W., & Jimenez-Builes, J. A. (2022). Intelligent process automation: An application in manufacturing industry. *Sustainability (Basel), 14*(14), 8804. DOI: 10.3390/su14148804

Lindner, D., Fitzek, A., Bräuninger, H., Aleshcheva, G., Edler, C., Meissner, K., Scherschel, K., Kirchhof, P., Escher, F., Schultheiss, H.-P., Blankenberg, S., Püschel, K., & Westermann, D. (2020). Association of cardiac infection with SARS-CoV-2 in confirmed COVID-19 autopsy cases. *JAMA Cardiology, 5*(11), 1281–1285. DOI: 10.1001/jamacardio.2020.3551 PMID: 32730555

Liu, Y., Liu, J., Xia, H., Zhang, X., Fontes-Garfias, C. R., Swanson, K. A., Cai, H., Sarkar, R., Chen, W., Cutler, M., Cooper, D., Weaver, S. C., Muik, A., Sahin, U., Jansen, K. U., Xie, X., Dormitzer, P. R., & Shi, P.-Y. (2021). Neutralizing activity of BNT162b2-elicited serum. *The New England Journal of Medicine, 384*(15), 1466–1468. DOI: 10.1056/NEJMc2102017 PMID: 33684280

Lopes, R. D., Macedo, A. V. S., & de Barros, E. (2021). Effect of Discontinuing vs Continuing Angiotensin-Converting Enzyme Inhibitors and Angiotensin II Receptor Blockers on Days Alive and Out of the Hospital in Patients Admitted With COVID-19: A Randomized Clinical Trial. *Journal of the American Medical Association, 325*(3), 254–264. DOI: 10.1001/jama.2020.25864 PMID: 33464336

Manning, J., Billian, J., Matson, J., Allen, C., & Soares, N. (2021). Perceptions of families of individuals with autism spectrum disorder during the COVID-19 crisis. *Journal of Autism and Developmental Disorders, 51*(8), 2920–2928. DOI: 10.1007/s10803-020-04760-5 PMID: 33090358

Marr, B. (2019). *Artificial intelligence in practice: how 50 successful companies used AI and machine learning to solve problems*. John Wiley & Sons.

McAfee, A., & Brynjolfsson, E. (2017). *Machine, platform, crowd: Harnessing our digital future*. WW Norton & Company.

Meegahapola, L., Droz, W., Kun, P., De Götzen, A., Nutakki, C., Diwakar, S., Correa, S. R., Song, D., Xu, H., Bidoglia, M., Gaskell, G., Chagnaa, A., Ganbold, A., Zundui, T., Caprini, C., Miorandi, D., Hume, A., Zarza, J. L., Cernuzzi, L., & Gatica-Perez, D. (2023). Generalization and personalization of mobile sensing-based mood inference models: An analysis of college students in eight countries. *Proceedings of the ACM on Interactive, Mobile, Wearable and Ubiquitous Technologies*, *6*(4), 1–32. DOI: 10.1145/3569483

Mendling, J., Pentland, B. T., & Recker, J. (2020). Building a complementary agenda for business process management and digital innovation. [). Taylor & Francis.]. *European Journal of Information Systems*, *29*(3), 208–219. DOI: 10.1080/0960085X.2020.1755207

Moreira, S., Mamede, H. S., & Santos, A. (2023). Process automation using RPA–a literature review. *Procedia Computer Science*, *219*, 244–254. DOI: 10.1016/j.procs.2023.01.287

Moss, A., Robbins, S., Achanta, S., Kuttippurathu, L., Turick, S., Nieves, S., Hanna, P., Smith, E. H., Hoover, D. B., Chen, J., Cheng, Z. J., Ardell, J. L., Shivkumar, K., Schwaber, J. S., & Vadigepalli, R. (2021). A single cell transcriptomics map of paracrine networks in the intrinsic cardiac nervous system. *iScience*, *24*(7), 102713. DOI: 10.1016/j.isci.2021.102713 PMID: 34337356

Moulaï, K., Islam, G., Manning, S., & Terlinden, L. (2022). All too human" or the emergence of a techno-induced feeling of being less-able: Identity work, ableism and new service technologies. *International Journal of Human Resource Management*, *33*(22), 4499–4531. DOI: 10.1080/09585192.2022.2066982

Moulin, F., Jean, F., Melchior, M., Patanè, M., Pinucci, I., Sijbrandij, M., van der Waerden, J., & Galéra, C. (2023). Longitudinal impact of the COVID19 pandemic on mental health in a general population sample in France: Evidence from the COMET study. *Journal of Affective Disorders*, *320*, 275–283. DOI: 10.1016/j.jad.2022.09.142 PMID: 36191642

Murhekar, M. V., Bhatnagar, T., Selvaraju, S., Saravanakumar, V., Thangaraj, J. W. V., Shah, N., Kumar, M. S., Rade, K., Sabarinathan, R., Asthana, S., Balachandar, R., Bangar, S. D., Bansal, A. K., Bhat, J., Chopra, V., Das, D., Deb, A. K., Devi, K. R., Dwivedi, G. R., & Zaman, K. (2021). SARS-CoV-2 antibody seroprevalence in India, August–September, 2020: Findings from the second nationwide household serosurvey. *The Lancet. Global Health*, *9*(3), e257–e266. DOI: 10.1016/S2214-109X(20)30544-1 PMID: 33515512

Nasser, S. M. T., Rana, A. A., Doffinger, R., Kafizas, A., Khan, T. A., & Nasser, S. (2023). Elevated free interleukin-18 associated with severity and mortality in prospective cohort study of 206 hospitalised COVID-19 patients. *Intensive Care Medicine Experimental*, *11*(1), 9. DOI: 10.1186/s40635-022-00488-x PMID: 36823262

Ng, D. T. K., Leung, J. K. L., Chu, S. K. W., & Qiao, M. S. (2021). Conceptualizing AI literacy: An exploratory review. *Computers and Education: Artificial Intelligence*, *2*, 100041. https://doi.org/https://doi.org/10.1016/j.caeai.2021.100041

Ng, K. K. H., Chen, C.-H., Lee, C. K. M., Jiao, J. R., & Yang, Z.-X. (2021). A systematic literature review on intelligent automation: Aligning concepts from theory, practice, and future perspectives. *Advanced Engineering Informatics*, *47*, 101246. DOI: 10.1016/j.aei.2021.101246

Nguyen, T., Novak, R., Xiao, L., & Lee, J. (2021). Dataset distillation with infinitely wide convolutional networks. *Advances in Neural Information Processing Systems*, *34*, 5186–5198.

Nguyen, T. K., Nguyen, V. L., Nguyen, T. G., Mai, D. H., Nguyen, N. Q., Vu, T. A., Le, A. N., Nguyen, Q. H., Nguyen, C. T., & Nguyen, D. T. (2021). Lung-protective mechanical ventilation for patients undergoing abdominal laparoscopic surgeries: A randomized controlled trial. *BMC Anesthesiology*, *21*(1), 1–11. DOI: 10.1186/s12871-021-01318-5 PMID: 33784987

Nogueira-de-Almeida, C. A., Del Ciampo, L. A., Ferraz, I. S., Del Ciampo, I. R. L., Contini, A. A., & Ued, F. da V. (2020). COVID-19 and obesity in childhood and adolescence: A clinical review. *Jornal de Pediatria*, *96*(5), 546–558. DOI: 10.1016/j.jped.2020.07.001 PMID: 32768388

Nurk, S., Koren, S., Rhie, A., Rautiainen, M., Bzikadze, A. V., Mikheenko, A., Vollger, M. R., Altemose, N., Uralsky, L., Gershman, A., Aganezov, S., Hoyt, S. J., Diekhans, M., Logsdon, G. A., Alonge, M., Antonarakis, S. E., Borchers, M., Bouffard, G. G., Brooks, S. Y., & Phillippy, A. M. (2022). The complete sequence of a human genome. *Science*, *376*(6588), 44–53. DOI: 10.1126/science.abj6987 PMID: 35357919

Park, S., Lee, C.-W., Park, D.-I., Woo, H.-Y., Cheong, H. S., Shin, H. C., Ahn, K., Kwon, M.-J., & Joo, E.-J. (2021). Detection of SARS-CoV-2 in fecal samples from patients with asymptomatic and mild COVID-19 in Korea. *Clinical Gastroenterology and Hepatology*, *19*(7), 1387–1394. DOI: 10.1016/j.cgh.2020.06.005 PMID: 32534042

Parsons, S. K., Keegan, T. H. M., Kirchhoff, A. C., Parsons, H. M., Yabroff, K. R., & Davies, S. J. (2023). Cost of cancer in adolescents and young adults in the United States: Results of the 2021 report by Deloitte Access Economics, commissioned by Teen Cancer America. *Journal of Clinical Oncology*, *41*(17), 3260–3268. DOI: 10.1200/JCO.22.01985 PMID: 36827624

Power, D. J., & Phillips-Wren, G. (2011). Impact of social media and Web 2.0 on decision-making. *Journal of Decision Systems*, *20*(3), 249–261. DOI: 10.3166/jds.20.249-261

Rath, A., Das Gupta, A., Rohilla, V., Balyan, A., & Mann, S. (2022). Intelligent Smart Waste Management Using Regression Analysis: An Empirical Study. *International Conference on Emerging Technologies in Computer Engineering*, 138–148. DOI: 10.1007/978-3-031-07012-9_12

Richardson, S., Hirsch, J. S., Narasimhan, M., Crawford, J. M., McGinn, T., Davidson, K. W., Barnaby, D. P., Becker, L. B., Chelico, J. D., Cohen, S. L., Cookingham, J., Coppa, K., Diefenbach, M. A., Dominello, A. J., Duer-Hefele, J., Falzon, L., Gitlin, J., Hajizadeh, N., Harvin, T. G., & Zanos, T. P. (2020). Presenting Characteristics, Comorbidities, and Outcomes Among 5700 Patients Hospitalized With COVID-19 in the New York City Area. *Journal of the American Medical Association*, *323*(20), 2052–2059. DOI: 10.1001/jama.2020.6775 PMID: 32320003

Ruiz, P. L.-D., Gunnes, N., Gran, J. M., Karlstad, Ø., Selmer, R., Dahl, J., Bøås, H., White, R. A., Hofman, A. C., & Paulsen, T. H. (2023). Short-term safety of COVID-19 mRNA vaccines with respect to all-cause mortality in the older population in Norway. *Vaccine*, *41*(2), 323–332. DOI: 10.1016/j.vaccine.2022.10.085 PMID: 36376216

Salviotti, G., Gaur, A., & Pennarola, F. (2019). *Strategic factors enabling digital maturity: An extended survey.*

Sarker, I. H. (2022). AI-based modeling: Techniques, applications and research issues towards automation, intelligent and smart systems. *SN Computer Science*, *3*(2), 158. DOI: 10.1007/s42979-022-01043-x PMID: 35194580

Schulte, B., Richter, E., Büning, A., Baum, M., Breuer, A., Zorn, J., König, J., Geiger, M., Eschbach-Bludau, M., Heuser, J., Zölzer, D., Korencak, M., Hollstein, R., Beins, E., Emmert, D., Aldabbagh, S., Eis-Hübinger, A. M., & Streeck, H. (2024). A longitudinal study on SARS-CoV-2 seroconversion, reinfection and neutralisation spanning several variant waves and vaccination campaigns, Heinsberg, Germany, April 2020 to November 2022. *Eurosurveillance*, *29*(26), 2300659. DOI: 10.2807/1560-7917.ES.2024.29.26.2300659 PMID: 38940003

Sheth, J. (2020). Impact of Covid-19 on consumer behavior: Will the old habits return or die? *Journal of Business Research, 117*, 280–283. DOI: 10.1016/j.jbusres.2020.05.059 PMID: 32536735

Shmueli, G., & Koppius, O. R. (2011). Predictive analytics in information systems research. *Management Information Systems Quarterly*, *35*(3), 553–572. DOI: 10.2307/23042796

Siderska, J. (2020). Robotic Process Automation—A driver of digital transformation? *Engineering Management in Production and Services*, *12*(2), 21–31. DOI: 10.2478/emj-2020-0009

Siderska, J., Alsqour, M., & Alsaqoor, S. (2023). Employees' attitudes towards implementing robotic process automation technology at service companies. *Human Technology, 19*(1 SE-Articles), 23–40. DOI: 10.14254/1795-6889.2023.19-1.3

Süße, M., Prell, B., Stoldt, J., Strehle, H.-M., Richter, M., Schlegel, A., & Putz, M. (2019). Load and energy management for factories through multi-stage production planning and optimization. *Procedia CIRP*, *80*, 257–262. https://doi.org/https://doi.org/10.1016/j.procir.2019.01.076. DOI: 10.1016/j.procir.2019.01.076

Topol, E. J. (2020). COVID-19 can affect the heart. *Science*, *370*(6515), 408–409. DOI: 10.1126/science.abe2813 PMID: 32967937

Veale, M., & Zuiderveen Borgesius, F. (2021). Demystifying the Draft EU Artificial Intelligence Act—Analysing the good, the bad, and the unclear elements of the proposed approach. *Computer Law Review International*, *22*(4), 97–112. DOI: 10.9785/cri-2021-220402

Verhoef, P. C., Broekhuizen, T., Bart, Y., Bhattacharya, A., Dong, J. Q., Fabian, N., & Haenlein, M. (2021). Digital transformation: A multidisciplinary reflection and research agenda. *Journal of Business Research*, *122*, 889–901. DOI: 10.1016/j.jbusres.2019.09.022

Vial, G. (2021). Understanding digital transformation: A review and a research agenda. *Managing Digital Transformation*, 13–66.

Vlahovic, S., Suznjevic, M., Pavlin-Bernardic, N., & Skorin-Kapov, L. (2021). The effect of VR gaming on discomfort, cybersickness, and reaction time. *2021 13th International Conference on Quality of Multimedia Experience (QoMEX)*, 163–168.

von Mehren, M., Kane, J. M., Agulnik, M., Bui, M. M., Carr-Ascher, J., Choy, E., Connelly, M., Dry, S., Ganjoo, K. N., Gonzalez, R. J., Holder, A., Homsi, J., Keedy, V., Kelly, C. M., Kim, E., Liebner, D., McCarter, M., McGarry, S. V., Mesko, N. W., & Bergman, M. A. (2022). Soft tissue sarcoma, version 2.2022, NCCN clinical practice guidelines in oncology. *Journal of the National Comprehensive Cancer Network : JNCCN*, *20*(7), 815–833. DOI: 10.6004/jnccn.2022.0035 PMID: 35830886

Willcocks, L., Lacity, M., & Craig, A. (2017). Robotic process automation: Strategic transformation lever for global business services? *Journal of Information Technology Teaching Cases*, *7*(1), 17–28. DOI: 10.1057/s41266-016-0016-9

Willcocks, L. P., Lacity, M., & Craig, A. (2015). *The IT function and robotic process automation.*

Wilson, H. J., & Daugherty, P. R. (2018). Collaborative intelligence: Humans and AI are joining forces. *Harvard Business Review*, *96*(4), 114–123.

Wong, C. K. H., Au, I. C. H., Lau, K. T. K., Lau, E. H. Y., Cowling, B. J., & Leung, G. M. (2022). Real-world effectiveness of molnupiravir and nirmatrelvir plus ritonavir against mortality, hospitalisation, and in-hospital outcomes among community-dwelling, ambulatory patients with confirmed SARS-CoV-2 infection during the omicron wave in Hong Kong: An observational study. *Lancet*, *400*(10359), 1213–1222. DOI: 10.1016/S0140-6736(22)01586-0 PMID: 36216007

Xia, S., Zhang, Y., Wang, Y., Wang, H., Yang, Y., Gao, G. F., Tan, W., Wu, G., Xu, M., Lou, Z., Huang, W., Xu, W., Huang, B., Wang, H., Wang, W., Zhang, W., Li, N., Xie, Z., Ding, L., & Yang, X. (2021). Safety and immunogenicity of an inactivated SARS-CoV-2 vaccine, BBIBP-CorV: A randomised, double-blind, placebo-controlled, phase 1/2 trial. *The Lancet. Infectious Diseases*, *21*(1), 39–51. DOI: 10.1016/S1473-3099(20)30831-8 PMID: 33069281

Yao, M. (2024). RPA Technology Enables Highly Automated Development of Corporate Financial Accounting Processes. *Applied Mathematics and Nonlinear Sciences*, *9*(1), 20240541. Advance online publication. DOI: 10.2478/amns-2024-0541

Yapıcı, A., & Akcayol, M. A. (2022). Derin Öğrenme ile Görüntülerde Gürültü Azaltma Üzerine Kapsamlı Bir İnceleme. *International Journal of Advances in Engineering and Pure Sciences*, *34*(1), 65–90. DOI: 10.7240/jeps.938188

Zeiser, F. A., Donida, B., da Costa, C. A., de Oliveira Ramos, G., Scherer, J. N., Barcellos, N. T., Alegretti, A. P., Ikeda, M. L. R., Müller, A. P. W. C., & Bohn, H. C. (2022). First and second COVID-19 waves in Brazil: A cross-sectional study of patients' characteristics related to hospitalization and in-hospital mortality. *Lancet Regional Health. Americas*, *6*, 6. DOI: 10.1016/j.lana.2021.100107 PMID: 34746913

Zhao, S., Lin, Q., Ran, J., Musa, S. S., Yang, G., Wang, W., Lou, Y., Gao, D., Yang, L., He, D., & Wang, M. H. (2020). Preliminary estimation of the basic reproduction number of novel coronavirus (2019-nCoV) in China, from 2019 to 2020: A data-driven analysis in the early phase of the outbreak. *International Journal of Infectious Diseases*, *92*, 214–217. DOI: 10.1016/j.ijid.2020.01.050 PMID: 32007643

Compilation of References

Abdullahi, Z. H.. (2015). Mobile robot voice recognition in control movements. [IJCSEE]. *International Journal of Computer Science and Electronics Engineering*, *3*(1), 11–16.

Abo-Hammour, Z. E., Alsmadi, O. M., Bataineh, S. I., Al-Omari, M. A., & Affach, N. (2011, December 1). Continuous genetic algorithms for collision-free cartesian path planning of robot manipulators. *International Journal of Advanced Robotic Systems*, *8*(6), 74. DOI: 10.5772/50902

Abu-Dakka, F. J., Assad, I. F., Alkhdour, R. M., & Abderahim, M. (2017, March 1). Statistical evaluation of an evolutionary algorithm for minimum time trajectory planning problem for industrial robots. *International Journal of Advanced Manufacturing Technology*, *89*(1-4), 389–406. DOI: 10.1007/s00170-016-9050-1

Administrator. (2024, April 23). AI & RPA in Healthcare: Benefits, use cases, Trends 2024. AutomationEdge. https://automationedge.com/blogs/ai-and-rpa-in-healthcare/

Agostinelli, S., Marrella, A., & Mecella, M. (2019). Research challenges for intelligent robotic process automation. *Business Process Management Workshops: BPM 2019 International Workshops, Vienna, Austria, September 1–6, 2019, Revised Selected Papers 17*, 12–18.

Agostinelli, S., Marrella, A., & Mecella, M. (2019). Research challenges for intelligent robotic process automation. In *Business Process Management Workshops: BPM 2019 International Workshops, Vienna, Austria, September 1–6, 2019, Revised Selected Papers 17* (pp. 12-18). Springer International Publishing. DOI: 10.1007/978-3-030-37453-2_2

Agostinelli, S., Marrella, A., & Mecella, M. (2020). Towards intelligent robotic process automation for BPMers. arXiv preprint arXiv:2001.00804.

Agostinelli, S., Marrella, A., & Mecella, M. (2020). Towards intelligent robotic process automation for BPMers. *ArXiv Preprint ArXiv:2001.00804.*

Agostinelli, F., Doepke, M., Sorrenti, G., & Zilibotti, F. (2020). *It takes a village: The economics of parenting with neighborhood and peer effects*. National Bureau of Economic Research. DOI: 10.3386/w27050

Agostinelli, S., Lupia, M., Marrella, A., & Mecella, M. (2020). Automated generation of executable RPA scripts from user interface logs. In *Business Process Management: Blockchain and Robotic Process Automation Forum: BPM 2020 Blockchain and RPA Forum, Seville, Spain, September 13–18, 2020* [Springer International Publishing.]. *Proceedings*, *18*, 116–131.

Agostinelli, S., Marrella, A., & Mecella, M. (2021). Exploring the challenge of automated segmentation in robotic process automation. In *Proceedings of the Conference* (pp. 38-54). https://doi.org/DOI: 10.1007/978-3-030-75018-3_3

Agrawal, O. P. (2008, January 1). A quadratic numerical scheme for fractional optimal control problems. *Journal of Dynamic Systems, Measurement, and Control*, *130*(1), 011010. DOI: 10.1115/1.2814055

Aguirre, S., & Rodriguez, A. (2017). Automation of a business process using robotic process automation (RPA): A case study. *Applied Computer Sciences in Engineering: 4th Workshop on Engineering Applications, WEA 2017, Cartagena, Colombia, September 27-29, 2017*. *Proceedings*, *4*, 65–71.

Aguirre, S., & Rodriguez, A. (2017). Automation of a business process using robotic process automation (RPA): A case study. In *Applied Computer Sciences in Engineering: 4th Workshop on Engineering Applications, WEA 2017, Cartagena, Colombia, September 27-29, 2017* [Springer International Publishing.]. *Proceedings*, *4*, 65–71.

Ahmad, A. Y. A. B. (2024). Ethical implications of artificial intelligence in accounting: A framework for responsible ai adoption in multinational corporations in Jordan. *International Journal of Data and Network Science*, *8*(1), 401–414. DOI: 10.5267/j.ijdns.2023.9.014

Ahmad, H., Hanandeh, R., Alazzawi, F., Al-Daradkah, A., ElDmrat, A., Ghaith, Y., & Darawsheh, S. (2023). The effects of big data, artificial intelligence, and business intelligence on e-learning and business performance: Evidence from Jordanian telecommunication firms. *International Journal of Data and Network Science*, *7*(1), 35–40. DOI: 10.5267/j.ijdns.2022.12.009

Al Ayyubi, M. C., Mahmudah, H., Saleh, A., & Rachmadi, R. R. (2020, July). Implementation of poka-yoke system to prevent human error in material preparation for industry. In 2020 International Seminar on Intelligent Technology and Its Applications (ISITIA) (pp. 273-278). IEEE. DOI: 10.1109/ISITIA49792.2020.9163707

Al Zarooni, L., & El Khatib, M. (2023). Robotics Process Automation (RPA) and Project Risk Management. [IJBAS]. *International Journal of Business Analytics and Security, 3*(1), 75–91. DOI: 10.54489/ijbas.v3i1.198

AlbertElijah, R. R. (2023). RPA: Transforming Organizational Processes with Intelligent Automation. *International Journal of Advanced Engineering Technologies and Innovations, 1*(02), 201–218.

Alberth, M., & Mattern, M. (2017). Understanding robotic process automation (RPA). *Journal of Financial Transformation, 46*, 54–61.

Albus, J. S. (1994). *A reference model architecture for intelligent systems design*. Citeseer.

Alejo, D., Cobano, J. A., Heredia, G., & Ollero, A. (2015, June). Collision-free trajectory planning based on maneuver selection-particle swarm optimization. In 2015 international conference on unmanned aircraft systems (ICUAS) (pp. 72-81). IEEE.

Alexenko, T., Biondo, M., Banisakher, D., & Skubic, M. (2013, March). Android-based speech processing for eldercare robotics. In *Proceedings of the companion publication of the 2013 international conference on Intelligent user interfaces companion* (pp. 87-88).

Ali, M. A., Babu, N. R., & Varghese, K. (2005, April). Collision free path planning of cooperative crane manipulators using genetic algorithm. *Journal of Computing in Civil Engineering, 19*(2), 182–193. DOI: 10.1061/(ASCE)0887-3801(2005)19:2(182)

Alimoradi, R., Aliee, F. S., & Farahani, B. (2023, November). An Efficient Method For Selecting and Prioritizing Business Processes In Robotic Process Automation. In *2023 7th Iranian Conference on Advances in Enterprise Architecture (ICAEA)* (pp. 85-92). IEEE. DOI: 10.1109/ICAEA60387.2023.10414471

Alirezazadeh, S. (2020). Optimal algorithm allocation for single robot cloud systems. *arXiv*. https://doi.org//arxiv.2003.08683DOI: 10.48550

Allgaier, J., Mulansky, L., Draelos, R. L., & Pryss, R. (2023). How does the model make predictions? A systematic literature review on the explainability power of machine learning in healthcare. *Artificial Intelligence in Medicine, 143*, 102616. DOI: 10.1016/j.artmed.2023.102616 PMID: 37673561

Alvarez, V. H., Larico, R., Ianos, Y., & Aznar, M. (2008, June). Parameter estimation for VLE calculation by global minimization: The genetic algorithm. *Brazilian Journal of Chemical Engineering, 25*(2), 409–418. DOI: 10.1590/S0104-66322008000200018

Anagnoste, S. (2017, July). Robotic Automation Process-The next major revolution in terms of back office operations improvement. In *Proceedings of the International Conference on Business Excellence* (Vol. 11, No. 1, pp. 676-686). DOI: 10.1515/picbe-2017-0072

Anagnoste, S. (2018). The road to intelligent automation in the energy sector. *Management Dynamics in the Knowledge Economy*, *6*(3), 489–502. DOI: 10.25019/MDKE/6.3.08

Anagnoste, S. (2018, May). Robotic Automation Process–The operating system for the digital enterprise. In *Proceedings of the International Conference on Business Excellence* (Vol. 12, No. 1, pp. 54-69). DOI: 10.2478/picbe-2018-0007

Andrade, D. (2020). Challenges of automated software testing with robotic process automation RPA—A comparative analysis of UiPath and Automation Anywhere. [IJICR]. *International Journal of Intelligent Computing and Research*, *11*(1), 1066–1072. DOI: 10.20533/ijicr.2042.4655.2020.0129

Antoniadis, J., Arumugam, P., Arumugam, S., Babak, S., Bagchi, M., Nielsen, A.-S. B., Bassa, C. G., Bathula, A., Berthereau, A., & Bonetti, M. (2023). The second data release from the European Pulsar Timing Array-III. Search for gravitational wave signals. *Astronomy & Astrophysics*, *678*, A50. DOI: 10.1051/0004-6361/202346844

Antwiadjei, L. (2021). Evolution of Business Organizations: An Analysis of Robotic Process Automation. Eduzone: International Peer Reviewed/Refereed Multidisciplinary Journal, 10(2), 101-105.

Anywhere, A. (2022, April 11). Revenue Cycle Management-RCM Automation. Automation Anywhere. https://www.automationanywhere.com/solutions/healthcare/rcm-automation

Arif, S., & MacNeil, M. A. (2022). Predictive models aren't for causal inference. *Ecology Letters*, *25*(8), 1741–1745. DOI: 10.1111/ele.14033 PMID: 35672133

Aristidou, A., Lasenby, J., Chrysanthou, Y., & Shamir, A. (2018, September). Inverse kinematics techniques in computer graphics: A survey. *Computer Graphics Forum*, *37*(6), 35–58.

Arocena, I., Huegun-Burgos, A., & Rekalde-Rodriguez, I. (2022). Robotics and Education: A Systematic Review. *TEM Journal*, *11*, 1.

Aryawati, N. P. A., Harahap, T. K., Yanti, N. N. S. A., Mahardika, I. M. N. O., Widiniarsih, D. M., Ahmad, M. I. S., Mattunruang, A. A., & Amali, L. M. (2023). *Manajemen keuangan*. Penerbit Tahta Media.

Asadov, J. (2023). Azerbaijan's Multicultural Policy of Ethnic Diversity. *Państwa Postradzieckie w Procesie Modernizacji, April.*

Asatiani, A., Copeland, O., & Penttinen, E. (2023). Deciding on the robotic process automation operating model: A checklist for RPA managers. *Business Horizons, 66*(1), 109–121. DOI: 10.1016/j.bushor.2022.03.004

Ashok, K., & Gopikrishnan, S. (2023). Statistical analysis of remote health monitoring based IoT security models & deployments from a pragmatic perspective. *IEEE Access : Practical Innovations, Open Solutions, 11*, 2621–2651. DOI: 10.1109/ACCESS.2023.3234632

Ata, A. A. (2007). Optimal trajectory planning of manipulators: a review. Journal of Engineering Science and technology, 2(1), 32-54.

Atiyah, A. N., Adzhar, N., & Jaini, N. I. (2021, July). An overview: On path planning optimization criteria and mobile robot navigation. [). IOP Publishing.]. *Journal of Physics: Conference Series, 1988*(1), 012036.

Attaran, M. (2023). The impact of 5G on the evolution of intelligent automation and industry digitization. *Journal of Ambient Intelligence and Humanized Computing, 14*(5), 5977–5993. DOI: 10.1007/s12652-020-02521-x PMID: 33643481

Axmann, B., & Harmoko, H. (2020, September). Robotic process automation: An overview and comparison to other technology in industry 4.0. In *2020 10th International Conference on Advanced Computer Information Technologies (ACIT)* (pp. 559-562). IEEE.

Axmann, B., & Harmoko, H. (2022). Process & software selection for robotic process automation (RPA). *Tehnički Glasnik, 16*(3), 412–419. DOI: 10.31803/tg-20220417182552

Ayoobi, H., Kasaei, H., Cao, M., Verbrugge, R., & Verheij, B. (2022). Local-HDP: Interactive open-ended 3D object category recognition in real-time robotic scenarios. *Robotics and Autonomous Systems, 147*, 103911. DOI: 10.1016/j.robot.2021.103911

Ayvaz, S., & Alpay, K. (2021). Predictive maintenance system for production lines in manufacturing: A machine learning approach using IoT data in real-time. *Expert Systems with Applications, 173*, 114598. DOI: 10.1016/j.eswa.2021.114598

Badmus, A. D. (2023). Implementing Technology Manufacturing with Robotics Process Automation (RPA). *Journal of Knowledge Learning and Science Technology ISSN: 2959-6386 (online),* 2(2), 96-103.

Bakker, P., & Kuniyoshi, Y. (1996, April). Robot see, robot do: An overview of robot imitation. In AISB96 Workshop on Learning in Robots and Animals (Vol. 5, pp. 3-11).

Balaguru, S. (n.d.). *Beyond automation: Redefining healthcare revenue cycles through RPA*. NLP and Gen AI.

Bamshad, A., & Cho, H. J. (2021). Laserjet printed micro/nano sensors and microfluidic systems: A simple and facile digital platform for inexpensive, flexible, and low-volume devices. *Advanced Materials Technologies*, *6*(12), 2100401. DOI: 10.1002/admt.202100401

Banala, S. (2024). The future of IT operations: Harnessing cloud automation for enhanced efficiency and the role of generative AI operational excellence. *International Journal of Machine Learning and Artificial Intelligence*, *5*(5), 1–15.

Banks, A., Vincent, J., & Anyakoha, C. (2008, March 1). A review of particle swarm optimization. Part II: Hybridisation, combinatorial, multicriteria and constrained optimization, and indicative applications. *Natural Computing*, *7*(1), 109–124. DOI: 10.1007/s11047-007-9050-z

Bartz-Beielstein, T., & Zaefferer, M. (2017, June 1). Model-based methods for continuous and discrete global optimization. *Applied Soft Computing*, *55*, 154–167. DOI: 10.1016/j.asoc.2017.01.039

Battling, S., Pastoors, J., Deitert, A., Götzen, T., Hartmann, L., Schröder, E., Yordanov, S., & Büchs, J. (2022). Development of a novel defined minimal medium for Gluconobacter oxydans 621H by systematic investigation of metabolic demands. *Journal of Biological Engineering*, *16*(1), 31. DOI: 10.1186/s13036-022-00310-y PMID: 36414992

Bédard, M., Leshob, A., Benzarti, I., Mili, H., Rab, R., & Hussain, O. (2024). A rule-based method to effectively adopt robotic process automation. *Journal of Software (Malden, MA)*, *36*(11), 2709. DOI: 10.1002/smr.2709

Beer Mohamed, M. I., & Hassan, M. F. (2022, November). Adaptive security for cognitive robotic process automation in enterprise computing using AI-powered analytical engine. In *International Conference on Artificial Intelligence for Smart Community: AISC 2020, 17–18 December, Universiti Teknologi Petronas, Malaysia* (pp. 825-835). Singapore: Springer Nature Singapore. DOI: 10.1007/978-981-16-2183-3_78

Beerbaum, D. (2023). Generative artificial intelligence (GAI) ethics taxonomy-applying ChatGPT for robotic process automation (GAI-RPA) as business case. *Available atSSRN* 4385025.

Beerbaum, D. (2023). Generative artificial intelligence (GAI) with ChatGPT for accounting–a business case. *Available atSSRN* 4385651.

Beithou, N., Mansour, M. A., Abdellatif, N., Alsaqoor, S., Tarawneh, S., Jaber, A. H., Andruszkiewicz, A., Alsqour, M., Borowski, G., & Alahmer, A. (2023). Effect of the Residential Photovoltaic Systems Evolution on Electricity and Thermal Energy Usage in Jordan. *Advances in Science and Technology. Research Journal, 17*(3).

Bellman, M., & Göransson, G. (2019). *Intelligent process automation: building the bridge between Robotic Process Automation and artificial intelligence*.

Bender, E. M., Gebru, T., McMillan-Major, A., & Shmitchell, S. (2021). On the dangers of stochastic parrots: Can language models be too big? . *Proceedings of the 2021 ACM Conference on Fairness, Accountability, and Transparency*, 610–623. DOI: 10.1145/3442188.3445922

Benko Loknar, M., Blažič, S., & Klančar, G. (2021, October 19). Minimum-time velocity profile planning for planar motion considering velocity, acceleration and jerk constraints. *International Journal of Control*, •••, 1–5.

Berruti, A., Grisanti, S., Pulzer, A., Claps, M., Daffara, F., Loli, P., Mannelli, M., Boscaro, M., Arvat, E., Tiberio, G., Hahner, S., Zaggia, B., Porpiglia, F., Volante, M., Fassnacht, M., & Terzolo, M. (2017). Long-Term Outcomes of Adjuvant Mitotane Therapy in Patients With Radically Resected Adrenocortical Carcinoma. *The Journal of Clinical Endocrinology and Metabolism, 102*(4), 1358–1365. DOI: 10.1210/jc.2016-2894 PMID: 28324035

Berti, A., Merlet, J. P., & Carricato, M. (2016, May). Solving the direct geometrico-static problem of underconstrained cable-driven parallel robots by interval analysis. *The International Journal of Robotics Research, 35*(6), 723–739. DOI: 10.1177/0278364915595277

Bestaoui, Y. (1992, February). On-line reference trajectory definition with joint torque and velocity constraints. *The International Journal of Robotics Research, 11*(1), 75–85. DOI: 10.1177/027836499201100106

Betts, J. T. (1998, March). Survey of numerical methods for trajectory optimization. *Journal of Guidance, Control, and Dynamics, 21*(2), 193–207. DOI: 10.2514/2.4231

Beudaert, X., Lavernhe, S., & Tournier, C. (2012, June 1). Feedrate interpolation with axis jerk constraints on 5- axis NURBS and G1 tool path. *International Journal of Machine Tools & Manufacture*, *57*, 73–82. DOI: 10.1016/j.ijmachtools.2012.02.005

Beuerle, M. (2019). *Development of low-scaling methods to calculate ground state energies and analytical gradients based on the adiabatic-connection fluctuation-dissipation theorem* (Doctoral dissertation, lmu).

Bharathi, A., & Dong, J. (2016, February 1). Feedrate optimization for smooth minimum-time trajectory generation with higher order constraints. *International Journal of Advanced Manufacturing Technology*, *82*(5-8), 1029–1040. DOI: 10.1007/s00170-015-7447-x

Bhattacharyya, S., Banerjee, J. S., & De, D. (Eds.). (2023). *Confluence of Artificial Intelligence and Robotic Process Automation* (Vol. 335). Springer Nature. DOI: 10.1007/978-981-19-8296-5

Binns, R. (2018). Fairness in machine learning: Lessons from political philosophy. *Conference on Fairness, Accountability and Transparency*, 149–159.

Bi, Q. (2023). Analysis of the application of generative AI in business management. *Advances in Economics and Management Research*, *6*(1), 36–36. DOI: 10.56028/aemr.6.1.36.2023

Boryga, M. (2020). The use of higher-degree polynomials for trajectory planning with jerk, acceleration and velocity constraints. *International Journal of Computer Applications in Technology*, *63*(4), 337–347. DOI: 10.1504/IJCAT.2020.110414

Boscariol, P., & Gasparetto, A. (2013, August 1). Model-based trajectory planning for flexible-link mechanisms with bounded jerk. *Robotics and Computer-integrated Manufacturing*, *29*(4), 90–99. DOI: 10.1016/j.rcim.2012.11.003

Bosco, A., Augusto, A., Dumas, M., Rosa, M., & Fortino, G. (2019). Discovering automatable routines from user interaction logs. *In Advances in Intelligent Systems and Computing* (pp. 144-162). https://doi.org/DOI: 10.1007/978-3-030-26643-1_9

Busscher, P., Williams, R. L.II, Bryson, L. S., & Castro-Lacouture, D. (2007, November 1). Cable-suspended robotic contour crafting system. *Automation in Construction*, *17*(1), 45–55. DOI: 10.1016/j.autcon.2007.02.011

Bower, D. C., & Walker, D. H. (2007). Planning knowledge for phased rollout projects. *Project Management Journal*, *38*(3), 45–60. DOI: 10.1002/pmj.20005

Brito, T. Intelligent collision avoidance system for industrial manipulators, Doctoral Dissertation), http://hdl.handle.net/10198/19319

Brynjolfsson, E., Li, D., & Raymond, L. R. (2023). *Generative AI at work (No. w31161)*. National Bureau of Economic Research. DOI: 10.3386/w31161

Brynjolfsson, E., & Mcafee, A. (2017). Artificial intelligence, for real. *Harvard Business Review*, *1*, 1–31.

Bui, D. P., Bajema, K. L., Huang, Y., Yan, L., Li, Y., Rajeevan, N., Berry, K., Rowneki, M., Argraves, S., & Hynes, D. (2023). Development of a prediction model for 30-day COVID-19 hospitalization and death in a national cohort of Veterans Health Administration patients-March 2022-April 2023. MedRxiv, 2011–2023. DOI: 10.1101/2023.11.17.23298653

Business Process 101 : Definition, Steps and example [Guide for 2024]. (2024, May 15). https://kissflow.com/workflow/bpm/business-process/

Bygstad, B. (2017). Generative innovation: A comparison of lightweight and heavyweight IT. *Journal of Information Technology*, *32*(2), 180–193. DOI: 10.1057/jit.2016.15

Cakir, M., Guvenc, M. A., & Mistikoglu, S. (2021). The experimental application of popular machine learning algorithms on predictive maintenance and the design of IIoT based condition monitoring system. *Computers & Industrial Engineering*, *151*, 106948. DOI: 10.1016/j.cie.2020.106948

Campbell, S., Naeem, W., & Irwin, G. W. (2012, December 1). A review on improving the autonomy of unmanned surface vehicles through intelligent collision avoidance manoeuvres. *Annual Reviews in Control*, *36*(2), 267–283. DOI: 10.1016/j.arcontrol.2012.09.008

Canali, F., Guarino Lo Bianco, C., & Locatelli, M. (2014, June 3). Minimum-jerk online planning by a mathematical programming approach. *Engineering Optimization*, *46*(6), 763–783. DOI: 10.1080/0305215X.2013.806916

Carabin, G., Scalera, L., Wongratanaphisan, T., & Vidoni, R. (2021, February 1). An energy-efficient approach for 3D printing with a Linear Delta Robot equipped with optimal springs. *Robotics and Computer-integrated Manufacturing*, *67*, 102045. DOI: 10.1016/j.rcim.2020.102045

Cernat, M., Staicu, A. N., & Stefanescu, A. (2020, November). Towards automated testing of RPA implementations. In *Proceedings of the 11th ACM SIGSOFT International Workshop on Automating TEST Case Design, Selection, and Evaluation* (pp. 21-24).

Chakraborti, T., Isahagian, V., Khalaf, R., Khazaeni, Y., Muthusamy, V., Rizk, Y., & Unuvar, M. (2020). From Robotic Process Automation to Intelligent Process Automation: –Emerging Trends–. In Business Process Management: Blockchain and Robotic Process Automation Forum: BPM 2020 Blockchain and RPA Forum, Seville, Spain, September 13–18, 2020, Proceedings 18 (pp. 215-228). Springer International Publishing.0

Chakraborti, T., Sreedharan, S., Grover, S., & Kambhampati, S. (2019, March). Plan explanations as model reconciliation--an empirical study. In *2019 14th ACM/IEEE International Conference on Human-Robot Interaction (HRI)* (pp. 258-266). Ieee.

Chakraborti, T., Isahagian, V., Khalaf, R., Khazaeni, Y., Muthusamy, V., Rizk, Y., & Unuvar, M. (2020). From Robotic Process Automation to Intelligent Process Automation: –Emerging Trends–. *Business Process Management: Blockchain and Robotic Process Automation Forum: BPM 2020 Blockchain and RPA Forum, Seville, Spain, September 13–18, 2020. Proceedings*, *18*, 215–228.

Chakraborti, T., Rizk, Y., Isahagian, V., Aksar, B., & Fuggitti, F. (2022). From natural language to workflows: Towards emergent intelligence in robotic process automation. *International Conference on Business Process Management*, 123–137. DOI: 10.1007/978-3-031-16168-1_8

Chakraborty, R., Asthana, A., Singh, A. K., Jain, B., & Susan, A. B. H. (2022). Adsorption of heavy metal ions by various low-cost adsorbents: A review. *International Journal of Environmental Analytical Chemistry*, *102*(2), 342–379. DOI: 10.1080/03067319.2020.1722811

Chanopas, A., Krairit, D., & Ba Khang, D. (2006). Managing information technology infrastructure: A new flexibility framework. *Management Research News*, *29*(10), 632–651. DOI: 10.1108/01409170610712335

Chavan, P., & Chavan, P. (2024, June). Data migration between on-prem to cloud using generative AI to reduce costing and overheads. In *2024 International Conference on Innovations and Challenges in Emerging Technologies (ICICET)* (pp. 1-7). IEEE.

Cheligeer, C., Huang, J., Wu, G., Bhuiyan, N., Xu, Y., & Zeng, Y. (2022). Machine learning in requirements elicitation: A literature review. *Artificial Intelligence for Engineering Design, Analysis and Manufacturing*, *36*, e32. Advance online publication. DOI: 10.1017/S0890060422000166

Chen, C. T., & Pham, H. V. (2012, January). Trajectory planning in parallel kinematic manipulators using a constrained multi-objective evolutionary algorithm. *Nonlinear Dynamics*, *67*(2), 1669–1681. DOI: 10.1007/s11071-011-0095-2

Chen, M. F., Chen, Y. P., & Hsiao, W. T. (2009). Correction of field distortion of laser marking systems using surface compensation function. *Optics and Lasers in Engineering*, *47*(1), 84–89. DOI: 10.1016/j.optlaseng.2008.07.017

Chen, N., Zhou, M., Dong, X., Qu, J., Gong, F., Han, Y., Qiu, Y., Wang, J., Liu, Y., Wei, Y., Xia, J., Yu, T., Zhang, X., & Zhang, L. (2020). Epidemiological and clinical characteristics of 99 cases of 2019 novel coronavirus pneumonia in Wuhan, China: A descriptive study. *Lancet*, *395*(10223), 507–513. DOI: 10.1016/S0140-6736(20)30211-7 PMID: 32007143

Chen, Y. C. (1991, October). Solving robot trajectory planning problems with uniform cubic B- splines. *Optimal Control Applications & Methods*, *12*(4), 247–262. DOI: 10.1002/oca.4660120404

Chen, Y., Agostini, L., Moretti, G., Fontana, M., & Vertechy, R. (2019, October 1). Dielectric elastomer materials for large-strain actuation and energy harvesting: A comparison between styrenic rubber, natural rubber and acrylic elastomer. *Smart Materials and Structures*, *28*(11), 114001. DOI: 10.1088/1361-665X/ab3b32

Cheong, T.-S., & Yoon, C.-H. (1999). A memory based classifier using the recursive partition averaging. *Proceedings of IEEE. IEEE Region 10 Conference. TENCON 99.'Multimedia Technology for Asia-Pacific Information Infrastructure'(Cat. No. 99CH37030)*, *2*, 1038–1041. DOI: 10.1109/TENCON.1999.818599

Chettibi, T. (2019, March 1). Smooth point-to-point trajectory planning for robot manipulators by using radial basis functions. *Robotica*, *37*(3), 539–559. DOI: 10.1017/S0263574718001169

Chettibi, T., Lehtihet, H. E., Haddad, M., & Hanchi, S. (2004, July 1). Minimum cost trajectory planning for industrial robots. *European Journal of Mechanics. A, Solids*, *23*(4), 703–715. DOI: 10.1016/j.euromechsol.2004.02.006

Chiddarwar, S. S., & Babu, N. R. (2012, January 1). Optimal trajectory planning for industrial robot along a specified path with payload constraint using trigonometric splines. *International Journal of Automation and Control.*, *6*(1), 39–65. DOI: 10.1504/IJAAC.2012.045439

Choi, D., R'bigui, H., & Cho, C. (2022). Enabling the gap between RPA and process mining: User interface interactions recorder. *IEEE Access : Practical Innovations, Open Solutions*, *10*, 39604–39612. DOI: 10.1109/ACCESS.2022.3165797

Chou LS, Song SM. Geometric work of manipulators and path planning based on minimum energy consumption.

Chowdhury, P., Paul, S. K., Kaisar, S., & Moktadir, M. A. (2021). COVID-19 pandemic related supply chain studies: A systematic review. *Transportation Research Part E, Logistics and Transportation Review*, *148*, 102271. DOI: 10.1016/j.tre.2021.102271 PMID: 33613082

Chugh, R., Macht, S., & Hossain, R. (2022). Robotic Process Automation: A review of organizational grey literature. *International Journal of Information Systems and Project Management*, *10*(1), 5–26. DOI: 10.12821/ijispm100101

Clodic, A., Fleury, S., Alami, R., Chatila, R., Bailly, G., Brethes, L., . . . Montreuil, V. (2006, September). Rackham: An interactive robot-guide. In ROMAN 2006-the 15th IEEE International Symposium on Robot and Human Interactive Communication (pp. 502-509). IEEE.

Codourey, A. (1998, December). Dynamic modeling of parallel robots for computed-torque control implementation. *The International Journal of Robotics Research*, *17*(12), 1325–1336. DOI: 10.1177/027836499801701205

Colbert, A., Yee, N., & George, G. (2016). The digital workforce and the workplace of the future. []). Academy of Management Briarcliff Manor, NY.]. *Academy of Management Journal*, *59*(3), 731–739. DOI: 10.5465/amj.2016.4003

Colombo, C., Burgel, P.-R., Gartner, S., van Koningsbruggen-Rietschel, S., Naehrlich, L., Sermet-Gaudelus, I., & Southern, K. W. (2020). Impact of COVID-19 on people with cystic fibrosis. *The Lancet. Respiratory Medicine*, *8*(5), e35–e36. DOI: 10.1016/S2213-2600(20)30177-6 PMID: 32304639

Connolly, F., Walsh, C. J., & Bertoldi, K. (2017, January 3). Automatic design of fiber-reinforced soft actuators for trajectory matching. *Proceedings of the National Academy of Sciences of the United States of America*, *114*(1), 51–56. DOI: 10.1073/pnas.1615140114 PMID: 27994133

Consolini, L., Locatelli, M., & Minari, A. (2021). A sequential approach for speed planning under jerk constraints. arXiv preprint arXiv:2105.15095.

Consolini, L., Locatelli, M., Minari, A., & Piazzi, A. (2017, May 1). An optimal complexity algorithm for minimumtime velocity planning. *Systems & Control Letters*, *103*, 50–57. DOI: 10.1016/j.sysconle.2017.02.001

Cooper, L. A., Holderness, D. K.Jr, Sorensen, T. L., & Wood, D. A. (2019). Robotic process automation in public accounting. *Accounting Horizons*, *33*(4), 15–35. DOI: 10.2308/acch-52466

Costa, D. A. D. S., Mamede, H. S., & Silva, M. M. D. (2022). Robotic Process Automation (RPA) adoption: a systematic literature review. *Engineering Management in Production and Services, 14*(2).

Czarnecki, C., & Fettke, P. (2021). *Robotic process automation: Management, technology, applications*. De Gruyter. DOI: 10.1515/9783110676693

Dahiya, R. S., & Valle, M. (2008). *Tactile sensing for robotic applications*. Sensors, Focus on Tactile, Force and Stress Sensors.

Damian, C. S., Devarajan, Y., & Jayabal, R. (2024). Biodiesel production in India: Prospects, challenges, and sustainable directions. *Biotechnology and Bioengineering, 121*(3), 894–902. DOI: 10.1002/bit.28643 PMID: 38164703

Davenport, T. H., & Ronanki, R. (2018). Artificial intelligence for the real world. *Harvard Business Review, 96*(1), 108–116.

de Almeida, P. G. R., dos Santos, C. D., & Farias, J. S. (2021). Artificial intelligence regulation: A framework for governance. *Ethics and Information Technology, 23*(3), 505–525. DOI: 10.1007/s10676-021-09593-z

De Barrie, D., Pandya, M., Pandya, H., Hanheide, M., & Elgeneidy, K. (2021). A deep learning method for vision based force prediction of a soft fin ray gripper using simulation data. *Frontiers in Robotics and AI, 8*, 631371. DOI: 10.3389/frobt.2021.631371 PMID: 34113655

De Simone, V., Di Pasquale, V., & Miranda, S. (2023). An overview on the use of AI/ML in Manufacturing MSMEs: Solved issues, limits, and challenges. *Procedia Computer Science, 217*, 1820–1829. DOI: 10.1016/j.procs.2022.12.382

Dechamma, P. R., & Shobha, N. S. (2020). A review on robotic process automation. *International Journal of Research in Engineering. Science and Management, 3*(5), 237–244.

Deepak, A.. (2024). Image Processing based Robotic Car for Agricultural Ploughing using Machine Learning Approach. *International Journal of Intelligent Systems and Applications in Engineering, 12*(2s), 718–724.

Deloitte. (2017,). Reimagining business process reengineering in the age of automation.https://www2.deloitte.com/content/dam/Deloitte/us/Documents/public-sector/us-fed-reimagining-business-process-reengineering-in-the-age-of-automation.pdf

Devarajan, Y. (2018). A study of robotic process automation use cases today for tomorrow's business. *International Journal of Computers and Technology, 5*(6), 12–18.

Devarajan, Y. (2018). A study of robotic process automation uses cases today for tomorrow's business. *International Journal of Computers and Technology*, *5*(6), 12–18.

Devlin, J. (2018). Bert: Pre-training of deep bidirectional transformers for language understanding. *ArXiv Preprint ArXiv:1810.04805*.

Diao, Q., Zeng, Y., & Chen, J. (2024). The Applications and Latest Progress of Ceramic 3D Printing. Additive Manufacturing Frontiers, 200113.

Dietmair, A., & Verl, A. (2009, June 1). A generic energy consumption model for decision making and energy efficiency optimisation in manufacturing. *International Journal of Sustainable Engineering*, *2*(2), 123–133. DOI: 10.1080/19397030902947041

Dietrich, A., Wu, X., Bussmann, K., Harder, M., Iskandar, M., Englsberger, J., Ott, C., & Albu-Schäffer, A. (2021, September 1). Albu- chäffer A. Practical consequences of inertia shaping for interaction and tracking in robot control. *Control Engineering Practice*, *114*, 104875. DOI: 10.1016/j.conengprac.2021.104875

Dinçer, U., & Çevik, M. (2019, February 1). Improved trajectory planning of an industrial parallel mechanism by a composite polynomial consisting of Bezier curves and cubic polynomials. *Mechanism and Machine Theory*, *132*, 248–263. DOI: 10.1016/j.mechmachtheory.2018.11.009

Dong, J., Wang, T., Li, B., & Ding, Y. (2014, January 1). Smooth feedrate planning for continuous short line tool path with contour error constraint. *International Journal of Machine Tools & Manufacture*, *76*, 1–2. DOI: 10.1016/j.ijmachtools.2013.09.009

Douglas, R., & Roger, J. (2024). The era of intelligent automation: Cloud computing, AI, and beyond. *International Journal of Advanced Engineering Technologies and Innovations*, *1*(1), 390–414.

Duan, Y., Edwards, J. S., & Dwivedi, Y. K. (2019). Artificial intelligence for decision making in the era of Big Data–evolution, challenges and research agenda. *International Journal of Information Management*, *48*, 63–71. DOI: 10.1016/j.ijinfomgt.2019.01.021

Dubowsky, S., Norris, M., & Shiller, Z. (1986, April). Time optimal trajectory planning for robotic manipulators with obstacle avoidance: A CAD approach. In Proceedings. 1986 []. IEEE.]. *IEEE International Conference on Robotics and Automation*, *3*, 1906–1912.

Dubowsky, S., & Papadopoulos, E. (1993, October). The kinematics, dynamics, and control of free-flying and freefloating space robotic systems. *IEEE Transactions on Robotics and Automation*, *9*(5), 531–543. DOI: 10.1109/70.258046

Duindam, V., Alterovitz, R., Sastry, S., & Goldberg, K. (2008, May). Screw-based motion planning for bevel-tip flexible needles in 3D environments with obstacles. In 2008 IEEE international conference on robotics and automation (pp. 2483-2488). IEEE.

Dumas, M., Rosa, L. M., Mendling, J., & Reijers, A. H. (2018). Fundamentals of business process management. Springer-Verlag.L

Durna, A., Fries, J., Hrabovsky, L., Sliva, A., & Zarnovsky, J. (2020). Research and development of laser engraving and material cutting machine from 3D printer. *Management Systems in Production Engineering*, *28*(1), 47–52. DOI: 10.2478/mspe-2020-0008

El-Attar, M., & Miller, J. (2010). Developing comprehensive acceptance tests from use cases and robustness diagrams. *Requirements Engineering*, *15*(3), 285–306. DOI: 10.1007/s00766-009-0088-6

Elbanhawi, M., Simic, M., & Jazar, R. N. (2015, December 1). Continuous path smoothing for car-like robots using Bspline curves. *Journal of Intelligent & Robotic Systems*, *80*(1), 23–56. DOI: 10.1007/s10846-014-0172-0

Ella, R., Reddy, S., Blackwelder, W., Potdar, V., Yadav, P., Sarangi, V., Aileni, V. K., Kanungo, S., Rai, S., Reddy, P., Verma, S., Singh, C., Redkar, S., Mohapatra, S., Pandey, A., Ranganadin, P., Gumashta, R., Multani, M., Mohammad, S., & Waghmare, S. (2021). Efficacy, safety, and lot-to-lot immunogenicity of an inactivated SARS-CoV-2 vaccine (BBV152): Interim results of a randomised, double-blind, controlled, phase 3 trial. *Lancet*, *398*(10317), 2173–2184. DOI: 10.1016/S0140-6736(21)02000-6 PMID: 34774196

Elliott, C. (2010). Botnets: To what extent are they a threat to information security? *Information Security Technical Report*, *15*(3), 79–103. DOI: 10.1016/j.istr.2010.11.003

Enríquez, J. G., Jiménez-Ramírez, A., Domínguez-Mayo, F. J., & García-García, J. A. (2020). Robotic process automation: A scientific and industrial systematic mapping study. *IEEE Access : Practical Innovations, Open Solutions*, *8*, 39113–39129. DOI: 10.1109/ACCESS.2020.2974934

Ens, B., Bach, B., Cordeil, M., Engelke, U., Serrano, M., Willett, W., Prouzeau, A., Anthes, C., Büschel, W., & Dunne, C. (2021). Grand challenges in immersive analytics. *Proceedings of the 2021 CHI Conference on Human Factors in Computing Systems*, 1–17.

Erdogan, R., Saglam, Z., Cetintav, G., & Karaoglan Yilmaz, F. G. (2023). Examination of the usability of Tinkercad application in educational robotics teaching by eye tracking technique. *Smart Learning Environments, 10*(1), 27.

Esteva, A., Chou, K., Yeung, S., Naik, N., Madani, A., Mottaghi, A., Liu, Y., Topol, E., Dean, J., & Socher, R. (2021). Deep learning-enabled medical computer vision. *NPJ Digital Medicine, 4*(1), 5. DOI: 10.1038/s41746-020-00376-2 PMID: 33420381

Ezair, B., Tassa, T., & Shiller, Z. (2014, May). Planning high order trajectories with general initial and final conditions and asymmetric bounds. *The International Journal of Robotics Research, 33*(6), 898–916. DOI: 10.1177/0278364913517148

Fan, C., Miller, K., & Mitra, S. (2020, July). Fast and guaranteed safe controller synthesis for nonlinear vehicle models. In *International Conference on Computer Aided Verification* (pp. 629-652). Cham: Springer International Publishing.

Fang, Y., Hu, J., Liu, W., Shao, Q., Qi, J., & Peng, Y. (2019, July 1). Smooth and time-optimal S-curve trajectory planning for automated robots and machines. *Mechanism and Machine Theory, 137*, 127–153. DOI: 10.1016/j.mechmachtheory.2019.03.019

Fang, Y., Qi, J., Hu, J., Wang, W., & Peng, Y. (2020, November 1). An approach for jerk-continuous trajectory generation of robotic manipulators with kinematical constraints. *Mechanism and Machine Theory, 153*, 103957. DOI: 10.1016/j.mechmachtheory.2020.103957

Fan, W., Gao, X. S., Yan, W., & Yuan, C. M. (2012, September). Interpolation of parametric CNC machining path under confined jounce. *International Journal of Advanced Manufacturing Technology, 62*(5), 719–739. DOI: 10.1007/s00170-011-3842-0

Fan, W., Lee, C. H., & Chen, J. H. (2015, September 1). A realtime curvature-smooth interpolation scheme and motion planning for CNC machining of short line segments. *International Journal of Machine Tools & Manufacture, 96*, 27–46. DOI: 10.1016/j.ijmachtools.2015.04.009

Farhadi, M., & Zhao, L. (2024). Exploring the Impact of Iran-China Trade on Environmental Sustainability. *Journal of Energy and Environmental Policy Options, 7*(1), 1–8.

Farid, Y., & Ruggiero, F. (2022, January 1). Finite-time extended state observer and fractional-order sliding mode controller for impulsive hybrid port-Hamiltonian systems with input delay and actuators saturation: Application to ball-juggler robots. *Mechanism and Machine Theory, 167*, 04577. DOI: 10.1016/j.mechmachtheory.2021.104577

Feddema, J. T. (1996, April). Kinematically optimal robot placement for minimum time coordinated motion. In proceedings of IEEE international conference on robotics and automation (Vol. 4, pp. 3395-3400). IEEE.

Feng, Y., Yao-nan, W., & Yi-min, Y. (2014, September 18). Inverse kinematics solution for robot manipulator based on neural network under joint subspace. *International Journal of Computers, Communications & Control*, *7*(3), 459–472. DOI: 10.15837/ijccc.2012.3.1387

Fernandez, D., & Aman, A. (2021). Planning for a successful Robotic Process Automation (RPA) project: A case study. [JIKM]. *Journal of Information and Knowledge Management*, *11*(1), 103–117.

Fernández-Loría, C., & Loría, J. (2022). Learning to Rank by Causal Effects Without Data to Accurately Estimate Causal Effects. *ArXiv Preprint ArXiv:2206.12532*.

Fettke, P., & Strohmeier, S. (2022). HR robotic process automation. In *Handbook of Research on Artificial Intelligence in Human Resource Management* (pp. 187–206). Edward Elgar Publishing.

Feuerriegel, S., Hartmann, J., Janiesch, C., & Zschech, P. (2024). Generative AI. *Business & Information Systems Engineering*, *66*(1), 111–126. DOI: 10.1007/s12599-023-00834-7

Fivesdigital. (n.d.). How the healthcare industry is benefitting from Robotic Process Automation. Copyright 2023 @ fivesdigital.com. https://www.fivesdigital.com/blog/how-the-healthcare-industry-is-benefitting-from-robotic-process-automation

Flash, T., Meirovitch, Y., & Barliya, A. (2013, April 1). Models of human movement: Trajectory planning and inverse kinematics studies. *Robotics and Autonomous Systems*, *61*(4), 330–339. DOI: 10.1016/j.robot.2012.09.020

Flechsig, C. (2021). The impact of intelligent process automation on purchasing and supply management–Initial insights from a multiple case study. *Logistics Management: Contributions of the Section Logistics of the German Academic Association for Business Research, 2021, Dresden, Germany*, 67–89.

Foehn, P., Romero, A., & Scaramuzza, D. (2021, July 21). Time-optimal planning for quadrotor waypoint flight. *Science Robotics*, *6*(56), eabh1221. DOI: 10.1126/scirobotics.abh1221 PMID: 34290102

Fonseca, C. M., & Fleming, P. J. (1995, March). An overview of evolutionary algorithms in multiobjective optimization. *Evolutionary Computation*, *3*(1), 1–6. DOI: 10.1162/evco.1995.3.1.1

Fraichard, T. (1998, January 1). Trajectory planning in a dynamic workspace: A′ state-time space' approach. *Advanced Robotics, 13*(1), 75–94. DOI: 10.1163/156855399X00928

Furman, J., & Seamans, R. (2019). AI and the Economy. *Innovation Policy and the Economy, 19*(1), 161–191. DOI: 10.1086/699936

Furnham, A., & Chamorro-Premuzic, T. (2004). Estimating one's own personality and intelligence scores. *British Journal of Psychology, 95*(2), 149–160. DOI: 10.1348/000712604773952395 PMID: 15142299

Gangolli, L. V., Duggal, R., & Shukla, A. (2005). Review of healthcare in India. Mumbai: Centre for enquiry into health and allied themes.

Garcia, R. R., Bittencourt, A. C., & Villani, E. (2018, September). Relevant factors for the energy consumption of industrial robots. *Journal of the Brazilian Society of Mechanical Sciences and Engineering, 40*(9), 1–5. DOI: 10.1007/s40430-018-1376-1

Garg, D. P., & Kumar, M. (2002, June 1). Optimization techniques applied to multiple manipulators for path planning and torque minimization. *Engineering Applications of Artificial Intelligence, 15*(3-4), 241–252. DOI: 10.1016/S0952-1976(02)00067-2

Gasparetto, A., & Zanotto, V. (2007, April 1). A new method for smooth trajectory planning of robot manipulators. *Mechanism and Machine Theory, 42*(4), 455–471. DOI: 10.1016/j.mechmachtheory.2006.04.002

Gasparetto, A., & Zanotto, V. (2008, June 1). A technique for time-jerk optimal planning of robot trajectories. *Robotics and Computer-integrated Manufacturing, 24*(3), 415–426. DOI: 10.1016/j.rcim.2007.04.001

Gasparetto, A., & Zanotto, V. (2010, April 1). Optimal trajectory planning for industrial robots. *Advances in Engineering Software, 41*(4), 548–556. DOI: 10.1016/j.advengsoft.2009.11.001

Geiger, M., Jacot, A., Spigler, S., Gabriel, F., Sagun, L., d'Ascoli, S., Biroli, G., Hongler, C., & Wyart, M. (2020). Scaling description of generalization with number of parameters in deep learning. *Journal of Statistical Mechanics, 2020*(2), 23401. DOI: 10.1088/1742-5468/ab633c

Geyer-Klingeberg, J., Nakladal, J., Baldauf, F., & Veit, F. (2018). Process Mining and Robotic Process Automation: A Perfect Match. *BPM (Dissertation/Demos/Industry), 2196*, 124-131.

Geyer-Klingeberg, J., Nakladal, J., Baldauf, F., & Veit, F. (2018). Process Mining and Robotic Process Automation: A Perfect Match. *BPM (Dissertation/Demos/ Industry), 2196*, 124–131.

Ghaisas, S., & Singhal, A. (2024). Dealing with data for RE: Mitigating challenges using NLP and generative AI. *arXiv preprint arXiv:2402.16977.*

Gianni, M., & Gotzamani, K. (2015). Management systems integration: Lessons from an abandonment case. *Journal of Cleaner Production, 86*, 265–276. DOI: 10.1016/j.jclepro.2014.08.023

Gill, S. S., Xu, M., Ottaviani, C., Patros, P., Bahsoon, R., Shaghaghi, A., Golec, M., Stankovski, V., Wu, H., Abraham, A., Singh, M., Mehta, H., Ghosh, S. K., Baker, T., Parlikad, A. K., Lutfiyya, H., Kanhere, S. S., Sakellariou, R., Dustdar, S., & Uhlig, S. (2022). AI for next generation computing: Emerging trends and future directions. *Internet of Things : Engineering Cyber Physical Human Systems, 19*, 100514. DOI: 10.1016/j.iot.2022.100514

Glohr, C., Kellermann, J., & Dörnemann, H. (2013). The IT factory: a vision of standardization and automation. In *The Road to a Modern IT Factory: Industrialization–Automation–Optimization* (pp. 101–109). Springer Berlin Heidelberg.

Goerzen, C., Kong, Z., & Mettler, B. (2010, January). A survey of motion planning algorithms from the perspective of autonomous UAV guidance. *Journal of Intelligent & Robotic Systems, 57*(1), 65–100. DOI: 10.1007/s10846-009-9383-1

Gonzalez, F. (2023). Learning robot differential movements using a new educational robotics simulation tool. *Education and Information Technologies, 28*(9), 1–24. DOI: 10.1007/s10639-022-11433-6 PMID: 36846490

Goodfellow, I., Pouget-Abadie, J., Mirza, M., Xu, B., Warde-Farley, D., Ozair, S., Courville, A., & Bengio, Y. (2020). Generative adversarial networks. *Communications of the ACM, 63*(11), 139–144. DOI: 10.1145/3422622

Gordon, N. J., Salmond, D. J., & Smith, A. F. (1993, April). Novel approach to nonlinear/non-Gaussian Bayesian state estimation. In IEE proceedings F (radar and signal processing) (Vol. 140, No. 2, pp. 107-113). IEE.

Gosen, N. (2019). *How Robotic Process Automation (RPA) influences firm financial performance in The Netherlands* (Master's thesis, University of Twente).

Gotthardt, M., Koivulaakso, D., Paksoy, O., Saramo, C., Martikainen, M., & Lehner, O. (2020). *Current state and challenges in the implementation of smart robotic process automation in accounting and auditing*. ACRN Journal of Finance and Risk Perspectives. DOI: 10.35944/jofrp.2020.9.1.007

Grella, F., Baldini, G., Canale, R., Sagar, K., Wang, S. A., Albini, A., . . . Zoppi, M. (2021, December). A tactile sensor-based architecture for collaborative assembly tasks with heavy-duty robots. In 2021 20th International Conference on Advanced Robotics (ICAR) (pp. 1030-1035). IEEE.

Grigori, D., Casati, F., Castellanos, M., Dayal, U., Sayal, M., & Shan, M. C. (2004). Business process intelligence. *Computers in Industry*, *53*(3), 321–343. DOI: 10.1016/j.compind.2003.10.007

Guan, Y., Yokoi, K., Stasse, O., & Kheddar, A. (2005, July). On robotic trajectory planning using polynomial interpolations. In 2005 IEEE international conference on robotics and biomimetics-ROBIO (pp. 111-116). IEEE.

Guo, C. Y., Hsieh, T. L., Chang, C. C., & Perng, J. W. (2023). A novel smart photoelectric lock system: Speech transmitted by laser and speech to text. *Heliyon*, *9*(3), e14510. DOI: 10.1016/j.heliyon.2023.e14510 PMID: 36967961

Gupta, S., Rani, S., & Dixit, A. (2019, October). Recent trends in automation—a study of RPA development tools. In *2019 3rd International Conference on Recent Developments in Control, Automation & Power Engineering (RDCAPE)* (pp. 159-163). IEEE.

Gupta, A., Gupta, S., Memoria, M., Kumar, R., Kumar, S., Singh, D., Tyagi, S., & Ansari, N. (2022). Artificial Intelligence and Smart Cities: A Bibliometric Analysis. *2022 International Conference on Machine Learning, Big Data, Cloud and Parallel Computing, COM-IT-CON 2022*. DOI: 10.1109/COM-IT-CON54601.2022.9850656

Haban, D., & Shin, K. G. (1990). Application of real-time monitoring to scheduling tasks with random execution times. *IEEE Transactions on Software Engineering*, *16*(12), 1374–1389. DOI: 10.1109/32.62446

Habibi, Y., Srifariyati, S., Hasan, H., & Subhi, M. R. (2017). Strategi pembelajaran anak usia dini berbasis multiple intelligence. *Madaniyah*, *7*(2), 237–260.

Haleem, A., Javaid, M., Singh, R. P., Rab, S., & Suman, R. (2021). Hyperautomation for the enhancement of automation in industries. *Sensors International*, *2*, 100124. DOI: 10.1016/j.sintl.2021.100124

Hallikainen, P., Bekkhus, R., & Pan, S. L. (2018). How OpusCapita Used Internal RPA Capabilities to Offer Services to Clients. *MIS Quarterly Executive*, *17*(1).

Hall, J. L., & McGraw, D. (2014). For telehealth to succeed, privacy and security risks must be identified and addressed. *Health Affairs*, *33*(2), 216–221. DOI: 10.1377/hlthaff.2013.0997 PMID: 24493763

Hamet, P., & Tremblay, J. (2017). Artificial intelligence in medicine. *Metabolism: Clinical and Experimental*, *69*, S36–S40. https://doi.org/https://doi.org/10.1016/j.metabol.2017.01.011. DOI: 10.1016/j.metabol.2017.01.011 PMID: 28126242

Han, W. G., Baek, S. M., & Kuc, T. Y. (1997, October). Genetic algorithm based path planning and dynamic obstacle avoidance of mobile robots. In 1997 IEEE International Conference on Systems, Man, and Cybernetics. Computational Cybernetics and Simulation (Vol. 3, pp. 2747-2751). IEEE.

Hannonen, A. (2020). Automated testing for software process automation.

Hansen, C., Öltjen, J., Meike, D., & Ortmaier, T. (2012, August). Enhanced approach for energy-efficient trajectory generation of industrial robots. In 2012 [IEEE.]. *IEEE International Conference on Automation Science and Engineering (CASE)*, •••, 1–7.

Hashemi, A. (2008). *The relationship between multiple intelligences and reading comprehension.*

Hashemian, A., Hosseini, S. F., & Nabavi, S. N. (2017, December 17). Kinematically smoothing trajectories by NURBS reparameterization–an innovative approach. *Advanced Robotics*, *31*(23-24), 1296–1312. DOI: 10.1080/01691864.2017.1396923

Hathaway, A. P. (1982). Intelligence and non-intelligence factors contributing to scores on the Rorschach Prognostic Rating Scale. *Journal of Personality Assessment*, *46*(1), 8–11. DOI: 10.1207/s15327752jpa4601_3 PMID: 7069572

Held, M., & Kaaser, D. (2014, March 4). C2 approximation of planar curvilinear profiles by cubic B-splines. *Computer-Aided Design and Applications*, *11*(2), 206–219. DOI: 10.1080/16864360.2014.846092

Herm, L., Janiesch, C., Steinbach, T., & Wüllner, D. (2021). Managing RPA implementation projects. *Robotic Process Automation*, 27-46.

Herm, L.-V., Janiesch, C., Reijers, H. A., & Seubert, F. (2021). From symbolic RPA to intelligent RPA: challenges for developing and operating intelligent software robots. *Business Process Management: 19th International Conference, BPM 2021, Rome, Italy, September 06–10, 2021. Proceedings*, *19*, 289–305.

He, S., Hu, C., Zhu, Y., & Tomizuka, M. (2020, December 1). Time optimal control of triple integrator with input saturation and full state constraints. *Automatica*, *122*, 109240. DOI: 10.1016/j.automatica.2020.109240

Hester, A. (2024, May 3). Automated prior authorization with intelligent automation. Nividous Intelligent Automation Company.https://nividous.com/blogs/automated-prior-authorization

Hirakawa, A. R., & Kawamura, A. (1997, April). Trajectory planning of redundant manipulators for minimum energy consumption without matrix inversion. In *Proceedings of international conference on robotics and automation* (Vol. 3, pp. 2415-2420). IEEE.

Hlásny, T., König, L., Krokene, P., Lindner, M., Montagné-Huck, C., Müller, J., Qin, H., Raffa, K. F., Schelhaas, M.-J., Svoboda, M., Viiri, H., & Seidl, R. (2021). Bark beetle outbreaks in Europe: State of knowledge and ways forward for management. *Current Forestry Reports*, *7*(3), 138–165. DOI: 10.1007/s40725-021-00142-x

Hofmann, P., Samp, C., & Urbach, N. (2020). Robotic process automation. *Electronic Markets*, *30*(1), 99–106. DOI: 10.1007/s12525-019-00365-8

Hogan, N. (1984, November 1). An organizing principle for a class of voluntary movements. *The Journal of Neuroscience : The Official Journal of the Society for Neuroscience*, *4*(11), 2745–2754. DOI: 10.1523/JNEUROSCI.04-11-02745.1984 PMID: 6502203

Hollerbach, J. O., & Suh, K. (1987, August). Redundancy resolution of manipulators through torque optimization. *IEEE Journal on Robotics and Automation*, *3*(4), 308–316. DOI: 10.1109/JRA.1987.1087111

Hong, B., Ly, M., & Lin, H. (2023). Robotic process automation risk management: Points to consider. *Journal of Emerging Technologies in Accounting*, *20*(1), 125–145. DOI: 10.2308/JETA-2022-004

Huang, J., Hu, P., Wu, K., & Zeng, M. (2018, March 1). Optimal time-jerk trajectory planning for industrial robots. *Mechanism and Machine Theory*, *121*, 530–544. DOI: 10.1016/j.mechmachtheory.2017.11.006

Hutton, R. J. B., Warwick, W., Stanard, T., McDermott, P. L., & McIlwaine, S. (2001). Computational model of recognition-primed decisions (RPD): Improving realism in computer-generated forces (CGF). *Proceedings of the Human Factors and Ergonomics Society Annual Meeting*, *45*(26), 1833–1837. DOI: 10.1177/154193120104502607

Hu, Y.-H., Fu, J. S., & Yeh, H.-C. (2023). Developing an early-warning system through robotic process automation: Are intelligent tutoring robots as effective as human teachers? *Interactive Learning Environments*, •••, 1–14. DOI: 10.1080/10494820.2022.2160467

Huypens, V. (2020, November). Constant speed lines–curves—NURBS reference pulse IPOs (part I). *International Journal of Advanced Manufacturing Technology*, *111*(5), 1247–1275. DOI: 10.1007/s00170-020-05339-1

Hwangbo, J., Lee, J., Dosovitskiy, A., Bellicoso, D., Tsounis, V., Koltun, V., & Hutter, M. (2019, January 16). Learning agile and dynamic motor skills for legged robots. *Science Robotics*, *4*(26), eaau5872. DOI: 10.1126/scirobotics.aau5872 PMID: 33137755

Ibrahim, S. S. N. (2018). The Role of RPA in Shaping the Future of Enterprise Systems.

Ichter, B., Harrison, J., & Pavone, M. (2018, May). Learning sampling distributions for robot motion planning. In *2018 IEEE International Conference on Robotics and Automation (ICRA)* (pp. 7087-7094). IEEE.

Inam, A., Joseph, S., & Chemingui, H. (2024). Beyond the Screen. *Hyperautomation in Business and Society*, 266.

Intelligent automation in healthcare & cognitive computing. (n.d.).https://www.optum.in/thought-leadership/library/intelligent-automation.html

Ivančić, L., Suša Vugec, D., & Bosilj Vukšić, V. (2019). Robotic process automation: systematic literature review. In *Business Process Management: Blockchain and Central and Eastern Europe Forum: BPM 2019 Blockchain and CEE Forum, Vienna, Austria, September 1–6, 2019* [Springer International Publishing.]. *Proceedings*, *17*, 280–295.

Ivanov, D. (2021). *Introduction to supply chain resilience: Management, modelling, technology*. Springer Nature.

Jackson, C. B., Farzan, M., Chen, B., & Choe, H. (2022). Mechanisms of SARS-CoV-2 entry into cells. *Nature Reviews. Molecular Cell Biology*, *23*(1), 3–20. DOI: 10.1038/s41580-021-00418-x PMID: 34611326

Jain, A., Paliwal, S., Sharma, M., Vig, L., & Shroff, G. (2024). SmartFlow: Robotic process automation using LLMs. *arXiv preprint arXiv:2405.12842*.

Jain, R., & Bhatnagar, R. (2019). Robotic process automation in healthcare-a review. *International Robotics & Automation Journal*, *5*(1), 12–14. DOI: 10.15406/iratj.2019.05.00164

Jandyal, A., Chaturvedi, I., Wazir, I., Raina, A., & Haq, M. I. U. (2022). 3D printing–A review of processes, materials and applications in industry 4.0. *Sustainable Operations and Computers*, *3*, 33–42. DOI: 10.1016/j.susoc.2021.09.004

Janjoš, F., Reichart, R., & Niermeyer, P. (2017, July 1). Smooth path-generation around obstacles using quartic splines and rrts. *IFAC-PapersOnLine*, *50*(1), 9108–9113. DOI: 10.1016/j.ifacol.2017.08.1708

Jarrahi, M. H. (2018). Artificial intelligence and the future of work: Human-AI symbiosis in organizational decision making. *Business Horizons*, *61*(4), 577–586. DOI: 10.1016/j.bushor.2018.03.007

Jaunat, J., Garel, E., Huneau, F., Erostate, M., Santoni, S., Robert, S., Fox, D., & Pasqualini, V. (2019). Combinations of geoenvironmental data underline coastal aquifer anthropogenic nitrate legacy through groundwater vulnerability mapping methods. *The Science of the Total Environment*, *658*, 1390–1403. DOI: 10.1016/j.scitotenv.2018.12.249 PMID: 30677999

Jędrzejka, D. (2019). Robotic process automation and its impact on accounting. *Zeszyty Teoretyczne Rachunkowości*, *2019*(105), 137–166. DOI: 10.5604/01.3001.0013.6061

Jha, N., Prashar, D., & Nagpal, A. (2021). Combining artificial intelligence with robotic process automation—an intelligent automation approach. *Deep Learning and Big Data for Intelligent Transportation: Enabling Technologies and Future Trends*, 245–264.

Jha, N., Prashar, D., & Nagpal, A. (2021). Combining artificial intelligence with robotic process automation—An intelligent automation approach. In *Deep learning and big data for intelligent transportation: Enabling technologies and future trends* (pp. 245-264).

Jha, P., Deshmukh, Y., Tumbe, C., Suraweera, W., Bhowmick, A., Sharma, S., Novosad, P., Fu, S. H., Newcombe, L., Gelband, H., & Brown, P. (2022). COVID mortality in India: National survey data and health facility deaths. *Science*, *375*(6581), 667–671. DOI: 10.1126/science.abm5154 PMID: 34990216

Jiang, C., Han, X., Guan, F. J., & Li, Y. H. (2007, November 1). An uncertain structural optimization method based on nonlinear interval number programming and interval analysis method. *Engineering Structures*, *29*(11), 3168–3177. DOI: 10.1016/j.engstruct.2007.01.020

Jimenez-Ramirez, A., Reijers, H. A., Barba, I., & Del Valle, C. (2019). A method to improve the early stages of the robotic process automation lifecycle. In *Advanced Information Systems Engineering: 31st International Conference, CAiSE 2019, Rome, Italy, June 3–7, 2019* [Springer International Publishing.]. *Proceedings*, *31*, 446–461.

Jin, R., Rocco, P., & Geng, Y. (2021, January 1). Cartesian trajectory planning of space robots using a multi-objective optimization. *Aerospace Science and Technology*, *108*, 106360. DOI: 10.1016/j.ast.2020.106360

Jolad, B., Arora, M., Ganu, R., & Bhatia, C. (2017). Voice controlled robotic vehicle. International Research Journal of Engineering and Technology (IRJET), 4(06).

Jurdi, A. P. S., Queiróz, M., Feio, C. B. A. P. R., Barros, N. A., Cagnin, G. M., Iassia, A. M., Santos, L. C., & Pinheiro, A. C. (2022). PET–Saúde Interprofissionalidade CER II: Narrativas de um grupo sobre ações cotidianas de trabalho frente às alterações provocadas pela Covid-19. *Interface: Comunicacao, Saude, Educacao, 26*, e210597. DOI: 10.1590/interface.210597

Kaidarova, A., Alsharif, N., Oliveira, B. N. M., Marengo, M., Geraldi, N. R., Duarte, C. M., & Kosel, J. (2020). Laser-printed, flexible graphene pressure sensors. *Global Challenges (Hoboken, NJ), 4*(4), 2000001. DOI: 10.1002/gch2.202000001 PMID: 32257383

Kaidarova, A., & Kosel, J. (2020). Physical sensors based on laser-induced graphene: A review. *IEEE Sensors Journal, 21*(11), 12426–12443. DOI: 10.1109/JSEN.2020.3034845

Kala, R. (2018, January). On repelling robotic trajectories: Coordination in navigation of multiple mobile robots. *Intelligent Service Robotics, 11*(1), 79–95. DOI: 10.1007/s11370-017-0238-5

Kamat, A. (2019). Challenges of robotic process automation adoption in banking and financial services. *International Journal of Research and Analytical Reviews, 6*(2), 597–606.

Kannan, K., & Selvakumar, J. (2015). Arduino based voice controlled robot. [IRJET]. *International Research Journal of Engineering and Technology, 2*(01), 49–55.

Kant, R., Dixit, A. K., Memoria, M., & Chaudhary, T. (2023). Artificial Intelligence Based Transformation in the Field of Legal Profession. *2023 3rd International Conference on Innovative Sustainable Computational Technologies (CISCT)*, 1–6.

Kant, R., Dixit, A., & Memoria, M. (2023). Robots, Revolutionizing the Practice of Law: A Bright Future Ahead. *2023 International Conference on IoT, Communication and Automation Technology (ICICAT)*, 1–6. DOI: 10.1109/ICICAT57735.2023.10263664

Kapania, N. R. (2016). *Trajectory planning and control for an autonomous race vehicle*. Stanford Digital University.

Kar, N., Kar, B., & Kar, S. (2021). Stress and coping during COVID-19 pandemic: Result of an online survey. *Psychiatry Research, 295*, 113598. DOI: 10.1016/j.psychres.2020.113598 PMID: 33264677

Kashyap, A. K., Parhi, D. R., Muni, M. K., & Pandey, K. K. (2020, November 1). A hybrid technique for path planning of humanoid robot NAO in static and dynamic terrains. *Applied Soft Computing, 96*, 106581. DOI: 10.1016/j.asoc.2020.106581

Katrakazas, C., Quddus, M., Chen, W. H., & Deka, L. (2015, November 1). Real-time motion planning methods for autonomous on-road driving: State-of-the-art and future research directions. *Transportation Research Part C, Emerging Technologies*, *60*, 416–442. DOI: 10.1016/j.trc.2015.09.011

Katritsis, D. G. (2021). Artificial intelligence, superintelligence and intelligence. *Arrhythmia & Electrophysiology Review*, *10*(4), 223–224. DOI: 10.15420/aer.2021.61 PMID: 35106171

Kaur, B., Kumar, S., & Kaushik, B. K. (2022). Recent advancements in optical biosensors for cancer detection. *Biosensors & Bioelectronics*, *197*, 113805. DOI: 10.1016/j.bios.2021.113805 PMID: 34801795

Kaya, C. T., Türkyılmaz, M., & Birol, B. (2019). Impact of RPA technologies on accounting systems. *Muhasebe ve Finansman Dergisi*, (82).

Kazakovtsev, L. A., Orlov, V., Stupina, A. A., & Kazakovtsev, V. (2015, February 17). Modied Genetic Algorithm with Greedy Heuristic for Continuous and Discrete p-Median Problems. Facta Universitatis, Series. *Mathematics and Informatics.*, *30*(1), 89–106.

Kazem, B. I., Mahdi, A. I., & Oudah, A. T. (2008, September). Motion planning for a robot arm by using genetic algorithm. *Jjmie.*, *2*(3), 131–136.

Kedziora, D. (2022). Botsourcing, Roboshoring or Virtual Backoffice? Perspectives on Implementing Robotic Process Automation (RPA) and Artificial Intelligence (AI). *Human Technology*, *18*(2), 92–97. DOI: 10.14254/1795-6889.2022.18-2.1

Kedziora, D., & Hyrynsalmi, S. (2023). Turning robotic process automation onto intelligent automation with machine learning. *Proceedings of the 11th International Conference on Communities and Technologies*, 1–5. DOI: 10.1145/3593743.3593746

Kedziora, D., & Smolander, K. (2022). *Responding to healthcare emergency outbreak of COVID-19 pandemic with Robotic Process Automation*. RPA. DOI: 10.24251/HICSS.2022.752

Kehoe, B., Patil, S., Abbeel, P., & Goldberg, K. (2015). A survey of research on cloud robotics and automation. *IEEE Transactions on Automation Science and Engineering*, *12*(2), 398–409. DOI: 10.1109/TASE.2014.2376492

Kelly, A., & Nagy, B. (2003, July). Reactive nonholonomic trajectory generation via parametric optimal control. *The International Journal of Robotics Research*, *22*(7-8), 583–601. DOI: 10.1177/02783649030227008

Kewlani, G., Ishigami, G., & Iagnemma, K. (2009, October). Stochastic mobility-based path planning in uncertain environments. In *2009 IEEE/RSJ International Conference on Intelligent Robots and Systems* (pp. 1183-1189). IEEE.

Khan, S. (2020). Comparative analysis of RPA tools—UiPath, Automation Anywhere, and Blue Prism. *International Journal of Computer Science and Mobile Applications*, *8*(11), 1–6. DOI: 10.47760/ijcsma.2020.v08i11.001

Khan, S. A. R., Yu, Z., Belhadi, A., & Mardani, A. (2020). Investigating the effects of renewable energy on international trade and environmental quality. *Journal of Environmental Management*, *272*, 111089. DOI: 10.1016/j.jenvman.2020.111089 PMID: 32854892

Khan, S., & Tailor, R. K. (2023). Application of RPA in Human Security Systems in Smart Cities. In *Application and Adoption of Robotic Process Automation for Smart Cities* (pp. 28–46). IGI Global. DOI: 10.4018/978-1-6684-7193-7.ch002

Khan, S., Tailor, R. K., Uygun, H., & Gujrati, R. (2022). Application of robotic process automation (RPA) for supply chain management, smart transportation and logistics. *International Journal of Health Sciences*, *6*(S3), 11051–11063. DOI: 10.53730/ijhs.v6nS3.8554

Khogali, H. O., & Mekid, S. (2023). The blended future of automation and AI: Examining some long-term societal and ethical impact features. *Technology in Society*, *73*, 102232. DOI: 10.1016/j.techsoc.2023.102232

Kholiya, N., & Kumar, A. (2023a). Phytotoxicity of Chromium (VI) on Germination. *Growth Attributes and Pigmentation in Cluster Bean.*, *19*(3), 94–101.

Kholiya, N., & Kumar, A. (2023b). Phytotoxicity of chromium (VI) on germination, growth attributes and pigmentation in cluster bean. *Journal of Stress Physiology & Biochemistry*, *19*(3), 94–101.

Khoukhi, A., Baron, L., & Balazinski, M. (2009, August 1). Constrained multi-objective trajectory planning of parallel kinematic machines. *Robotics and Computer-integrated Manufacturing*, *25*(4-5), 756–769. DOI: 10.1016/j.rcim.2008.09.002

Kim, J., Kim, S. R., Kim, S. J., & Kim, D. H. (2010, January 12). A practical approach for minimum- time trajectory planning for industrial robots. Industrial Robot. *International Journal (Toronto, Ont.)*.

Kim, K. T., Jeong, S. N., & Lee, S. H. (2021). Development of RPA with information extraction module. In *Proceedings of the Korean Society of Computer Information Conference* (pp. 435-436). Korean Society of Computer Information.

Kim, N., Cha, S., & Peng, H. (2010, August 23). Optimal control of hybrid electric vehicles based on Pontryagin's minimum principle. *IEEE Transactions on Control Systems Technology*, *19*(5), 1279–1287.

Kim, S. H. (2023). Development of evaluation criteria for Robotic Process Automation (RPA) solution selection. *Electronics (Basel)*, *12*(4), 986. DOI: 10.3390/electronics12040986

Kim, S., Chen, J., Cheng, T., Gindulyte, A., He, J., He, S., Li, Q., Shoemaker, B. A., Thiessen, P. A., Yu, B., Zaslavsky, L., Zhang, J., & Bolton, E. E. (2021). PubChem in 2021: New data content and improved web interfaces. *Nucleic Acids Research*, *49*(D1), D1388–D1395. DOI: 10.1093/nar/gkaa971 PMID: 33151290

Kirchmer, M. (2017). Robotic process automation–pragmatic solution or dangerous illusion? *BTOES Insights*, (June), 17.

Kirkegaard, E. O. W., & Nyborg, H. (2021). Intelligence and general psychopathology in the vietnam experience study: A closer look. *The Mankind Quarterly*, *61*(4), 792–819. DOI: 10.46469/mq.2021.61.4.2

Kochhar, K., Purohit, H., & Chutani, R. (2019). The rise of artificial intelligence in banking sector. In *The 5th International Conference on Educational Research and Practice (icerp)* (Vol. 127).

Köker, R.KöKer R. (2013, February 10). A genetic algorithm approach to a neural-network-based inverse kinematics solution of robotic manipulators based on error minimization. *Information Sciences*, *222*, 528–543. DOI: 10.1016/j.ins.2012.07.051

Kokina, J., & Blanchette, S. (2019). Early evidence of digital labor in accounting: Innovation with Robotic Process Automation. *International Journal of Accounting Information Systems*, *35*, 100431. DOI: 10.1016/j.accinf.2019.100431

Kong, K., Bae, J., & Tomizuka, M. (2009, February 13). Control of rotary series elastic actuator for ideal force-mode actuation in human–robot interaction applications. *IEEE/ASME Transactions on Mechatronics*, *14*(1), 105–118. DOI: 10.1109/TMECH.2008.2004561

Koppisetti, V. S. K. (2024). Robotic process automation: Streamlining operations in the digital era. *ESP International Journal of Advancements in Computational Technology (ESP-IJACT)*, *2*(2), 74-81.

Korayem, M. H., Nikoobin, A., & Azimirad, V. (2009, October 1). Trajectory optimization of flexible link manipulators in point-to-point motion. *Robotica*, *27*(6), 825–840. DOI: 10.1017/S0263574708005183

Kotagiri, A., & Yada, A. (2024). Crafting a strong anti-fraud defense: RPA, ML, and NLP collaboration for resilience in US finance's. *International Journal of Management Education for Sustainable Development*, *7*(7), 1–15.

Kota, S., Hetrick, J., Li, Z., & Saggere, L. (1999, December). Tailoring unconventional actuators using compliant transmissions: Design methods and applications. *IEEE/ASME Transactions on Mechatronics*, *4*(4), 396–408. DOI: 10.1109/3516.809518

Kothandapani, H. P. (2021). Integrating robotic process automation and machine learning in data lakes for automated model deployment, retraining, and data-driven decision making. *Sage Science Review of Applied Machine Learning*, *4*(2), 16–30.

Kreuzer, M. P. (2014). *Remotely piloted aircraft: Evolution, diffusion, and the future of air warfare* (Doctoral dissertation, Princeton University).

Krishnan, M. G., & Sankar, A. (2022). Image space trajectory tracking of 6-DOF robot manipulator in assisting visual servoing. *Automatika (Zagreb)*, *63*(2), 199–215. DOI: 10.1080/00051144.2021.2022889

Kronkvist, J. (2020). Development of robotic process automation enhancement tools: Platform centralization of RPA key aspects.

Kucuk, S. (2017, December 1). Optimal trajectory generation algorithm for serial and parallel manipulators. *Robotics and Computer-integrated Manufacturing*, *48*, 219–232. DOI: 10.1016/j.rcim.2017.04.006

Kumar, A. (2023). The transformation of the Indian healthcare system. *Cureus*, *15*(5). Advance online publication. DOI: 10.7759/cureus.39079 PMID: 37378105

Kumar, K. N., & Balaramachandran, P. R. (2018). Robotic process automation-a study of the impact on customer experience in retail banking industry. *Journal of Internet Banking and Commerce*, *23*(3), 1–27.

Kumar, V., Verma, R., Kango, S., & Sharma, V. S. (2021). Recent progresses and applications in laser-based surface texturing systems. *Materials Today. Communications*, *26*, 101736. DOI: 10.1016/j.mtcomm.2020.101736

Kunduru, A. R. (2023). Cloud BPM Application (Appian) Robotic Process Automation Capabilities. *Asian Journal of Research in Computer Science*, *16*(3), 267–280. DOI: 10.9734/ajrcos/2023/v16i3361

Kurylets, A., & Goranin, N. (2023). Security ontology OntoSecRPA for robotic process automation domain. *Applied Sciences (Basel, Switzerland)*, *13*(9), 5568. DOI: 10.3390/app13095568

Kurylets, A., & Goranin, N. (2024). Design and implementation of a UMLRPAsec-extension for robotic process automation. *Electronics (Basel), 13*(4), 769. DOI: 10.3390/electronics13040769

Kyheröinen, T. (2018). *Implementation of robotic process automation to a target process–a case study* (Master's thesis).

Lacity, M. C., & Willcocks, L. P. (2021). Becoming strategic with intelligent automation. *MIS Quarterly Executive*, *20*(2), 7.

Lacity, M., & Willcocks, L. P. (2018). *Robotic process and cognitive automation: The next phase*. SB Publishing.

Lacity, M., Willcocks, L., & Gozman, D. (2021). Influencing information systems practice: The action principles approach applied to robotic process and cognitive automation. *Journal of Information Technology*, *36*(3), 216–240. DOI: 10.1177/0268396221990778

Lamberton, C., Brigo, D., & Hoy, D. (2017). Impact of Robotics, RPA and AI on the insurance industry: Challenges and opportunities. *Journal of Financial Perspectives*, *4*(1).

Lampariello, R., Nguyen-Tuong, D., Castellini, C., Hirzinger, G., & Peters, J. (2011, May). Trajectory planning for optimal robot catching in real-time. In *2011 IEEE International Conference on Robotics and Automation* (pp. 3719-3726). IEEE.

Landers, R. G., Barton, K., Devasia, S., Kurfess, T., Pagilla, P., & Tomizuka, M. (2020, November 1). A Review of Manufacturing Process Control. *Journal of Manufacturing Science and Engineering*, *142*(11), 110814. DOI: 10.1115/1.4048111

Leahy, M. B., & Valavanis, K. P. (2020, August). Dynamics-based control of robotic manipulators. In *Intelligent Robotic Systems* (pp. 245–282). CRC Press.

LeCun, Y., Bengio, Y., & Hinton, G. (2015). Deep learning. *Nature*, *521*(7553), 436–444. DOI: 10.1038/nature14539 PMID: 26017442

Lee, S. A. (2020). How much “Thinking” about COVID-19 is clinically dysfunctional? *Brain, Behavior, and Immunity*, *87*, 97–98. DOI: 10.1016/j.bbi.2020.04.067 PMID: 32353520

Lengyel, J., Reichert, M., Donald, B. R., & Greenberg, D. P. (1990, September 1). Real-time robot motion planning using rasterizing computer graphics hardware. *Computer Graphics*, *24*(4), 327–335. DOI: 10.1145/97880.97915

Leno, V., Polyvyanyy, A., Dumas, M., La Rosa, M., & Maggi, F. M. (2021). Robotic process mining: Vision and challenges. *Business & Information Systems Engineering*, *63*(3), 301–314. DOI: 10.1007/s12599-020-00641-4

Leno, V., Polyvyanyy, A., La Rosa, M., Dumas, M., & Maggi, F. M. (2021). Action-oriented process mining: Unveiling the control flow of actions in process models. *Information Systems*, *102*, 101782.

Leopold, H., van Der Aa, H., & Reijers, H. A. (2018). Identifying candidate tasks for robotic process automation in textual process descriptions. In *Enterprise, Business-Process and Information Systems Modeling: 19th International Conference, BPMDS 2018, 23rd International Conference, EMMSAD 2018, Held at CAiSE 2018, Tallinn, Estonia, June 11-12, 2018* [Springer International Publishing.]. *Proceedings*, *19*, 67–81.

Letizia, N. A., Salamat, B., & Tonello, A. M. (2021, February 22). A Novel Recursive Smooth Trajectory Generation Method for Unmanned Vehicles. *IEEE Transactions on Robotics*, *37*(5), 1792–1805. DOI: 10.1109/TRO.2021.3053649

Liang, J. H., & Lee, C. H. (2015, January 1). Efficient collision-free path-planning of multiple mobile robots system using efficient artificial bee colony algorithm. *Advances in Engineering Software*, *79*, 47–56. DOI: 10.1016/j.advengsoft.2014.09.006

Li, D., & Du, L. (2021, September). AUV Trajectory Tracking Models and Control Strategies: A Review. *Journal of Marine Science and Engineering*, *9*(9), 1020. DOI: 10.3390/jmse9091020

Lievano-Martínez, F. A., Fernández-Ledesma, J. D., Burgos, D., Branch-Bedoya, J. W., & Jimenez-Builes, J. A. (2022). Intelligent process automation: An application in manufacturing industry. *Sustainability (Basel)*, *14*(14), 8804. DOI: 10.3390/su14148804

Li, H., Wu, W. J., Rastegar, J., & Guo, A. (2019, July). A real-time and look-ahead interpolation algorithm with axial jerk-smooth transition scheme for computer numerical control machining of micro-line segments. *Proceedings of the Institution of Mechanical Engineers. Part B, Journal of Engineering Manufacture*, *233*(9), 2007–2019. DOI: 10.1177/0954405418809768

Lillis, L., Siverson, J., Lee, A., Cantera, J., Parker, M., Piepenburg, O., Lehman, D. A., & Boyle, D. S. (2016). Factors influencing recombinase polymerase amplification (RPA) assay outcomes at point of care. *Molecular and Cellular Probes*, *30*(2), 74–78. DOI: 10.1016/j.mcp.2016.01.009 PMID: 26854117

Lin, C., Chang, P., & Luh, J. (1983, December). Formulation and optimization of cubic polynomial joint trajectories for industrial robots. *IEEE Transactions on Automatic Control*, *28*(12), 1066–1074. DOI: 10.1109/TAC.1983.1103181

Lindner, D., Fitzek, A., Bräuninger, H., Aleshcheva, G., Edler, C., Meissner, K., Scherschel, K., Kirchhof, P., Escher, F., Schultheiss, H.-P., Blankenberg, S., Püschel, K., & Westermann, D. (2020). Association of cardiac infection with SARS-CoV-2 in confirmed COVID-19 autopsy cases. *JAMA Cardiology*, *5*(11), 1281–1285. DOI: 10.1001/jamacardio.2020.3551 PMID: 32730555

Ling, X., Gao, M., & Wang, D. (2020, November). Intelligent document processing based on RPA and machine learning. In *2020 Chinese Automation Congress (CAC)* (pp. 1349-1353). IEEE. DOI: 10.1109/CAC51589.2020.9326579

Lin, J., Somani, N., Hu, B., Rickert, M., & Knoll, A. (2018, October). An efficient and time-optimal trajectory generation approach for waypoints under kinematic constraints and error bounds. In *2018 IEEE/RSJ International Conference on Intelligent Robots and Systems (IROS)* (pp. 5869-5876). IEEE.

Liu, H., Lai, X., & Wu, W. (2013, April 1). Time-optimal and jerk-continuous trajectory planning for robot manipulators with kinematic constraints. *Robotics and Computer-integrated Manufacturing*, *29*(2), 309–317. DOI: 10.1016/j.rcim.2012.08.002

Liu, Y., Liu, J., Xia, H., Zhang, X., Fontes-Garfias, C. R., Swanson, K. A., Cai, H., Sarkar, R., Chen, W., Cutler, M., Cooper, D., Weaver, S. C., Muik, A., Sahin, U., Jansen, K. U., Xie, X., Dormitzer, P. R., & Shi, P.-Y. (2021). Neutralizing activity of BNT162b2-elicited serum. *The New England Journal of Medicine*, *384*(15), 1466–1468. DOI: 10.1056/NEJMc2102017 PMID: 33684280

Liu, Y., Xu, H., & Zhang, X. (2018, January 1). Dynamic Characteristics and Restraint Mechanism of Elastically Restrained Connecting Joints on the Flexible Robotic Manipulator. *Shock and Vibration*, *2018*(1), 2018. DOI: 10.1155/2018/3217106

Liu, Y., Yang, Z., Yu, Z., Liu, Z., Liu, D., Lin, H., Li, M., Ma, S., Avdeev, M., & Shi, S. (2023). Generative artificial intelligence and its applications in materials science: Current situation and future perspectives. *Journal of Materiomics*, *9*(4), 798–816. DOI: 10.1016/j.jmat.2023.05.001

Lopes, R. D., Macedo, A. V. S., & de Barros, E. (2021). Effect of Discontinuing vs Continuing Angiotensin-Converting Enzyme Inhibitors and Angiotensin II Receptor Blockers on Days Alive and Out of the Hospital in Patients Admitted With COVID-19: A Randomized Clinical Trial. *Journal of the American Medical Association*, *325*(3), 254–264. DOI: 10.1001/jama.2020.25864 PMID: 33464336

López-Carnicer, J. M., Del Valle, C., & Enríquez, J. G. (2020). Towards an open-source logger for the analysis of RPA projects. In *Business Process Management: Blockchain and Robotic Process Automation Forum: BPM 2020 Blockchain and RPA Forum, Seville, Spain, September 13–18, 2020* [Springer International Publishing.]. *Proceedings*, *18*, 176–184.

Lo, W., Yang, C.-M., Zhang, Q., & Li, M. (2024). Increased productivity and reduced waste with robotic process automation and generative AI-powered IoE services. *Journal of Web Engineering*, •••, 53–88. DOI: 10.13052/jwe1540-9589.2313

Luo, H., & Luo, H. (2021). RPA and Artificial Intelligence in Budget Management Based on Multiperspective Recognition Based on Network Communication Integration. *Wireless Communications and Mobile Computing*, *2021*(1), 9723379. DOI: 10.1155/2021/9723379

Luo, L. P., Yuan, C., Yan, R. J., Yuan, Q., Wu, J., Shin, K. S., & Han, C. S. (2015, May 1). Trajectory planning for energy minimization of industry robotic manipulators using the Lagrange interpolation method. *International Journal of Precision Engineering and Manufacturing*, *16*(5), 911–917. DOI: 10.1007/s12541-015-0119-9

Lu, S., Ding, B., & Li, Y. (2020, March). Minimum-jerk trajectory planning pertaining to a translational 3- degree of freedom parallel manipulator through piecewise quintic polynomials interpolation. *Advances in Mechanical Engineering*, *12*(3), 1687814020913667. DOI: 10.1177/1687814020913667

Ma, S. (1995, November). Real-time algorithm for quasi-minimum energy control of robotic manipulators. In Proceedings of IECON'95-21st Annual Conference on IEEE Industrial Electronics (Vol. 2, pp. 1324-1329). IEEE.

Maalla, A. (2019, December). Development prospect and application feasibility analysis of robotic process automation. In *2019 IEEE 4th Advanced Information Technology,Electronic and Automation Control Conference (IAEAC)* (Vol. 1, pp. 2714-2717). IEEE. DOI: 10.1109/IAEAC47372.2019.8997983

Macfarlane, S., & Croft, E. A. (2003, February 19). Jerk-bounded manipulator trajectory planning: Design for real- time applications. *IEEE Transactions on Robotics and Automation*, *19*(1), 42–52. DOI: 10.1109/TRA.2002.807548

Madakam, S., Holmukhe, R. M., & Jaiswal, D. K. (2019). The future digital work force: Robotic process automation (RPA). *Journal of Information Systems and Technology Management*, *16*, e201916001. DOI: 10.4301/S1807-1775201916001

Maddipatla, D., Narakathu, B. B., & Atashbar, M. (2020). Recent progress in manufacturing techniques of printed and flexible sensors: A review. *Biosensors (Basel), 10*(12), 199. DOI: 10.3390/bios10120199 PMID: 33287324

Mahmood, F., Arshad, J., Ben Othman, M. T., Hayat, M. F., Bhatti, N., Jaffery, M. H., Rehman, A. U., & Hamam, H. (2022). Implementation of an Intelligent Exam Supervision System Using Deep Learning Algorithms. *Sensors (Basel), 22*(17), 6389. DOI: 10.3390/s22176389 PMID: 36080848

Ma, J. W., Gao, S., Yan, H. T., Lv, Q., & Hu, G. Q. (2021, February 12). A new approach to time-optimal trajectory planning with torque and jerk limits for robot. *Robotics and Autonomous Systems, 140*, 103744. DOI: 10.1016/j.robot.2021.103744

Malyuta, D., Yu, Y., Elango, P., & Açıkmeşe, B. (2021, November 2). Advances in trajectory optimization for space vehicle control. *Annual Reviews in Control, 52*, 282–315. DOI: 10.1016/j.arcontrol.2021.04.013

Mandvikar, S., & Achanta, A. (2023). Process automation 2.0 with generative AI framework. *International Journal of Science and Research (Raipur, India), 12*(10), 1614–1619. DOI: 10.21275/SR231021200626

Manning, J., Billian, J., Matson, J., Allen, C., & Soares, N. (2021). Perceptions of families of individuals with autism spectrum disorder during the COVID-19 crisis. *Journal of Autism and Developmental Disorders, 51*(8), 2920–2928. DOI: 10.1007/s10803-020-04760-5 PMID: 33090358

Manukonda, K. R. R. (2024). Leveraging robotic process automation (RPA) for end-to-end testing in agile and DevOps environments: A comparative study. *Journal of Artificial Intelligence & Cloud Computing, SRC/JAICC-334*(3), 315, 2-5. https://doi.org/DOI: 10.47363/JAICC/2024

Marr, B. (2019). *Artificial intelligence in practice: how 50 successful companies used AI and machine learning to solve problems*. John Wiley & Sons.

Martínez-Rojas, A., Barba, I., & Enríquez, J. G. (2020). Towards a taxonomy of cognitive RPA components. *Business Process Management: Blockchain and Robotic Process Automation Forum: BPM 2020 Blockchain and RPA Forum, Seville, Spain, September 13–18, 2020. Proceedings, 18*, 161–175.

Martinho, R., Rijo, R., & Nunes, A. (2015). Complexity analysis of a business process automation: Case study on a healthcare organization. *Procedia Computer Science, 64*, 1226–1231. DOI: 10.1016/j.procs.2015.08.510

Martins, J. M., Mohamed, Z., Tokhi, M. O., Da Costa, J. S., & Botto, M. A. (2003, July 1). Approaches for dynamic modelling of flexible manipulator systems. *IEE Proceedings. Control Theory and Applications, 150*(4), 401–411. DOI: 10.1049/ip-cta:20030496

Masehian, E., & Katebi, Y. (2007, May 26). Robot motion planning in dynamic environments with moving obstacles and target. *International Journal of Mechanical Systems Science and Engineering., 1*(1), 20–25.

Mazumdar, A. C., & Jyoti, A. (2019). Automation of Financial Services Using Artificial Intelligence with Human Touch. *International Journal of Modern Engineering & Management Research.*

McAfee, A., & Brynjolfsson, E. (2017). *Machine, platform, crowd: Harnessing our digital future.* WW Norton & Company.

Medeiros, V. S., Jelavic, E., Bjelonic, M., Siegwart, R., Meggiolaro, M. A., & Hutter, M. (2020, April 27). Trajectory optimization for wheeled-legged quadrupedal robots driving in challenging terrain. *IEEE Robotics and Automation Letters, 5*(3), 4172–4179. DOI: 10.1109/LRA.2020.2990720

Meegahapola, L., Droz, W., Kun, P., De Götzen, A., Nutakki, C., Diwakar, S., Correa, S. R., Song, D., Xu, H., Bidoglia, M., Gaskell, G., Chagnaa, A., Ganbold, A., Zundui, T., Caprini, C., Miorandi, D., Hume, A., Zarza, J. L., Cernuzzi, L., & Gatica-Perez, D. (2023). Generalization and personalization of mobile sensing-based mood inference models: An analysis of college students in eight countries. *Proceedings of the ACM on Interactive, Mobile, Wearable and Ubiquitous Technologies, 6*(4), 1–32. DOI: 10.1145/3569483

Mehmet, Ç. A. V. A. Ş., & Emin, A. Ğ. R. A. L. İ. (2020). V-REP robotik simülatör ile robot kol simülasyonu. *Fırat Üniversitesi Mühendislik Bilimleri Dergisi, 32*(2), 435–444.

Mei, Y., Lu, Y. H., Hu, Y. C., & Lee, C. G. (2006, June 5). Deployment of mobile robots with energy and timing constraints. *IEEE Transactions on Robotics, 22*(3), 507–522. DOI: 10.1109/TRO.2006.875494

Mellado, J., & Núñez, F. (2022). Design of an IoT-PLC: A containerized programmable logical controller for the industry 4.0. *Journal of Industrial Information Integration, 25*, 100250. DOI: 10.1016/j.jii.2021.100250

Mendelson, D., & Mendelson, D. (2017). Legal protections for personal health information in the age of Big Data–a proposal for regulatory framework. *Ethics, Medicine, and Public Health, 3*(1), 37–55. DOI: 10.1016/j.jemep.2017.02.005

Mendling, J., Pentland, B. T., & Recker, J. (2020). Building a complementary agenda for business process management and digital innovation. []. Taylor & Francis.]. *European Journal of Information Systems*, *29*(3), 208–219. DOI: 10.1080/0960085X.2020.1755207

Menolfi, D., Jiang, W., Lee, B. J., Moiseeva, T., Shao, Z., Estes, V., Frattini, M. G., Bakkenist, C. J., & Zha, S. (2018). Kinase-dead ATR differs from ATR loss by limiting the dynamic exchange of ATR and RPA. *Nature Communications*, *9*(1), 5351. DOI: 10.1038/s41467-018-07798-3 PMID: 30559436

Met, İ., Kabukçu, D., Uzunoğulları, G., Soyalp, Ü., & Dakdevir, T. (2020). Transformation of business model in finance sector with artificial intelligence and robotic process automation. *Digital Business Strategies in Blockchain Ecosystems: Transformational Design and Future of Global Business*, 3–29.

Mettin, U., La Hera, P. X., Morales, D. O., Shiriaev, A. S., Freidovich, L. B., & Westerberg, S. (2009, June). Trajectory planning and time-independent motion control for a kinematically redundant hydraulic manipulator. In *2009 International Conference on Advanced Robotics* (pp. 1-6). IEEE.

Micle, D. E., Deiac, F., Olar, A., Dren a, R. F., Florean, C., Coman, I. G., & Arion, F. H. (2021). Research on innovative business plan. Smart cattle farming using artificial intelligent robotic process automation. *Agriculture*, *11*(5), 430. DOI: 10.3390/agriculture11050430

Mishra, U. A., Chawla, I., & Pathak, P. M. (2020, September). On determining shortest path in joint space of a cable-driven parallel robot for point-to-point motion. In 2020 28th Mediterranean Conference on Control and Automation (MED) (pp. 984-989). IEEE.

Misra, G., & Bai, X. (2017, November). Task-constrained trajectory planning of free-floating space-robotic systems using convex optimization. *Journal of Guidance, Control, and Dynamics*, *40*(11), 2857–2870. DOI: 10.2514/1.G002405

Moderno, O. B. D. S., Braz, A. C., & Nascimento, P. T. D. S. (2024). Robotic process automation and artificial intelligence capabilities driving digital strategy: A resource-based view. *Business Process Management Journal*, *30*(1), 105–134. DOI: 10.1108/BPMJ-08-2022-0409

Moffitt, K. C., Rozario, A. M., & Vasarhelyi, M. A. (2018). Robotic process automation for auditing. *Journal of Emerging Technologies in Accounting*, *15*(1), 1–10. DOI: 10.2308/jeta-10589

Mohamed, S. A., Mahmoud, M. A., Mahdi, M. N., & Mostafa, S. A. (2022). Improving efficiency and effectiveness of robotic process automation in human resource management. *Sustainability (Basel), 14*(7), 3920. DOI: 10.3390/su14073920

Mohanarajah, G., Hunziker, D., D'Andrea, R., & Waibel, M. (2015). Rapyuta: A cloud robotics platform. *IEEE Transactions on Automation Science and Engineering, 12*(2), 481–493. DOI: 10.1109/TASE.2014.2329556

Molla, M., Ahmad, J., & Wan Kadir, W. M. N. (2024). A comparison of transforming the user stories and functional requirements into UML use case diagram. *International Journal of Innovative Computing, 14*(1), 29–36. DOI: 10.11113/ijic.v14n1.463

Monteiro, G., Camelo, L., Aquino, G., Fernandes, R. D. A., Gomes, R., Printes, A., Torné, I., Silva, H., Oliveira, J., & Figueiredo, C. (2023). A Comprehensive Framework for Industrial Sticker Information Recognition Using Advanced OCR and Object Detection Techniques. *Applied Sciences (Basel, Switzerland), 13*(12), 7320. DOI: 10.3390/app13127320

Moreira, S., Mamede, H. S., & Santos, A. (2023). Process automation using RPA–a literature review. *Procedia Computer Science, 219*, 244–254. DOI: 10.1016/j.procs.2023.01.287

Moshayedi, A. J., Roy, A. S., Liao, L., Lan, H., Gheisari, M., Abbasi, A., & Bamakan, S. M. (2021). Automation attendance systems approaches: A practical review. *BOHR Int. J. Internet Things Artif. Intell. Mach. Learn, 1*(1), 23–31. DOI: 10.54646/BIJIOTR.003

Moss, A., Robbins, S., Achanta, S., Kuttippurathu, L., Turick, S., Nieves, S., Hanna, P., Smith, E. H., Hoover, D. B., Chen, J., Cheng, Z. J., Ardell, J. L., Shivkumar, K., Schwaber, J. S., & Vadigepalli, R. (2021). A single cell transcriptomics map of paracrine networks in the intrinsic cardiac nervous system. *iScience, 24*(7), 102713. DOI: 10.1016/j.isci.2021.102713 PMID: 34337356

Moulaï, K., Islam, G., Manning, S., & Terlinden, L. (2022). All too human" or the emergence of a techno-induced feeling of being less-able: Identity work, ableism and new service technologies. *International Journal of Human Resource Management, 33*(22), 4499–4531. DOI: 10.1080/09585192.2022.2066982

Moulin, F., Jean, F., Melchior, M., Patanè, M., Pinucci, I., Sijbrandij, M., van der Waerden, J., & Galéra, C. (2023). Longitudinal impact of the COVID19 pandemic on mental health in a general population sample in France: Evidence from the COMET study. *Journal of Affective Disorders, 320*, 275–283. DOI: 10.1016/j.jad.2022.09.142 PMID: 36191642

Muenchhof, M., & Singh, T. (2002, May). Desensitized jerk limited-time optimal control of multi-input systems. *Journal of Guidance, Control, and Dynamics*, *25*(3), 474–481. DOI: 10.2514/2.4933

Mullakara, N., & Asokan, A. K. (2020). *Robotic process automation projects: Build real-world RPA solutions using UiPath and Automation Anywhere*. Packt Publishing Ltd.

Mulukuntla, S. (2022). Generative AI: Benefits, limitations, potential risks, and challenges in healthcare industry. *EPH-International Journal of Medical and Health Science*, *8*(4), 1–9.

Murhekar, M. V., Bhatnagar, T., Selvaraju, S., Saravanakumar, V., Thangaraj, J. W. V., Shah, N., Kumar, M. S., Rade, K., Sabarinathan, R., Asthana, S., Balachandar, R., Bangar, S. D., Bansal, A. K., Bhat, J., Chopra, V., Das, D., Deb, A. K., Devi, K. R., Dwivedi, G. R., & Zaman, K. (2021). SARS-CoV-2 antibody seroprevalence in India, August–September, 2020: Findings from the second nationwide household serosurvey. *The Lancet. Global Health*, *9*(3), e257–e266. DOI: 10.1016/S2214-109X(20)30544-1 PMID: 33515512

Murugamani, C., Sahoo, S. K., Kshirsagar, P. R., Prathap, B. R., Islam, S., Noorulhasan Naveed, Q., Hussain, M. R., Hung, B. T., & Teressa, D. M. (2022). Wireless communication for robotic process automation using machine learning technique. *Wireless Communications and Mobile Computing*, *2022*, 2022. DOI: 10.1155/2022/4723138

Murugappan, K., & Sree Kala, T. (2022). An enhanced security framework for robotic process automation. In *Cyber Security and Digital Forensics* [Springer Singapore.]. *Proceedings of ICCSDF*, *2021*, 231–238.

Muzikante, I., Kumar Agarwal, M., & Turner, J. (2018). Robotic process automation risk analysis in circle K business centre. *Bachelor thesis, Riga Technical University, Riga*.

Naglieri, J. A., & Otero, T. (2011). Cognitive Assessment System: Redefining intelligence from a neuropsychological perspective. *Handbook of Pediatric Neuropsychology*, 320–333.

Naithani, C., Sood, S. P., & Agrahari, A. (2023). The Indian healthcare system turns to digital health: eSanjeevaniOPD as a national telemedicine service. *Journal of Information Technology Teaching Cases*, *13*(1), 67–76. DOI: 10.1177/20438869211061575

Nakić, J., & Boban, D. (2024). Enhancing informal education with robotic process automation: A method for automating course search on e-learning platforms. *Journal of Communications Software and Systems*, *20*(2), 215–225. DOI: 10.24138/jcomss-2023-0178

Nalini, M., Dhanraj, R. K., Balusamy, B., Abirami, V., & Kavya, K. (2024). Intelligent assistants using natural language processing for hyperautomation. In *Hyperautomation for Next-Generation Industries* (pp. 91-126). DOI: 10.1002/9781394186518.ch4

Nasser, S. M. T., Rana, A. A., Doffinger, R., Kafizas, A., Khan, T. A., & Nasser, S. (2023). Elevated free interleukin-18 associated with severity and mortality in prospective cohort study of 206 hospitalised COVID-19 patients. *Intensive Care Medicine Experimental*, *11*(1), 9. DOI: 10.1186/s40635-022-00488-x PMID: 36823262

Negoiţă, R., & Borangiu, T. (2023). Robotic process automation of inventory demand with intelligent reservation. *Studies in Informatics and Control*, *32*(2), 5–14. DOI: 10.24846/v32i2y202301

Ng, D. T. K., Leung, J. K. L., Chu, S. K. W., & Qiao, M. S. (2021). Conceptualizing AI literacy: An exploratory review. *Computers and Education: Artificial Intelligence*, *2*, 100041. https://doi.org/https://doi.org/10.1016/j.caeai.2021.100041

Ng, K. K. H., Chen, C.-H., Lee, C. K. M., Jiao, J. R., & Yang, Z.-X. (2021). A systematic literature review on intelligent automation: Aligning concepts from theory, practice, and future perspectives. *Advanced Engineering Informatics*, *47*, 101246. DOI: 10.1016/j.aei.2021.101246

Nguyen, T. K., Nguyen, V. L., Nguyen, T. G., Mai, D. H., Nguyen, N. Q., Vu, T. A., Le, A. N., Nguyen, Q. H., Nguyen, C. T., & Nguyen, D. T. (2021). Lung-protective mechanical ventilation for patients undergoing abdominal laparoscopic surgeries: A randomized controlled trial. *BMC Anesthesiology*, *21*(1), 1–11. DOI: 10.1186/s12871-021-01318-5 PMID: 33784987

Nguyen, T., Novak, R., Xiao, L., & Lee, J. (2021). Dataset distillation with infinitely wide convolutional networks. *Advances in Neural Information Processing Systems*, *34*, 5186–5198.

Nieuwenhuisen, M., & Behnke, S. (2015, September). 3D planning and trajectory optimization for real-time generation of smooth MAV trajectories. In *2015 European Conference on Mobile Robots (ECMR)* (pp. 1-7). IEEE.

Nikolos, I. K., Valavanis, K. P., Tsourveloudis, N. C., & Kostaras, A. N. (2003, November 17). Evolutionary algorithm based offline/online path planner for UAV navigation. *IEEE Transactions on Systems, Man, and Cybernetics. Part B, Cybernetics, 33*(6), 898–912. DOI: 10.1109/TSMCB.2002.804370 PMID: 18238242

Nogueira-de-Almeida, C. A., Del Ciampo, L. A., Ferraz, I. S., Del Ciampo, I. R. L., Contini, A. A., & Ued, F. da V. (2020). COVID-19 and obesity in childhood and adolescence: A clinical review. *Jornal de Pediatria, 96*(5), 546–558. DOI: 10.1016/j.jped.2020.07.001 PMID: 32768388

Noppen, P., Beerepoot, I., Van De Weerd, I., Jonker, M., & Reijers, H. A. (2020). How to keep RPA maintainable? In *Business Process Management: 18th International Conference, BPM 2020, Seville, Spain, September 13–18, 2020* [Springer International Publishing.]. *Proceedings, 18*, 453–470.

Noy, S., & Zhang, W. (2023). Experimental evidence on the productivity effects of generative artificial intelligence. *Science, 381*(6654), 187–192. DOI: 10.1126/science.adh2586 PMID: 37440646

Nurk, S., Koren, S., Rhie, A., Rautiainen, M., Bzikadze, A. V., Mikheenko, A., Vollger, M. R., Altemose, N., Uralsky, L., Gershman, A., Aganezov, S., Hoyt, S. J., Diekhans, M., Logsdon, G. A., Alonge, M., Antonarakis, S. E., Borchers, M., Bouffard, G. G., Brooks, S. Y., & Phillippy, A. M. (2022). The complete sequence of a human genome. *Science, 376*(6588), 44–53. DOI: 10.1126/science.abj6987 PMID: 35357919

Oliveira, P. W., Barreto, G. A., & Thé, G. A. P. (2020, January 9). A General Framework for Optimal Tuning of PID-like Controllers for Minimum Jerk Robotic Trajectories. *Journal of Intelligent & Robotic Systems, 99*(3-4), 1–20. DOI: 10.1007/s10846-019-01121-y

Osmundsen, K., Iden, J., & Bygstad, B. (2019). Organizing robotic process automation: balancing loose and tight coupling.

Oyewole, A. T., Adeoye, O. B., Addy, W. A., Okoye, C. C., Ofodile, O. C., & Ugochukwu, C. E. (2024). Automating financial reporting with natural language processing: A review and case analysis. *World Journal of Advanced Research and Reviews, 21*(3), 575–589.

Ozaki, H., & Lin, C. J. (1996, April). Optimal B-spline joint trajectory generation for collision-free movements of a manipulator under dynamic constraints. In *Proceedings of IEEE International Conference on Robotics and Automation* (Vol. 4, pp. 3592-3597). IEEE.

Pahune, S., & Rewatkar, N. (2024). Cognitive automation in the supply chain: Unleashing the power of RPA vs. GEN AI.

Paik, J., Pirolli, P., Dong, W., Lebiere, C., & Thomson, R. (2013). An ACT-R Model of Sensemaking in Geospatial Intelligence Tasks. *Proceedings of the 22nd Annual Conference on Behavior Representation in Modeling and Simulation.*

Paing, M. S., Nshama, E. W., & Uchiyama, N. (2021, July 9). A Kinematically Constrained Reparameterization Approach to Optimal Time and Jerk Motion of Industrial Machines. *IEEE Access : Practical Innovations, Open Solutions, 9*, 97843–97854. DOI: 10.1109/ACCESS.2021.3095847

Paletz, S. B. F., Hughes, M. M., Sumer, A., Brugman, C. M., & Bunting, M. (2018). A preliminary validation of the Regional Proficiency Assessment Tool (RPAT). *Military Psychology : the Official Journal of the Division of Military Psychology, American Psychological Association, 30*(6), 528–546. DOI: 10.1080/08995605.2018.1506196

Parikh, D. P., Sattigeri, B. M., & Kumar, A. (2014). An update on growth and development of telemedicine with pharmacological implications. *International Journal of Medical Science and Public Health, 3*(5), 527–531. DOI: 10.5455/ijmsph.2014.020320141

Parker, J. K., Khoogar, A. R., & Goldberg, D. E. (1989, January). Inverse kinematics of redundant robots using genetic algorithms. In *1989 IEEE International Conference on Robotics and Automation* (pp. 271-272). IEEE Computer Society.

Park, S., Lee, C.-W., Park, D.-I., Woo, H.-Y., Cheong, H. S., Shin, H. C., Ahn, K., Kwon, M.-J., & Joo, E.-J. (2021). Detection of SARS-CoV-2 in fecal samples from patients with asymptomatic and mild COVID-19 in Korea. *Clinical Gastroenterology and Hepatology, 19*(7), 1387–1394. DOI: 10.1016/j.cgh.2020.06.005 PMID: 32534042

Parsons, S. K., Keegan, T. H. M., Kirchhoff, A. C., Parsons, H. M., Yabroff, K. R., & Davies, S. J. (2023). Cost of cancer in adolescents and young adults in the United States: Results of the 2021 report by Deloitte Access Economics, commissioned by Teen Cancer America. *Journal of Clinical Oncology, 41*(17), 3260–3268. DOI: 10.1200/JCO.22.01985 PMID: 36827624

Pedretti, A., Santini, M., Scolimoski, J., Queiroz, M., Toshioka, F., Rocha, E., & Ramos, M. (2021). Robotic process automation extended with artificial intelligence techniques in power distribution utilities. *Brazilian Archives of Biology and Technology, 64*(spe). https://doi.org/DOI: 10.1590/1678-4324-75years-2021210217

Pekarovskiy, A., Nierhoff, T., Hirche, S., & Buss, M. (2017, November 14). Dynamically consistent online adaptation of fast motions for robotic manipulators. *IEEE Transactions on Robotics*, *34*(1), 166–182. DOI: 10.1109/TRO.2017.2765666

Piazzi, A., & Visioli, A. (1998, January 1). Global minimum-time trajectory planning of mechanical manipulators using interval analysis. *International Journal of Control*, *71*(4), 631–652. DOI: 10.1080/002071798221713

Plattfaut, R., Borghoff, V., Godefroid, M., Koch, J., Trampler, M., & Coners, A. (2022). The critical success factors for robotic process automation. *Computers in Industry*, *138*, 103646. DOI: 10.1016/j.compind.2022.103646

Pott, A. (2018). Cable-driven parallel robots: theory and application, New York, Springer; NY, USA: Springer; 120: 52-56.

Power, D. J., & Phillips-Wren, G. (2011). Impact of social media and Web 2.0 on decision-making. *Journal of Decision Systems*, *20*(3), 249–261. DOI: 10.3166/jds.20.249-261

Pramanik, H. S., Kirtania, M., & Pani, A. K. (2019). Essence of digital transformation—Manifestations at large financial institutions from North America. *Future Generation Computer Systems*, *95*, 323–343. DOI: 10.1016/j.future.2018.12.003

Pramod, D. (2022). Robotic process automation for industry: Adoption status, benefits, challenges and research agenda. *Benchmarking*, *29*(5), 1562–1586. DOI: 10.1108/BIJ-01-2021-0033

Preauthorizations Intake - RPA Component | UIPATH Marketplace | Overview. (n.d.). UiPath Marketplace. https://marketplace.uipath.com/listings/solution-accelerator-preauthorizations-intake4369

Premavathi, T., Shekhar, A., Raj, A., Mohan, K., Palaniappan, D., & Shukla, M. (2024). The Utilization of Internet of Things (IoT) in the Field of Robotics Process Automation. In *Applications of New Technology in Operations and Supply Chain Management* (pp. 337–359). IGI Global. DOI: 10.4018/979-8-3693-1578-1.ch018

Prism, S. B. (2024, April 24). What is Intelligent Automation? SS&C Blue Prism. https://www.blueprism.com/guides/intelligent-automation/

Qi, Y. (2024, August). Application research of RPA technology in the construction of financial robot business automation processing platform. In *2024 3rd International Conference on Artificial Intelligence, Internet and Digital Economy (ICAID 2024)* (pp. 340-345). Atlantis Press. DOI: 10.2991/978-94-6463-490-7_37

Qian, S., Bao, K., Zi, B., & Zhu, W. D. (2020, July 1). Dynamic Trajectory Planning for a Three Degrees-of- Freedom Cable-Driven Parallel Robot Using Quintic B-Splines. *Journal of Mechanical Design, 142*(7), 073301. DOI: 10.1115/1.4045723

Rababah, K., Mohd, H., & Ibrahim, H. (2011). Customer relationship management (CRM) processes from theory to practice: The pre-implementation plan of CRM system. *International Journal of e-Education, e-Business, e-. Management Learning, 1*(1), 22–27.

Radmanesh, M., Kumar, M., & Sarim, M. (2018, June 1). Grey wolf optimization based sense and avoid algorithm in a Bayesian framework for multiple UAV path planning in an uncertain environment. *Aerospace Science and Technology, 77*, 168–179. DOI: 10.1016/j.ast.2018.02.031

Rai, D., Siddiqui, S., Pawar, M., & Goyal, S. (2019). Robotic process automation: The virtual workforce. *International Journal on Future Revolution in Computer Science & Communication Engineering*, *5*(2), 28–32.

Rainey, S. K., Brown, B., & Kirk, D. B. (2017). Bots, natural language processing, and machine learning. *Tax Executive*, *69*, 39.

Rane, N. L., Choudhary, S. P., & Rane, J. (2024). Artificial intelligence-driven corporate finance: Enhancing efficiency and decision-making through machine learning, natural language processing, and robotic process automation in corporate governance and sustainability. *Studies in Economics and Business Relations*, *5*(2), 1–22. DOI: 10.48185/sebr.v5i2.1050

Rashid, H., Ahmed, I. U., Osman, S. B., Newaz, Q., Rasheduzzaman, M., & Reza, S. T. (2017, January). Design and implementation of a voice controlled robot with human interaction ability. In *Proceedings of the International Conference on Computer, Communication, Chemical, Materials and Electronic Engineering*, Rajshahi, Bangladesh (pp. 26-27).

Rath, A., Das Gupta, A., Rohilla, V., Balyan, A., & Mann, S. (2022). Intelligent Smart Waste Management Using Regression Analysis: An Empirical Study. *International Conference on Emerging Technologies in Computer Engineering*, 138–148. DOI: 10.1007/978-3-031-07012-9_12

Ratliff, N., Zucker, M., Bagnell, J. A., & Srinivasa, S. (2009, May 12). CHOMP: Gradient optimization techniques for efficient motion planning. In2009 [IEEE.]. *IEEE International Conference on Robotics and Automation*, •••, 489–494. DOI: 10.1109/ROBOT.2009.5152817

Rawat, R., Chakrawarti, R. K., Sarangi, S. K., Choudhary, R., Gadwal, A. S., & Bhardwaj, V. (Eds.). (2023). *Robotic Process Automation*. John Wiley & Sons. DOI: 10.1002/9781394166954

Ray, J. K., Sultana, R., Bera, R., Sil, S., & Alfred, Q. M. (2023). A Comprehensive Review on Artificial Intelligence (AI) and Robotic Process Automation (RPA) for the Development of Smart Cities. *Confluence of Artificial Intelligence and Robotic Process Automation*, 289-311.

Ray, J. K., Sultana, R., Bera, R., Sil, S., & Alfred, Q. M. (2023). A Comprehensive Review on Artificial Intelligence (AI) and Robotic Process Automation (RPA) for the Development of Smart Cities. *Confluence of Artificial Intelligence and Robotic Process Automation*, 289–311.

Reddy, K. N., Harichandana, U., Alekhya, T., & Rajesh, S. M. (2019). A study of robotic process automation among artificial intelligence. *International Journal of Scientific and Research Publications*, *9*(2), 392–397.

Renner, M. (2019). Software robot-based automation of financial administration's processes.

Ribeiro, J., Lima, R., Eckhardt, T., & Paiva, S. (2021). Robotic process automation and artificial intelligence in industry 4.0–a literature review. *Procedia Computer Science*, *181*, 51–58. DOI: 10.1016/j.procs.2021.01.104

Richardson, S., Hirsch, J. S., Narasimhan, M., Crawford, J. M., McGinn, T., Davidson, K. W., Barnaby, D. P., Becker, L. B., Chelico, J. D., Cohen, S. L., Cookingham, J., Coppa, K., Diefenbach, M. A., Dominello, A. J., Duer-Hefele, J., Falzon, L., Gitlin, J., Hajizadeh, N., Harvin, T. G., & Zanos, T. P. (2020). Presenting Characteristics, Comorbidities, and Outcomes Among 5700 Patients Hospitalized With COVID-19 in the New York City Area. *Journal of the American Medical Association*, *323*(20), 2052–2059. DOI: 10.1001/jama.2020.6775 PMID: 32320003

Rod Koch, C. M. A., & CSCA, P. (2019). Can RPA improve agility? *Strategic Finance*, *100*(9), 68–69.

Romao, M., Costa, J., & Costa, C. J. (2019). Robotic process automation: A case study in the banking industry. *2019 14th Iberian Conference on Information Systems and Technologies (CISTI)*, 1–6.

Romao, M., Costa, J., & Costa, C. J. (2019, June). Robotic process automation: A case study in the banking industry. In *2019 14th Iberian Conference on information systems and technologies (CISTI)* (pp. 1-6). IEEE. DOI: 10.23919/CISTI.2019.8760733

Rout, A., Mahanta, G. B., Bbvl, D., & Biswal, B. B. (2020). Kinematic and dynamic optimal trajectory planning of industrial robot using improved multi-objective ant lion optimizer. Journal of The Institution of Engineers (India): Series C, 101(3), 559-569.

Rubio, F., Llopis-Albert, C., Valero, F., & Suñer, J. L. (2016, December 1). Industrial robot efficient trajectory generation without collision through the evolution of the optimal trajectory. *Robotics and Autonomous Systems*, *86*, 106–112. DOI: 10.1016/j.robot.2016.09.008

Rubio, F., Valero, F., Sunyer, J., & Cuadrado, J. (2012, January 6). Optimal time trajectories for industrial robots with torque, power, jerk and energy consumed constraints. Industrial Robot. *International Journal (Toronto, Ont.)*.

Ruiz, P. L.-D., Gunnes, N., Gran, J. M., Karlstad, Ø., Selmer, R., Dahl, J., Bøås, H., White, R. A., Hofman, A. C., & Paulsen, T. H. (2023). Short-term safety of COVID-19 mRNA vaccines with respect to all-cause mortality in the older population in Norway. *Vaccine*, *41*(2), 323–332. DOI: 10.1016/j.vaccine.2022.10.085 PMID: 36376216

Sahar, G., & Hollerbach, J. M. (1986, September). Planning of minimum-time trajectories for robot arms. *The International Journal of Robotics Research*, *5*(3), 90–100. DOI: 10.1177/027836498600500305

Sahay, R., Kumar, A., Udai, A. D., & Sinha, A. N. (2006). *An imitated human motion optimal b-spline path of a biped robot using genetic algorithm. InProceedings of All India Seminar on Advances in Product Development (APD-2006)*. New Age International.

Salomon, R. (1996, January 1). Re-evaluating genetic algorithm performance under coordinate rotation of benchmark functions. A survey of some theoretical and practical aspects of genetic algorithms. *Bio Systems*, *39*(3), 263–278. DOI: 10.1016/0303-2647(96)01621-8 PMID: 8894127

Saltzman, E., & Kelso, J. A. (1987, January). Skilled actions: A task-dynamic approach. *Psychological Review*, *94*(1), 84–106. DOI: 10.1037/0033-295X.94.1.84 PMID: 3823306

Salviotti, G., Gaur, A., & Pennarola, F. (2019). *Strategic factors enabling digital maturity: An extended survey.*

Samokhvalov, I. (2024). *Transforming Management Reporting with Intelligent Process Automation (IPA)*. Enhancing Business Analytics, Forecasting, and Decision-Making in Organizations.

Santos, F., Pereira, R., & Vasconcelos, J. B. (2020). Toward robotic process automation implementation: An end-to-end perspective. *Business Process Management Journal, 26*(2), 405–420. DOI: 10.1108/BPMJ-12-2018-0380

Sarker, I. H. (2022). AI-based modeling: Techniques, applications and research issues towards automation, intelligent and smart systems. *SN Computer Science, 3*(2), 158. DOI: 10.1007/s42979-022-01043-x PMID: 35194580

Sathya, V., Jayashree, K., & Malathi, S. (2023). Robotic Process Automation (RPA) Applications and Tools for the Workforce Management System. In *Designing Workforce Management Systems for Industry 4.0* (pp. 251-264). CRC Press.

Scalera, L., Carabin, G., Vidoni, R., & Gasparetto, A. (2020). Minimum-energy trajectory planning for industrial robotic applications: Analytical model and experimental results. In *Advances in Service and Industrial Robotics: Results of RAAD* (pp. 334–342). Springer International Publishing.

Schatsky, D., Muraskin, C., & Iyengar, K. (2017). *Robotic process automation. A Path to the Cognitive Enterprise*. Deloitte Consulting.

Schmidt, A. (2020). Regulatory Challenges in Healthcare IT: Ensuring Compliance with HIPAA and GDPR. *Academic Journal of Science and Technology, 3*(1), 1–7.

Schmitz, M., Stummer, C., & Gerke, M. (2019). Smart automation as enabler of digitalization? A review of RPA/AI potential and barriers to its realization. In *Future Telco* (pp. 349–358). Successful Positioning of Network Operators in the Digital Age. DOI: 10.1007/978-3-319-77724-5_31

Schulte, B., Richter, E., Büning, A., Baum, M., Breuer, A., Zorn, J., König, J., Geiger, M., Eschbach-Bludau, M., Heuser, J., Zölzer, D., Korencak, M., Hollstein, R., Beins, E., Emmert, D., Aldabbagh, S., Eis-Hübinger, A. M., & Streeck, H. (2024). A longitudinal study on SARS-CoV-2 seroconversion, reinfection and neutralisation spanning several variant waves and vaccination campaigns, Heinsberg, Germany, April 2020 to November 2022. *Eurosurveillance, 29*(26), 2300659. DOI: 10.2807/1560-7917.ES.2024.29.26.2300659 PMID: 38940003

Seereeram, S., & Wen, J. T. (1995, February). A global approach to path planning for redundant manipulators. *IEEE Transactions on Robotics and Automation, 11*(1), 152–160. DOI: 10.1109/70.345948

Sehr, M. A., Lohstroh, M., Weber, M., Ugalde, I., Witte, M., Neidig, J., Hoeme, S., Niknami, M., & Lee, E. A. (2020). Programmable logic controllers in the context of industry 4.0. *IEEE Transactions on Industrial Informatics, 17*(5), 3523–3533. DOI: 10.1109/TII.2020.3007764

Sethy, S., & Dadhich, M. (2023). *Adoption of QR Code Technology for Easy Access to Library and Management Information Services: A Comprehensive Analysis*. Library Philosophy & Practice.

Sharma, A. (2021). Artificial intelligence in health care. *International Journal of Humanities, Arts. Medicine Science*, *5*(1), 106–109.

Sharma, C., Bharadwaj, S. S., Gupta, N., & Jain, H. (2023). Robotic process automation adoption: Contextual factors from service sectors in an emerging economy. *Journal of Enterprise Information Management*, *36*(1), 252–274. DOI: 10.1108/JEIM-06-2021-0276

Sheth, J. (2020). Impact of Covid-19 on consumer behavior: Will the old habits return or die? *Journal of Business Research*, *117*, 280–283. DOI: 10.1016/j.jbusres.2020.05.059 PMID: 32536735

Shidaganti, G., Salil, S., Anand, P., & Jadhav, V. (2021). Robotic process automation with AI and OCR to improve business process. *2021 Second International Conference on Electronics and Sustainable Communication Systems (ICESC)*, 1612–1618. DOI: 10.1109/ICESC51422.2021.9532902

Shiller Z. Time-energy optimal control of articulated systems with geometric path constraints.

Shin, K., & McKay, N. (1986, June). A dynamic programming approach to trajectory planning of robotic manipulators. *IEEE Transactions on Automatic Control*, *31*(6), 491–500. DOI: 10.1109/TAC.1986.1104317

Shmueli, G., & Koppius, O. R. (2011). Predictive analytics in information systems research. *Management Information Systems Quarterly*, *35*(3), 553–572. DOI: 10.2307/23042796

Shojaedini, E., Majd, M., & Safabakhsh, R. (2019, April 1). Novel adaptive genetic algorithm sample consensus. *Applied Soft Computing*, *77*, 635–642. DOI: 10.1016/j.asoc.2019.01.052

Sibanyoni, N. A. (2021). A Blockchain-Based Robotic Process Automation Mechanism in Educational Setting. In *Industry Use Cases on Blockchain Technology Applications in IoT and the Financial Sector* (pp. 17–41). IGI Global. DOI: 10.4018/978-1-7998-6650-3.ch002

Siddiqi, S. S., Akram, G., & Iftikhar, M. (2012). Solution of seventh order boundary value problem by differential transformation method. *World Applied Sciences Journal*, *16*(11), 1521–1526.

Siderska, J., Aini, S. N. B. M., & Kedziora, D. (2024, March). Complementing robotic process automation with generative artificial intelligence (ChatGPT), case of Robocorp. In *Future of Information and Communication Conference* (pp. 37-53). Cham: Springer Nature Switzerland.

Siderska, J., Alsqour, M., & Alsaqoor, S. (2023). Employees' attitudes towards implementing robotic process automation technology at service companies. *Human Technology, 19*(1 SE-Articles), 23–40. DOI: 10.14254/1795-6889.2023.19-1.3

Siderska, J. (2020). Robotic Process Automation—A driver of digital transformation? *Engineering Management in Production and Services, 12*(2), 21–31. DOI: 10.2478/emj-2020-0009

Siderska, J., Aunimo, L., Süße, T., Stamm, J., Kedziora, D., & Aini, S. N. B. M. (2023). Towards intelligent automation (IA): Literature review on the evolution of robotic process automation (RPA), its challenges, and future trends. *Engineering Management in Production and Services, 15*(4), 90–103. DOI: 10.2478/emj-2023-0030

Siderska, J., & Jadaan, K. S. (2018). Cloud manufacturing: A service-oriented manufacturing paradigm. A review paper. *Engineering Management in Production and Services, 10*(1), 22–31. DOI: 10.1515/emj-2018-0002

Siegert, S., & Völker, M. (2021). Towards Decision Management for Robotic Process Automation. *ZEUS*, 9–13.

Singh, K., & Singh, P. D. (2023). Machine learning in robotics with fog/cloud computing and IoT. *EAI Endorsed Transactions on AI and Robotics, 2*. Advance online publication. DOI: 10.4108/airo.3621

Sinha, S., & Harrold, M. J. (2000). Analysis and testing of programs with exception handling constructs. *IEEE Transactions on Software Engineering, 26*(9), 849–871. DOI: 10.1109/32.877846

Sisodiya, M., Prabhu, S., & Ashok, P. (2024, January). The transformative power of robotic process automation: Redefining business in industry 4.0. In *2024 14th International Conference on Cloud Computing, Data Science & Engineering (Confluence)* (pp. 498-503). IEEE.

Sivanandam, S. N., & Deepa, S. N. (2008). *Genetic algorithms. InIntroduction to genetic algorithms*. Springer.

Sivčev, S., Coleman, J., Omerdić, E., Dooly, G., & Toal, D. (2018, September 1). Underwater manipulators: A review. *Ocean Engineering, 163*, 431–450. DOI: 10.1016/j.oceaneng.2018.06.018

Smeets, M., Erhard, R., & Kaußler, T. (2021). *Robotic process automation (RPA) in the Financial Sector*. Springer Fachmedien Wiesbaden. DOI: 10.1007/978-3-658-32974-7

Smith, G., & Stanley, G. (1980). Relationships between measures of intelligence and choice reaction time. *Bulletin of the Psychonomic Society*, *16*(1), 8–10. DOI: 10.3758/BF03337079

Sobczak, A. (2021). Robotic Process Automation implementation, deployment approaches and success factors–an empirical study. *Entrepreneurship and Sustainability Issues*, *8*(4), 122–147. DOI: 10.9770/jesi.2021.8.4(7)

Sood, S. P., Negash, S., Mbarika, V. W., Kifle, M., & Prakash, N. (2007). Differences in public and private sector adoption of telemedicine: Indian case study for sectoral adoption. *Studies in Health Technology and Informatics*, *130*, 257. PMID: 17917199

Sorek, N., Gildin, E., Boukouvala, F., Beykal, B., & Floudas, C. A. (2017, April 1). Dimensionality reduction for production optimization using polynomial approximations. *Computational Geosciences*, *21*(2), 247–266. DOI: 10.1007/s10596-016-9610-3

Sowjanya, S., & Vijaya Chamundeeswari, V. (2024, February). Information extraction using RPA and generative AI from unstructured documents: A case of invoices. In *International Conference on Computational Intelligence in Data Science* (pp. 250-264). Cham: Springer Nature Switzerland. DOI: 10.1007/978-3-031-69986-3_19

Söylemez, I. (2024). Accelerating low-code automation development with generative artificial intelligence.

SS&C Blue Prism. (2023, September 21). Intelligent automation for provider prior authorization. https://www.blueprism.com/resources/blog/healthcare-prior-authorization-automation/

SS&C Blue Prism. (2024, April 3). Optimizing Revenue Cycle Performance with Automation | RCM Automation.https://www.blueprism.com/resources/blog/healthcare-revenue-cycle-automation-rcm/

Staicu, S. (2009, May 31). Inverse dynamics of the 3-PRR planar parallel robot. *Robotics and Autonomous Systems*, *57*(5), 556–563. DOI: 10.1016/j.robot.2008.09.005

Stephan Hayes, R., & Baker, R. (2014). A participant observation study of the resolution of audit engagement challenges in government tax compliance audits. *Qualitative Research in Accounting & Management*, *11*(4), 416–439. DOI: 10.1108/QRAM-02-2013-0003

Sternberg, R. J. (1999). Intelligence as developing expertise. *Contemporary Educational Psychology*, *24*(4), 359–375. DOI: 10.1006/ceps.1998.0998 PMID: 10508532

Strömberg, K. (2018). *Robotic Process Automation of office work: benefits, challenges and capability development* (Master's thesis).

Sufi, F. (2023). Algorithms in low-code-no-code for research applications: A practical review. *Algorithms*, *16*(2), 108. DOI: 10.3390/a16020108

Suke, S. G., Kosta, P., & Negi, H. (2015). Role of pharmacovigilance in India: An overview. Online journal of public health informatics, 7(2). Tak, A. ISAR Journal of Multidisciplinary Research and Studies.

Suman, M. C., & Prathyusha, K. (2012). Wireless camera based online examination security. [IJERA]. *International Journal of Engineering Research and Applications*, *2*, 1432–1435.

Sundaralingam, B., & Hermans, T. (2019, February). Relaxed-rigidity constraints: Kinematic trajectory optimization and collision avoidance for in-grasp manipulation. *Autonomous Robots*, *43*(2), 469–483. DOI: 10.1007/s10514-018-9772-z

Sun, S., Lin, H., Zheng, L., Yu, J., & Hu, Y. (2016, May). A real-time and look-ahead interpolation methodology with dynamic B-spline transition scheme for CNC machining of short line segments. *International Journal of Advanced Manufacturing Technology*, *84*(5), 1359–1370.

Sun, Y., Ho, K. C., & Wan, Q. (2018, November 5). Solution and analysis of TDOA localization of a near or distant source in closed form. *IEEE Transactions on Signal Processing*, *67*(2), 320–335. DOI: 10.1109/TSP.2018.2879622

Suri, V. K., Elia, M., & van Hillegersberg, J. (2017). Software bots-the next frontier for shared services and functional excellence. In *Global Sourcing of Digital Services: Micro and Macro Perspectives: 11th Global Sourcing Workshop 2017, La Thuile, Italy, February 22-25, 2017, Revised Selected Papers 11* (pp. 81-94). Springer International Publishing. DOI: 10.1007/978-3-319-70305-3_5

Suriano, L., Otero, A., Rodríguez, A., Sánchez-Renedo, M., & De La Torre, E. (2020, June 26). Exploiting multi- level parallelism for run-time adaptive inverse kinematics on heterogeneous masons. *IEEE Access : Practical Innovations, Open Solutions*, *8*, 118707–118724. DOI: 10.1109/ACCESS.2020.3005202

Süße, M., Prell, B., Stoldt, J., Strehle, H.-M., Richter, M., Schlegel, A., & Putz, M. (2019). Load and energy management for factories through multi-stage production planning and optimization. *Procedia CIRP*, *80*, 257–262. https://doi.org/https://doi.org/10.1016/j.procir.2019.01.076. DOI: 10.1016/j.procir.2019.01.076

Sweeney, D., Nair, S., & Cormican, K. (2023). Scaling AI-based industry 4.0 projects in the medical device industry: An exploratory analysis. *Procedia Computer Science*, *219*, 759–766. DOI: 10.1016/j.procs.2023.01.349

Syed, R., Suriadi, S., Adams, M., Bandara, W., Leemans, S. J., Ouyang, C., ter Hofstede, A. H. M., van de Weerd, I., Wynn, M. T., & Reijers, H. A. (2020). Robotic process automation: Contemporary themes and challenges. *Computers in Industry*, *115*, 103162. DOI: 10.1016/j.compind.2019.103162

Szabó, K. (2020). Bridging natural language processing AI techniques and corporate communications: Towards an integrative model. *Információs Társadalom*, *19*(4), 100. DOI: 10.22503/inftars.XIX.2019.4.7

Szuba, T. (1998). A molecular quasi-random model of computations applied to evaluate collective intelligence. *Future Generation Computer Systems*, *14*(5–6), 321–339. DOI: 10.1016/S0167-739X(98)00037-5

Szuba, T. (2001). Universal formal model of collective intelligence and its iq measure. *International Workshop of Central and Eastern Europe on Multi-Agent Systems*, 303–312.

Tailor, R. K., Pareek, R., & Khang, A. (2022). Robot process automation in blockchain. In *The Data-Driven Blockchain Ecosystem* (pp. 113–125). CRC Press. DOI: 10.1201/9781003269281-8

Talatahari, S., Azar, B. F., Sheikholeslami, R., & Gandomi, A. H. (2012, March 1). Imperialist competitive algorithm combined with chaos for global optimization. *Communications in Nonlinear Science and Numerical Simulation*, *17*(3), 1312–1319. DOI: 10.1016/j.cnsns.2011.08.021

Talib, F., & Rahman, Z. (2013). Current health of Indian healthcare and hospitality industries: a demographic study. Talib, F., and Rahman, (2013), 1-17.

Tang, H. (2023). Task scheduling strategy of logistics cloud robot based on edge computing. *Research Square*. https://doi.org/DOI: 10.21203/rs.3.rs-2423689/v1

Tatikonda, V., Venigandla, K., & Vemuri, N. (2022). Transforming customer banking experiences: Ai-driven rpa for customized service delivery. *International Journal of Developmental Research*, *12*, 60674–60677.

Thainimit, S., Chaipayom, P., Sa-arnwong, N., Gansawat, D., Petchyim, S., & Pongrujikorn, S. (2022). Robotic process automation support in telemedicine: Glaucoma screening usage case. *Informatics in Medicine Unlocked*, *31*, 101001. DOI: 10.1016/j.imu.2022.101001

Thekkethil, M. S., Shukla, V. K., Beena, F., & Chopra, A. (2021, September). Robotic process automation in banking and finance sector for loan processing and fraud detection. In *2021 9th international conference on reliability, infocom technologies and optimization (trends and future directions)(ICRITO)* (pp. 1-6). IEEE. DOI: 10.1109/ICRITO51393.2021.9596076

Thompson, S. E., & Patel, R. V. (1987, May). Formulation of joint trajectories for industrial robots using B-splines. *IEEE Transactions on Industrial Electronics, IE-34*(2), 192–199. DOI: 10.1109/TIE.1987.350954

Tian, L., & Collins, C. (2004, June 1). An effective robot trajectory planning method using a genetic algorithm. *Mechatronics, 14*(5), 455–470. DOI: 10.1016/j.mechatronics.2003.10.001

Tika, A., Gafur, N., Yfantis, V., & Bajcinca, N. (2020, January 1). Optimal scheduling and model predictive control for trajectory planning of cooperative robot manipulators. *IFAC-PapersOnLine, 53*(2), 9080–9086. DOI: 10.1016/j.ifacol.2020.12.2136

Tiron-Tudor, A., Lacurezeanu, R., Bresfelean, V. P., & Dontu, A. N. (2024). Perspectives on how robotic process automation is transforming accounting and auditing services. *Accounting Perspectives, 23*(1), 7–38. DOI: 10.1111/1911-3838.12351

Topol, E. J. (2020). COVID-19 can affect the heart. *Science, 370*(6515), 408–409. DOI: 10.1126/science.abe2813 PMID: 32967937

Toxtli, C. (2024). Human-centered automation. *arXiv preprint arXiv:2405.15960.*

Tripathi, A. M. (2018). *Learning robotic process automation: Create software robots and automate business processes with the leading RPA tool–UiPath*. Packt Publishing Ltd.

Tsai KY, Kohli D, Arnold J. Trajectory planning in joint space for mechanical manipulators.

Tuncer, A., & Yildirim, M. (2012, November 1). Dynamic path planning of mobile robots with improved genetic algorithm. *Computers & Electrical Engineering, 38*(6), 1564–1572. DOI: 10.1016/j.compeleceng.2012.06.016

Turrisi, G., Capotondi, M., Gaz, C. R., Modugno, V., Oriolo, G., & De Luca, A. (2021, November 13). On-Line Learning for Planning and Control of Underactuated Robots with Uncertain Dynamics. *IEEE Robotics and Automation Letters.*

Understanding the role of RPA in Healthcare | MuleSoft. (n.d.). MuleSoft. https://www.mulesoft.com/resources/api/rpa-in-healthcare

Ur-Rehman, R., Caro, S., Chablat, D., & Wenger, P. (2010, August 1). Multi-objective path placement optimization of parallel kinematics machines based on energy consumption, shaking forces and maximum actuator torques: Application to the Orthoglide. *Mechanism and Machine Theory*, *45*(8), 1125–1141. DOI: 10.1016/j.mechmachtheory.2010.03.008

Vajgel, B., Corrêa, P. L. P., De Sousa, T. T., Quille, R. V. E., Bedoya, J. A. R., De Almeida, G. M., Filgueiras, L. V. L., Demuner, V. R. S., & Mollica, D. (2021). Development of intelligent robotic process automation: A utility case study in Brazil. *IEEE Access : Practical Innovations, Open Solutions*, *9*, 71222–71235. DOI: 10.1109/ACCESS.2021.3075693

Valluri, D. (2024). Integrating augmented reality, gesture recognition, and NLP for enhancing underwater human-robot interaction. *International Journal of Science and Research Archive*, *11*(2), 956–968. DOI: 10.30574/ijsra.2024.11.2.0509

Van Chuong, L., Hung, P. D., & Diep, V. T. (2019, July). Robotic process automation and opportunities for Vietnamese market. In *Proceedings of the 7th International Conference on Computer and Communications Management* (pp. 86-90). DOI: 10.1145/3348445.3348458

van der Aa, H., & Leopold, H. (2021). Automatically identifying process automation candidates using natural language processing. In *Blockchain and Robotic Process Automation* (pp. 77-86). DOI: 10.1007/978-3-030-81409-0_7

van der Aa, H., & Leopold, H. (2021). Supporting RPA through natural language processing. In *Robotic Process Automation: Management, Technology* (p. 187). Applications.

Van der Aalst, W. M., Bichler, M., & Heinzl, A. (2018). Robotic process automation. *Business & Information Systems Engineering*, *60*(4), 269–272. DOI: 10.1007/s12599-018-0542-4

van Dijk, N. J. (2006, June 13). Generic trajectory generation for industrial manipulators. Eindhoven University of Technology, Department of Mechanical Engineering. *Dynamics and Control.*

Veale, M., & Zuiderveen Borgesius, F. (2021). Demystifying the Draft EU Artificial Intelligence Act—Analysing the good, the bad, and the unclear elements of the proposed approach. *Computer Law Review International*, *22*(4), 97–112. DOI: 10.9785/cri-2021-220402

Venigandla, K. (2022). Integrating RPA with AI and ML for Enhanced Diagnostic Accuracy in Healthcare. Power System Technology, 46(4).

Venigandla, K., Vemuri, N., Thaneeru, N., & Tatikonda, V. M. (2023). Leveraging AI-Enhanced Robotic Process Automation for Retail Pricing Optimization: A Comprehensive Analysis. *Journal of Knowledge Learning and Science Technology ISSN: 2959-6386 (Online), 2*(2), 361–370.

Venigandla, K., & Vemuri, N. (2022). Autonomous DevOps: Integrating RPA, AI, and ML for Self-Optimizing Development Pipelines. *Asian Journal of Multidisciplinary Research & Review, 3*(2), 214–231.

Venigandla, K., Vemuri, N., & Nnamere Aneke, E. (2024). Empowering smart cities with AI and RPA: Strategies for intelligent urban management and sustainable development. *International Journal of Scientific Research and Management, 12*(4), 1117–1125. DOI: 10.18535/ijsrm/v12i04.ec02

Venkateswaran, V. (2020,). Patient Experience in India.https://healthcare-in-india.net/wp-content/uploads/2020/08/Patient-Experience-in-India.pdf

Verhoef, P. C., Broekhuizen, T., Bart, Y., Bhattacharya, A., Dong, J. Q., Fabian, N., & Haenlein, M. (2021). Digital transformation: A multidisciplinary reflection and research agenda. *Journal of Business Research, 122*, 889–901. DOI: 10.1016/j.jbusres.2019.09.022

Verno, J. (2019). Risk management in corporates; can digitalization help improve risk management in cash management?.

Verscheure, D., Demeulenaere, B., Swevers, J., De Schutter, J., & Diehl, M. (2009, September 22). Time-optimal path tracking for robots: A convex optimization approach. *IEEE Transactions on Automatic Control, 54*(10), 2318–2327. DOI: 10.1109/TAC.2009.2028959

Vezvari, M. R., Nikoobin, A., & Ghoddosian, A. (2021, July). Perfect torque compensation of planar 5R parallel robot in point-to-point motions, optimal control approach. *Robotica, 39*(7), 1163–1180. DOI: 10.1017/S0263574720000971

Vial, G. (2021). Understanding digital transformation: A review and a research agenda. *Managing Digital Transformation*, 13–66.

Victoire, T. A., & Jeyakumar, A. E. (2004, September 1). Hybrid PSO–SQP for economic dispatch with valve-point effect. *Electric Power Systems Research, 71*(1), 51–59. DOI: 10.1016/j.epsr.2003.12.017

Viehhauser, J. (2020). Is robotic process automation becoming intelligent? Early evidence of influences of artificial intelligence on robotic process automation. *Business Process Management: Blockchain and Robotic Process Automation Forum: BPM 2020 Blockchain and RPA Forum, Seville, Spain, September 13–18, 2020. Proceedings, 18*, 101–115.

Villar, A. S., & Khan, N. (2021). Robotic process automation in banking industry: A case study on Deutsche Bank. *Journal of Banking and Financial Technology, 5*(1), 71–86. DOI: 10.1007/s42786-021-00030-9

Vlahovic, S., Suznjevic, M., Pavlin-Bernardic, N., & Skorin-Kapov, L. (2021). The effect of VR gaming on discomfort, cybersickness, and reaction time. *2021 13th International Conference on Quality of Multimedia Experience (QoMEX)*, 163–168.

von Mehren, M., Kane, J. M., Agulnik, M., Bui, M. M., Carr-Ascher, J., Choy, E., Connelly, M., Dry, S., Ganjoo, K. N., Gonzalez, R. J., Holder, A., Homsi, J., Keedy, V., Kelly, C. M., Kim, E., Liebner, D., McCarter, M., McGarry, S. V., Mesko, N. W., & Bergman, M. A. (2022). Soft tissue sarcoma, version 2.2022, NCCN clinical practice guidelines in oncology. *Journal of the National Comprehensive Cancer Network : JNCCN, 20*(7), 815–833. DOI: 10.6004/jnccn.2022.0035 PMID: 35830886

Wada, Y., & Kawato, M. (2004, April 1). A via-point time optimization algorithm for complex sequential trajectory formation. *Neural Networks, 17*(3), 353–364. DOI: 10.1016/j.neunet.2003.11.009 PMID: 15037353

Wallace, E. (2022). *The Role of Organisational Change Management in Robotic Process Automation Implementation Projects* (Doctoral dissertation, Auckland University of Technology).

Wang, G., Shu, Q., Wang, J., & Li, L. (2017, July). Research on adaptive non-uniform rational B-spline real- time interpolation technology based on acceleration constraints. *International Journal of Advanced Manufacturing Technology, 91*(5), 2089–2100. DOI: 10.1007/s00170-016-9914-4

Wang, Z., Zhang, L., Zhang, F., Li, W., & Yang, Z. (2021). Research and Exploration on The Application of RPA and Artificial Intelligence Technology in The Field of Electric Power Marketing. *Proceedings of the 4th International Conference on Information Technologies and Electrical Engineering*, 1–5. DOI: 10.1145/3513142.3513236

Webb, D. J., Van Den Berg, J., & Kinodynamic, R. R. T. (2013). asymptotically optimal motion planning for robots with linear dynamics, 2013 IEEE Int. Conf. Rob. Autom.

Wen, Y., & Pagilla, P. R. (2021, September). Path-constrained optimal trajectory planning for robot manipulators with obstacle avoidance. In *2021 IEEE/RSJ International Conference on Intelligent Robots and Systems (IROS)* (pp. 1421-1426). IEEE.

Wewerka, J., & Reichert, M. (2021). Robotic process automation—A systematic mapping study and classification framework. *Enterprise Information Systems, 17*(2), 1986862. Advance online publication. DOI: 10.1080/17517575.2021.1986862

Whitesmith, M. (2020). Experimental research in reducing the risk of cognitive bias in intelligence analysis. *International Journal of Intelligence and CounterIntelligence, 33*(2), 380–405. DOI: 10.1080/08850607.2019.1690329

Wie, B. (2005, September). Singularity escape/avoidance steering logic for control moment gyro systems. *Journal of Guidance, Control, and Dynamics, 28*(5), 948–956. DOI: 10.2514/1.10136

Willcocks, L. P., Lacity, M., & Craig, A. (2015). The IT function and robotic process automation.

Willcocks, L., Hindle, J., & Lacity, M. (2018). Keys to RPA success. *Executive Research Report, Knowledge Capital Partners, USA, Technical Report.*

Willcocks, L. P., Hindle, J., Stanton, M., & Smith, J. (2024). *Maximizing Value with Automation and Digital Transformation: A Realist's Guide*. Springer Nature. DOI: 10.1007/978-3-031-46569-7

Willcocks, L., Lacity, M., & Craig, A. (2017). Robotic process automation: Strategic transformation lever for global business services? *Journal of Information Technology Teaching Cases, 7*(1), 17–28. DOI: 10.1057/s41266-016-0016-9

William, P., Choubey, S., Choubey, A., & Chhabra, G. S. (2023). Evolutionary Survey on Robotic Process Automation and Artificial Intelligence: Industry 4.0. *Robotic Process Automation*, 315–327.

Williams, P. (2021, September 21). Amazing Ways That RPA Can be Used in Healthcare - IBM Blog. IBM Blog. https://www.ibm.com/blog/amazing-ways-that-rpa-can-be-used-in-healthcare/

Wilson, H. J., & Daugherty, P. R. (2018). Collaborative intelligence: Humans and AI are joining forces. *Harvard Business Review, 96*(4), 114–123.

Witty, P. A., & Lehman, H. C. (1931). A study of vocational attitude and intelligence. *The Elementary School Journal, 31*(10), 735–746. DOI: 10.1086/456653

Wong, C. K. H., Au, I. C. H., Lau, K. T. K., Lau, E. H. Y., Cowling, B. J., & Leung, G. M. (2022). Real-world effectiveness of molnupiravir and nirmatrelvir plus ritonavir against mortality, hospitalisation, and in-hospital outcomes among community-dwelling, ambulatory patients with confirmed SARS-CoV-2 infection during the omicron wave in Hong Kong: An observational study. *Lancet*, *400*(10359), 1213–1222. DOI: 10.1016/S0140-6736(22)01586-0 PMID: 36216007

Woolhandler, S., & Himmelstein, D. U. (2014). Administrative work consumes one-sixth of US physicians' working hours and lowers their career satisfaction. *International Journal of Health Services*, *44*(4), 635–642. DOI: 10.2190/HS.44.4.a PMID: 25626223

Xiao, Y., Du, Z., & Dong, W. (2012, March 2). Smooth and near time-optimal trajectory planning of industrial robots for online applications. Industrial Robot. *International Journal (Toronto, Ont.)*.

Xia, S., Zhang, Y., Wang, Y., Wang, H., Yang, Y., Gao, G. F., Tan, W., Wu, G., Xu, M., Lou, Z., Huang, W., Xu, W., Huang, B., Wang, H., Wang, W., Zhang, W., Li, N., Xie, Z., Ding, L., & Yang, X. (2021). Safety and immunogenicity of an inactivated SARS-CoV-2 vaccine, BBIBP-CorV: A randomised, double-blind, placebo-controlled, phase 1/2 trial. *The Lancet. Infectious Diseases*, *21*(1), 39–51. DOI: 10.1016/S1473-3099(20)30831-8 PMID: 33069281

Xie, L., Zhao, G., Lu, J., & Jiang, B. (2023). Research on marketing intelligence inspection based on RPA technology. *Third International Seminar on Artificial Intelligence, Networking, and Information Technology (AINIT 2022)*, *12587*, 170–177.

Xie, Z., Liang, X., & Roberto, C. (2023). Learning-based robotic grasping: A review. *Frontiers in Robotics and AI*, *10*, 1038658. DOI: 10.3389/frobt.2023.1038658 PMID: 37082744

Xu, X., & Chen, Y. A method for trajectory planning of robot manipulators in Cartesian space. InProceedings of the 3rd World Congress on Intelligent Control and Automation (Cat. No. 00EX393) 2000 Jun 26 (Vol. 2, pp. 1220-1225). IEEE.

Xue, J., Wu, W., & Cheng, Q. (2023). Intelligent invigilator system based on target detection. *Multimedia Tools and Applications*, *82*(29), 44673–44695. DOI: 10.1007/s11042-023-15474-y PMID: 37362652

Xu, H. (2008, June 17). Tracking Lagrangian trajectories in position–velocity space. *Measurement Science & Technology*, *19*(7), 075105. DOI: 10.1088/0957-0233/19/7/075105

Xu, Z., Wei, S., Wang, N., & Zhang, X. Trajectory planning with Bezier curve in Cartesian space for industrial gluing robot. In *International Conference on Intelligent Robotics and Applications* 2014 Dec 17 (pp. 146-154). Springer, Cham. DOI: 10.1007/978-3-319-13963-0_15

Yang, J., Yu, T., Jian, L. R., Qiu, J., & Li, Y. (2011). An extreme automation framework for scaling cloud applications. *IBM Journal of Research and Development, 55*(6), 8–1. DOI: 10.1147/JRD.2011.2170929

Yano, F., & Toyoda, Y. (1999, August). Preferable movement of a multijoint robot arm using a genetic algorithm. In *Intelligent robots and computer vision XVIII: algorithms, techniques, and active vision* (Vol. 3837, pp. 80–88). SPIE.

Yao, M. (2024). RPA Technology Enables Highly Automated Development of Corporate Financial Accounting Processes. *Applied Mathematics and Nonlinear Sciences, 9*(1), 20240541. Advance online publication. DOI: 10.2478/amns-2024-0541

Yapıcı, A., & Akcayol, M. A. (2022). Derin Öğrenme ile Görüntülerde Gürültü Azaltma Üzerine Kapsamlı Bir İnceleme. *International Journal of Advances in Engineering and Pure Sciences, 34*(1), 65–90. DOI: 10.7240/jeps.938188

Yarlagadda, R. T. (2018). The RPA and AI automation. *International Journal of Creative Research Thoughts (IJCRT), ISSN*, 2320-2882.

Yarlagadda, R. T. (2018). The RPA and AI automation. [IJCRT]. *International Journal of Creative Research Thoughts, 2320*(2882).

Yatskiv, N., Yatskiv, S., & Vasylyk, A. (2020). Method of robotic process automation in software testing using artificial intelligence. *2020 10th International Conference on Advanced Computer Information Technologies (ACIT)*, 501–504.

Ylä-Kujala, A., Kedziora, D., Metso, L., Kärri, T., Happonen, A., & Piotrowicz, W. (2023). Robotic process automation deployments: A step-by-step method to investment appraisal. *Business Process Management Journal, 29*(8), 163–187. DOI: 10.1108/BPMJ-08-2022-0418

Yuan, M., Chen, Z., Yao, B., & Liu, X. (2019, December 16). Fast and accurate motion tracking of a linear motor system under kinematic and dynamic constraints: An integrated planning and control approach. *IEEE Transactions on Control Systems Technology, 29*(2), 804–811. DOI: 10.1109/TCST.2019.2955658

Zalzala A, Morris AS. An Optimum Trajectory Planner for Robot Manipulators in Joint- pace and Under Physical Constraints.

Zanotto, V., Gasparetto, A., Lanzutti, A., Boscariol, P., & Vidoni, R. (2011, November). Experimental validation of minimum time-jerk algorithms for industrial robots. *Journal of Intelligent & Robotic Systems*, *64*(2), 197–219. DOI: 10.1007/s10846-010-9533-5

Zeiser, F. A., Donida, B., da Costa, C. A., de Oliveira Ramos, G., Scherer, J. N., Barcellos, N. T., Alegretti, A. P., Ikeda, M. L. R., Müller, A. P. W. C., & Bohn, H. C. (2022). First and second COVID-19 waves in Brazil: A cross-sectional study of patients' characteristics related to hospitalization and in-hospital mortality. *Lancet Regional Health. Americas*, *6*, 6. DOI: 10.1016/j.lana.2021.100107 PMID: 34746913

Zhang, S., & Luo, J. (2022, December). Construction of Autonomous System of Educational Robot Under the Background of Artificial Intelligence Taking the Mobile Educational Robots toward Performing Intelligent Invigilation as an Example. In 2022 3rd International Conference on Artificial Intelligence and Education (IC-ICAIE 2022) (pp. 810-816). Atlantis Press.

Zhang, D., & Wei, B. (Eds.). (2017 Feb 3). *Adaptive control for robotic manipulators*. CRC Press. DOI: 10.1201/9781315166056

Zhang, H., Qiu, H., Zhang, X., & Hu, P. (2020, November 1). Recent Advances on Manipulator Trajectory Planning Methods. *Recent Patents on Mechanical Engineering*, *13*(4), 303–327. DOI: 10.2174/2212797613666200319151513

Zhang, K., Gao, X. S., Li, H. B., & Yuan, C. M. (2012). A greedy algorithm for feed-rate planning of CNC machines along curved tool paths with confined jerk for each axis. *Robotics and Computer-integrated Manufacturing*, *28*(4), 472–483. DOI: 10.1016/j.rcim.2012.02.006

Zhang, X., & Shi, G. (2022, April). Multi-objective optimal trajectory planning for manipulators in the presence of obstacles. *Robotica*, *40*(4), 888–906. DOI: 10.1017/S0263574721000886

Zhang, X., & Wen, Z. (2021). Thoughts on the development of artificial intelligence combined with RPA. *Journal of Physics: Conference Series*, *1883*(1), 012151. DOI: 10.1088/1742-6596/1883/1/012151

Zhao, L., Jiang, Z., Sun, Y., Zhao, J., & Liu, H. (2021, February 28). Collision-free kinematics for hyper-redundant manipulators in dynamic scenes using optimal velocity obstacles. *International Journal of Advanced Robotic Systems*, *18*(1), 1729881421996148. DOI: 10.1177/1729881421996148

Zhao, S., Lin, Q., Ran, J., Musa, S. S., Yang, G., Wang, W., Lou, Y., Gao, D., Yang, L., He, D., & Wang, M. H. (2020). Preliminary estimation of the basic reproduction number of novel coronavirus (2019-nCoV) in China, from 2019 to 2020: A data-driven analysis in the early phase of the outbreak. *International Journal of Infectious Diseases*, *92*, 214–217. DOI: 10.1016/j.ijid.2020.01.050 PMID: 32007643

Zhao, X., Oseni, T., & Medishetty, B. T. (2022, October). Overview of Business Hyper-automation. In *2022 IEEE International Conference on e-Business Engineering (ICEBE)* (pp. 100-105). IEEE. DOI: 10.1109/ICEBE55470.2022.00026

Zha, X. F. (2002, October 1). Optimal pose trajectory planning for robot manipulators. *Mechanism and Machine Theory*, *37*(10), 1063–1086. DOI: 10.1016/S0094-114X(02)00053-8

Zhou, J., Cao, H., Jiang, P., Li, C., Yi, H., & Liu, M. (2021, March 24). Energy-Saving Trajectory Planning for Robotic High-Speed Milling of Sculptured Surfaces. *IEEE Transactions on Automation Science and Engineering*.

Zhu, R., Liu, H., Xu, X., Lin, L., Chen, Y., & Li, W. (2024). A-PGRD: Attention-based automatic business process model generation from RPA process description. *Concurrency and Computation*, *36*(8), e7940. DOI: 10.1002/cpe.7940

Zlajpah, L. (1996, April). On time optimal path control of manipulators with bounded joint velocities and torques. In *Proceedings of IEEE international conference on robotics and automation* (Vol. 2, pp. 1572-1577). IEEE.

About the Contributors

Sachi Nandan Mohanty, received his Postdoc from IIT Kanpur in the year 2019 and Ph.D. from IIT Kharagpur in the year 2015, with MHRD scholarship from Govt of India. He is currently working as an Associate Professor in the Department of Computer Science & Engineering at ICFAI Foundation for Higher Education Hyderabad. Prof. Mohanty research areas include Data mining, Big Data Analysis, Cognitive Science, Emotional Intellegnce, Fuzzy Decision Making, Brain-Computer Interface, and Computational Intelligence. Prof. S N Mohanty has received 3 Best Paper Awards during his Ph.D at IIT Kharagpur from International Conference at Benjing, China, and the other at International Conference on Soft Computing Applications organized by IIT Rookee in the year 2013. He has published 20 SCI Journals. As a Fellow on Indian Society Technical Education (ISTE), The Institute of Engineering and Technology (IET), Computer Society of India (CSI), Senior Member of Institute of Engineers and IEEE Computer Society, he is actively involved in the activities of the Professional Bodies/Societies.

***

Minakshi is an accomplished academician with 18+ years of experience in computer science and engineering domains. Currently an Assistant Professor at King Khalid University, she has taught at esteemed institutions like Uttaranchal University and Tula's Institute. A prolific researcher with 80 Scopus publications, 18 patents, and 3 software copyrights, she has received accolades like Outstanding Section Volunteer Award (IEEE UP Section), Research Excellence Award, and HOD of the Year Award. Adept at curriculum development, research guidance, project management, and team leadership, Dr. Minakshi brings a holistic approach to education, fostering student success through quality instruction and innovative pedagogical methods.

Rohit is a Research Associate at Thumbay College of Management and AI in Healthcare, Gulf Medical University, UAE. He is MBA from Kurukshetra University Kurukshetra (A state government public university), India. He has published research papers in international journals and participated in various international conferences, workshops, and seminars. He worked as a management trainee and intern in the Thumbay Hospital, United Arab Emirates (UAE); Decathlon, Zirakpur, India; and Bajaj Allianz General Insurance, India. He completed various MOOCs from Copenhagen Business School, University of New Mexico, University of California, Irvine, University of London, and many other. His areas of interest are Entrepreneurship, consumer behavior, marketing strategies and strategic management. He is inspired with thought leadership and a regular contributor to educational magazines, debates, and cultural activities.

Ajay Chandel is working as an Associate Professor at Mittal School of Business, Lovely Professional University, Punjab. He has 14 years of teaching and research experience. He has published papers in SCOPUS, WOS, and UGC listed Journals in areas like Social Media Marketing, E-Commerce, and Consumer Behaviour. He has published cases on SMEs and Social Entrepreneurship in The Case Centre, UK. He also reviews The Case Journal, Emerald Group Publishing, and International Journal of Business and Globalisation, Inderscience. He has authored and developed MOOCs on Tourism and Hospitality Marketing under Epg-Pathshala- A gateway to all postgraduate courses (a UGC-MHRD project under its National Mission on Education Through ICT (NME-ICT).

Swati Chowdhuri, Professor of Institute of Engineering& Management. She did Ph.D and M.Tech from Jadavpur University and B.E. from Burdwan University with Honours. Her research interests in the field of MIMO communication, Wireless Network, Deep Learning application in image processing, Machine Learning, Brain Computer Interface. She has more than 40 publications in international journal and conferences. She has almost 17 years teaching experience. She is Executive Committee member IEEE WIE, Kolkata Section and Life Member of IE(I) and ISTE.

Manju Dagar, an esteemed Assistant Professor in the Department of Management Studies, boasts an impressive academic and professional profile. She completed her Ph.D. on the topic of "Attrition of women employees in the IT Industry". She has 13 years of academic experience and has also served as the head of the department. Dr. Dagar is a valued member of the University Placement Cell and HR association. Her areas of specialization include Human Resource Management and Marketing Management, showcasing well-rounded expertise that spans crucial aspects of organizational dynamics. In terms of research, Dr. Manju Dagar has made significant contributions, and her work has been published in various prestigious platforms. Her research papers have found a place in UGC CARE-recognized journals, international publications, and the esteemed Scopus database, reflecting high quality and global relevance to her scholarly output. Additionally, she has made innovative contributions to patents further highlighting her multifaceted approach to academic and intellectual pursuits. She has also successfully completed various Faculty Development Programs (FDPs) and workshops. Most recently, she completed a diploma in Coaching & Mentoring from Skills and Education Group based in United Kingdom organized by CEAPS SGT University.

Vijay Kumar Dalla is Assistant Professor in the Department of Mechanical Engg. at NIT Jamshedpur since 23rd May 2018. . His areas of interest are robotics, dynamics, and control. Before joining NIT Jamshedpur, he has been associated with 6 various institutes in different states. He pursued his Ph.D. from IIT Roorkee in Robotics in the year 2016. He completed a Bachelor of Engineering (B.E.) in Mechanical Engineering from Govt. Engg. College Bilaspur, Chhattisgarh, in 2006. He has 7 years of learning exposure to two IITs and 10 years of teaching exposure. His teaching interests are robotics & automation; kinematics, dynamics, and control of mechanical systems, AI in robotics; machine learning in mechanical engg; smart materials. His research areas are space robotics, dynamics, and control. Other than his robotics learning, he has also practical exposure to fracture mechanics, materials science, and ceramic powder processing. He has published 25 research papers in reputed international journals and conference proceedings and 03 book chapters. He is also involved in various outreach activities. He has supervised one PhD student as sole supervisor. At present, he is guiding 4 Ph.D. students. He is a lifetime member of Machines and Mechanism (AMM) and The Robotics Society. He has received MHRD fellowships for his persuasion of post-graduation and doctorate degrees. He has a strong thrust and dedication to developing application-based robots. He has conducted four sponsored FDPs/Workshops. He was associated with two BRNS Projects during his Senior Research Fellowship at IIT Bombay after his MTech in Materials Science from IIT Bombay in 2008. At present, he has two ongoing funded robotics projects from TiHaN IITH and IIT Guwahati Technology Innovation and Development Foundation of 32 lacs.

Suman Devi is currently working as Assistant Professor in School of Computing Science & Engineering. She has total experience of more than 11 years of teaching and has been associated with Galgotias University since 2017. Along with she is a research scholar in the Department of CSE at Galgotias University, India. She is firm Believer of productivity and efficiency in work. She exhibits an honest work ethic with the ability to excel in fast-paced, time-sensitive environment. Being a passionate teacher, she believes that teaching is not merely restricted to making the students understand the underlying concepts of a course but also to develop critical thinking and evaluate alternate approaches for problem solving. He always put his efforts towards the overall developments of her students. Her research interest area is Wireless sensor networks, Security. She has published several papers in peer reviewed journal and international conferences. She also organized and attended several workshops.

Narayanage Jayantha Dewasiri, Ph.D., is a Chartered Manager at the Chartered Management Institute, UK. He conducts research at the Open University of Sri Lanka and the Sri Lanka Institute of Marketing. He is the founding Editor-in-Chief of the South Asian Journal of Marketing. His research interests are in corporate social responsibility, corporate governance, marketing and sales, management, and finance. He is working as the Head of Field Force at the HMD Global Inc., Sri Lanka. He also serves as a senior lecturer at University of Sabaragamuwa, Sri Lanka. He can be reached at dnjay@ou.ac.lk

Shivani Dhand is an Associate Professor (14+ years of experience),a Certified SHRM (Society for Human Resource Management)faculty in the department of Mittal School of Business of Lovely Professional University.She is NLP(Neuro Linguistics Programming) Practioner She is also a member of the Indian Society for Training and Development (ISTD) and the Indian Society of Labour Economics. She is also a core committee member and Mentor of an NGO(Shalini Fellowship) Shivani Dhand has worked in an IT company as an HR. She has published 15 research papers in various journals which include Scopus-indexed, UGC care and peer-reviewed journals. She is a certified POSH trainer. She has worked with CPSC (Columbo Plan Staff College) Manila Philippines, and trained the faculty and Principals of various Institutions in Bhutan (2015). In 2016, She worked on one of the projects," Austria as an Innovation leader in EU in Austria and was awarded as best presentation and Excellent paper. Her research is situated in the field of Startups, HR Practices Human Resource Management, Entrepreneurship, and social entrepreneurship with a special focus on academic entrepreneurship.

Damith Gangodawilage is a practicing scholar and professional with extensive experience in finance and accounting, business management, entrepreneurship, and innovation. Currently serving as the Deputy Dean and the Cluster Head of Accounting, Economics, and Finance of the Qasim Ibrahim School of Business (QISB) at Villa College, Maldives, he has consistently demonstrated leadership in fostering academic excellence and industry engagement. Damith is also the Director of the Centre for Innovation and Entrepreneurship Development (CIED) at Villa

College, where he spearheads initiatives such as the VIgnite Business Incubator, driving innovation among youth in the Maldives. With a strong academic foundation, Damith has made significant contributions to teaching, research, and curriculum development. His expertise spans strategic areas such as design thinking, sustainable business practices, and digital transformation. An active researcher, he has authored numerous studies and presentations on topics like corporate sustainability and tax compliance behavior, establishing himself as a thought leader. Dr. Damith is a dedicated mentor and educator, passionately committed to equipping students with critical thinking and problem-solving skills. His innovative approach combines academic rigor with practical insights, making him a distinguished figure in both academia and industry. As a global expert in entrepreneurship and compliance management, he bridges the gap between theory and practice, effectively bringing local solutions to the global stage.

Souvick Kumar Guha: Engineering Innovator & Researcher Souvick Kumar Guha is a passionate B.Tech student in Electronics and Communication Engineering at JIS College of Engineering. With a fervent drive for innovation and research, Souvick delves into coding algorithms and conducting experiments, constantly seeking to push the boundaries of knowledge. A dedicated student researcher, Souvick is committed to uncovering solutions that shape the future, tackling challenges with curiosity and determination. His journey is fueled by a belief in the transformative power of innovation to address pressing global issues. At JIS College of Engineering, Souvick thrives in a dynamic learning environment that nurtures creativity and growth. As an advocate for innovation, he embraces the thrill of discovery, aiming to turn ideas into reality through collaborative efforts. Join Souvick on his exciting journey of exploration and innovation. Let's connect and create a better world through research and creativity. #INNOVATION #Research

Sakshi is a Research Scholar in the department of Mittal School of Business of Lovely Professional University. She earned her master's degree in management from Maharshi Dayanand University, Rohtak, Haryana. She is pursuing her PhD in the field of human resource management.

Mohan has a total of eight years of professional experience in both the corporate sector and academia. He has worked in a Public Sector Bank and a State Government University, focusing on Marketing and Entrepreneurship Development. He completed his PhD on "Motivational Factors and Challenges of Start-Ups in NCR" at the Institute of Management Studies and Research, Maharshi Dayanand University, Rohtak-124001, Haryana. Dr. Mohan specializes in Marketing Management, Human Resource Management, Quantitative Analysis, and Entrepreneurship Development.

His research has been published in esteemed national and international journals, and he has actively participated in both national and international conferences, presenting his research findings. In March 2022, he received the "Best Research Paper Award" at an international conference conducted by IITM, GGSIU, Delhi, for his paper titled "Psychological Wellbeing Before & During the COVID-19 Pandemic: A Longitudinal Sample Study of Creative Writers and Directors." Since May 2023, Dr. Mohan has been working at SGT University, Gurugram. He continues to contribute to academia through research, conferences, and workshops, furthering his expertise in his specialized domains.

Tarun Kumar is currently working as Assistant Professor in School of Computing Science & Engineering. He has total experience of more than 17 years of teaching and has been associated with Galgotias University since 2017. Along with he is a research scholar in the Department of CSE at the National Institute of Technology Patna, Bihar, India. He is firm Believer of productivity and efficiency in work. He exhibits an honest work ethic with the ability to excel in fast-paced, time-sensitive environment. Being a passionate teacher, he believes that teaching is not merely restricted to making the students understand the underlying concepts of a course but also to develop critical thinking and evaluate alternate approaches for problem solving. He always put his efforts towards the overall developments of his students. His research interest area is Cloud Computing, DNA Computing. He has published several papers in peer reviewed journal and international conferences. He also organized and attended several workshops and conferences.

Thi Minh Ngoc Luu is currently the Head of the Department of Economics and Management at the International School, Vietnam National University, Hanoi. She has over 17 years of experience in teaching and researching human resource management, management control, corporate governance, individual and organizational behavior, leadership, and sustainable development.

Shashank Mittal has done his FPM in Organizational behavior and Human Resource from Indian Institute of Management Raipur. He holds B. Tech from I.E.T. Lucknow. His post FPM work experience includes almost five years of industry and academics exposures in multiple roles. Prior to joining FPM, he has over two years of industrial experience and three years of teaching experience in various organizations. He has published multiple papers in ABDC ranked and SSCI indexed journals of international repute such as Journal of Knowledge Management, Journal of Behavioral and Experimental Finance, International Journal of Conflict Management, Journal of Management and Organization and Current Psychology. His current research interest includes Social identity, Knowledge exchanges, Status,

Proactive helping, Employer branding, Organizational justice, and Humanitarian relief management. He enjoys badminton and cycling during leisure time.

Ali Najeeb is an Associate Professor at Qasim Ibrahim School of Business, Villa College, Maldives. His research interests include higher education leadership, governance and management, industrial relations, employment law and labour issues in small island states, migrant workers, HRM/ER in tourism and hospitality industry. He has published several research papers in refereed international journals, shared his experience in local and international panel discussions and presented his research work in international scholarly conferences. He is a recipient of the Emerald Literati Award for Excellence 2014 – the outstanding paper award.

Biswarup Neogi presently attached with JIS College of Engineering (JIS Group, India) as a Dean Research & Development and Professor of ECE Dept. in addition as a Precedent of Institutional Innovation Council-JISCE under Ministry of Education Innovation Cell (Govt. of India). As an Innovation Ambassador, Institutional Innovation Council – JISCE, he is enhancing the innovation activities in JISCE and propagate this activity in society. He is associated as a Mentor of 02 ATL Lab under Atal Innovation Mission, NITI Aayog . He also received Visiting Scientist Post from Institute of Coal, Federal Research Center Coal and Coal Chemistry Siberian Branch Russian Academy of Science. He is the Executive Council Member of Maulana Abul Kalam Azad University of Technology (MAKAUT), formerly West Bengal University of Technology, (Govt. of WB).

Moumita Pal, serves as the Head of the Department and an Associate Professor at JIS College of Engineering, Kalyani, where she has been employed since January 2012. Previously. Dr. Pal holds a Ph.D. from Mizoram Central University (2021), with her dissertation focused on the "Modeling and Simulation of Silicon Carbide Based Microelectromechanical Ultrasonic Transducer." Her research focuses on MEMS, steganography, and image processing. She has successfully managed a UGC-funded project on retinal image identification. Dr. Pal has organized numerous FDPs and conferences and has published more than 15 Nos. of research articles extensively in SCI and Scopus-indexed journals. She has several patents, including innovations in biometric verification systems and IoT-based health monitoring.

Majdi Anwar Quttainah earned his Ph.D. degree in Management from Rensselaer Polytechnic Institute, Lally School of Management & Technology (USA); MBA degree with merits from the Business School & Entrepreneurship at Newcastle University (UK), and bachelor's degree in International Business from the American International University in London (UK). Dr. Quttainah is an associate professor of management in the College of Business Administration at

Kuwait University. His research and teaching interests are corporate governance and strategies, entrepreneurship & small businesses, and organizational development. In addition, as a recognition of his research, he was awarded the Best Young Researcher academic year 2016/2017 Kuwait University. He is an active scholar, presently working as senior editor of FIIB Business Review (Scopus Q1, WoS, Published by the Sage Publishing). His research won several recognition and awards including the Best Paper Award at the 76th Annual Meeting of the Academy of Management Anaheim, CA, the Best Paper Award at the AGBA's 12th Annual World Congress, Kuantan City (State of Pahang), Malaysia, and the Best Paper Award of the Gorontalo International Conference of Project Management (ICPM) - Gorontalo/Indonesia. Also, at the home country level the best paper award for the Kuwait Economic Researcher Award by the Central Bank of Kuwait; May 2019 and 2022.

Sandeep Kumar Singh is a faculty in the area of Information Systems and Analytics at Jindal Global Business School, O.P. Jindal Global University, Sonipat. Prior to joining Jindal Global Business School, he worked as an Assistant Professor at JKLU Jaipur for two years. He has earned his Ph.D. in computational physics from Antwerpen University, Belgium. His PhD work focused on mathematical modeling and simulation of two-dimensional materials. Post to his PhD, he worked as a research scientist in Sweden for four years in energy storage systems, bioelectronics. He also did MTech in industrial mathematics and scientific computing from IIT Madras, Chennai. His current research interests are to apply complex mathematical modeling in the business and financial world.

Mohit Yadav is an Associate Professor in the area of Human Resource Management at Jindal Global Business School (JGBS). He has a rich blend of work experience from both Academics as well as Industry. Prof. Mohit holds a Ph.D. from Department of Management Studies, Indian Institute of Technology Roorkee (IIT Roorkee) and has completed Master of Human Resource and Organizational Development (MHROD) from prestigious Delhi School of Economics, University of Delhi. He also holds a B.Com (Hons.) degree from University of Delhi and UGC-JRF scholarship. He has published various research papers and book chapters with reputed publishers like Springer, Sage, Emerald, Elsevier, Inderscience etc. and presented research papers in national and International conferences both in India and abroad. He has many best paper awards on his credit too. He is reviewer of various international journals like Computers in Human Behavior, Policing etc. His areas of interest are Organizational Behavior, HRM, Recruitment and Selection, Organizational Citizenship Behavior, Quality of work life and role.

Index

A

B

H

I

J

L

M

N

O

P

R

S

T

U

W

www.ingramcontent.com/pod-product-compliance
Lightning Source LLC
LaVergne TN
LVHW080349011026
849161LV00009B/2242